MULTIPLIER CONVERGENT SERIES

MULTIPLIER CONVERGENT SERIES

Charles Swartz

New Mexico State University, USA

 **World Scientific**

NEW JERSEY · LONDON · SINGAPORE · BEIJING · SHANGHAI · HONG KONG · TAIPEI · CHENNAI

Published by

World Scientific Publishing Co. Pte. Ltd.

5 Toh Tuck Link, Singapore 596224

USA office: 27 Warren Street, Suite 401-402, Hackensack, NJ 07601

UK office: 57 Shelton Street, Covent Garden, London WC2H 9HE

British Library Cataloguing-in-Publication Data
A catalogue record for this book is available from the British Library.

MULTIPLIER CONVERGENT SERIES

ISBN-13 978-981-283-387-7
ISBN-10 981-283-387-0

Printed in Singapore.

To the Memory of My Mother

Preface

This monograph contains an exposition of the properties and applications of multiplier convergent series with values in a topological vector space. If λ is a space of scalar valued sequences and $\sum_j x_j$ is a (formal) series with values in a topological vector space X, the series $\sum_j x_j$ is λ multiplier convergent if the series $\sum_{j=1}^{\infty} t_j x_j$ converge in X for every $\{t_j\} \in \lambda$. For example, if $M_0 = \{\chi_\sigma : \sigma \subset \mathbb{N}\}$, where χ_σ is the characteristic function of σ, then M_0 multiplier convergence is just subseries convergence. Basic properties of multiplier convergent series are developed in Chapter 2 and applications of multiplier convergent series to topics in topological vector spaces and vector valued measures are given in Chapter 3. A classical result of Orlicz and Pettis states that if a series in a normed linear space is subseries convergent (M_0 multiplier convergent) in the weak topology of the space, then the series is actually subseries convergent (M_0 multiplier convergent) in the norm topology of the space. Generalizations of this theorem to λ multiplier convergent series with values in a locally convex space are given in Chapters 4, 5 and 6. Another classical theorem of Hahn and Schur asserts that if $\sum_j t_{ij}$ is absolutely convergent for every $i \in \mathbb{N}$ and if $\lim_i \sum_{j \in \sigma} t_{ij}$ exists for every $\sigma \subset \mathbb{N}$ with $t_j = \lim_i t_{ij}$, then the series $\sum_j t_j$ is absolutely convergent and

$$\lim_i \sum_{j=1}^{\infty} |t_{ij} - t_j| = 0.$$

In Chapter 7 we establish generalizations of the Hahn-Schur Theorem to λ multiplier convergent series with values in a topological vector space. Chapters 8, 9 and 10 contain applications of the Hahn-Schur Theorems to spaces of multiplier convergent series, double series and automatic continuity of matrix mappings between sequence spaces.

Chapter 11 extends the notion of multiplier convergent series to series with operator values and multiplier sequences with values in the domains of the operators. Chapters 12 and 13 extend the Orlicz-Pettis Theorem and Hahn-Schur Theorem to operator valued series and vector valued multipliers. Chapter 13 also contains applications to measures with values in a space of continuous linear operators. Chapter 14 considers automatic continuity results for operator valued matrices acting on vector valued sequence spaces.

Contents

Chapter 1

Introduction

One of the most interesting and useful theorems in the early history of functional analysis is a result now known as the Orlicz-Pettis Theorem. The result was originally proven by Orlicz for weakly sequentially complete normed spaces although the result in full generality for normed spaces was known by the Polish mathematicians and appears in Banach's book ([Or], [Ba]). The first version of the theorem available in English was proven by Pettis and was used to treat topics in vector valued measures and vector valued integrals ([Pe]; see [Ka3] and [FL] for discussions of the history of the theorem). If X is a topological vector space (TVS), a series $\sum_j x_j$ in X is subseries convergent in X if the subseries $\sum_{j=1}^{\infty} x_{n_j}$ converges in X for every subsequence $\{n_j\}$. The Orlicz-Pettis Theorem for normed spaces states that if the series $\sum_j x_j$ is subseries convergent in the weak topology of the space, then the series is actually subseries convergent in the norm topology of the space ([Or], [Pe]). The theorem was extended to locally convex spaces by McArthur ([Mc]). If σ is any subset of $\mathbb{N}$ and χ_σ is the characteristic function of σ, then a series $\sum_j x_j$ in a TVS X is subseries convergent iff the series $\sum_{j=1}^{\infty} \chi_\sigma(j) x_j = \sum_{j \in \sigma} x_j$ converges in X for every $\sigma \subset \mathbb{N}$. Thus, if $m_0 = span\{\chi_\sigma : \sigma \subset \mathbb{N}\}$, the sequence space of real valued sequences with finite range, a series $\sum_j x_j$ in a TVS X is subseries convergent iff the series $\sum_{j=1}^{\infty} t_j x_j$ converges for every $t = \{t_j\} \in m_0$. To obtain a generalization of the notion of subseries convergence, we may replace the space m_0 by a general vector space λ of real valued sequences. If λ is a vector space of real valued sequences and $\{x_j\}$ is a sequence in the TVS X, the (formal) series $\sum_j x_j$ is said to be λ multiplier convergent if the series $\sum_{j=1}^{\infty} t_j x_j$ converges in X for every $t = \{t_j\} \in \lambda$; the elements $t \in \lambda$ are called *multipliers*. This suggests that generalizations of the Orlicz-

Pettis Theorem might be obtained by replacing subseries convergent series by λ multiplier convergent series for certain sequence spaces λ. We show that such generalizations are possible in Chapters 4, 5, and 6.

Another classical result which involves subseries convergent series is a result which is often referred to as the Hahn-Schur Theorem. One version of the Hahn-Schur Theorem states that if $\sum_j x_{ij}$ is subseries convergent for every $i \in \mathbb{N}$, $\lim_i \sum_{j=1}^{\infty} x_{in_j}$ exists for every subsequence $\{n_j\}$ and if $x_j = \lim_i x_{ij}$, then the series $\sum_j x_j$ is subseries convergent and $\lim_i \sum_{j \in \sigma} x_{ij} = \sum_{j \in \sigma} x_j$ uniformly for $\sigma \subset \mathbb{N}$ ([Ha], [Sc], [Sw1]; this version of the theorem actually holds for series with values in an Abelian topological group). A series $\sum_j x_j$ in a TVS is said to be bounded multiplier convergent if the series $\sum_j x_j$ is l^{∞} multiplier convergent ([Day]). There is a version of the Hahn-Schur Theorem for bounded multiplier convergent series which states that if $\sum_j x_{ij}$ is bounded multiplier convergent for every $i \in \mathbb{N}$, $\lim_i \sum_{j=1}^{\infty} t_j x_{ij}$ exists for every $t = \{t_j\} \in l^{\infty}$ and if $x_j = \lim_i x_{ij}$, then the series $\sum_j x_j$ is bounded multiplier convergent and $lim_i \sum_{j=1}^{\infty} t_j x_{ij} = \sum_{j=1}^{\infty} t_j x_j$ uniformly for $t \in l^{\infty}, \|t\|_{\infty} \leq 1$ ([Sw2]). Again this suggest that one might obtain generalizations of both versions of the Hahn-Schur Theorem by replacing subseries and bounded multiplier convergent series by λ multiplier convergent series for certain sequence spaces λ. We show in Chapter 7 that versions of the Hahn-Schur Theorem are obtainable for λ multiplier convergent series if the sequence space λ satisfies sufficient conditions.

There are further applications of λ multiplier convergent series to topics in Banach space theory, sequence spaces and matrix mappings. For example, a result of Bessaga and Pelczynski states that a Banach space X contains no subspace isomorphic to c_0 iff every c_0 multiplier convergent series in X is subseries convergent (or bounded multiplier convergent) ([BP]). A generalization of this result to sequentially complete locally convex topological vector spaces (LCTVS) is given in Chapter 3.15. A characterization of dual spaces not containing c_0 is given in terms of subseries convergent series in 3.20, a characterization of locally complete LCTVS in terms of c_0 multiplier convergent series is given in 3.10 and a characterization of Banach-Mackey spaces in terms of l^1 multiplier convergent series is given in 3.23. In Chapter 3, we also give applications of multiplier convergent series to vector valued measures. We give a characterization of bounded vector measures in terms of c_0 multiplier convergent series in 3.33 and a characterization of strongly bounded (strongly additive) vector measures in terms of subseries convergence in 3.43.

Further applications to various topics in sequence spaces and matrix mappings between sequence spaces are given in later chapters.

Multiplier convergent series are interesting in their own right and we develop their basic properties in Chapter 2.

In the last four chapters we consider operator valued series and vector valued spaces of multipliers. Let X, Y be TVS, $L(X, Y)$ the space of all continuous linear operators from X into Y and E be a vector space of X valued sequences. A series $\sum_j T_j$ in $L(X, Y,)$ is E multiplier convergent if the series $\sum_{j=1}^{\infty} T_j x_j$ converges in Y for every sequence $\{x_j\} \in E$. The basic properties of operator valued series with vector valued multipliers sometimes closely parallels the properties of series with scalar multipliers but sometimes require additional assumptions. We present these properties in Chapter 11. Versions of the Orlicz-Pettis Theorem and the Hahn-Schur Theorem for operator valued series and vector valued multipliers are presented in Chapters 12 and 13.

The basic notations, definitions and terminology are presented in Appendices A, B and C. Appendices D and E contain material not easily accessible and which is used in the text.

Chapter 2

Basic Properties of Multiplier Convergent Series

In this chapter we will develop the basic properties of multiplier convergent series.

In what follows λ will denote a scalar sequence space which contains the subspace c_{00} of sequences which are eventually 0 and $\Lambda \subset \lambda$ will denote a subset while X will denote a Hausdorff topological vector space (TVS).

Definition 2.1. A (formal) series $\sum_j x_j$ in X is Λ multiplier convergent in X if the series $\sum_{j=1}^{\infty} t_j x_j$ converges in X for every $t = \{t_j\} \in \Lambda$. The series is Λ multiplier Cauchy in X if the series $\sum_{j=1}^{\infty} t_j x_j$ satisfies a Cauchy condition for every $t = \{t_j\} \in \Lambda$. The elements $t = \{t_j\} \in \Lambda$ are called multipliers.

A series $\sum_j x_j$ which is m_0 multiplier convergent is said to be *subseries convergent*; thus, a series is subseries convergent iff the subseries $\sum_j x_{n_j}$ is convergent for every subsequence $\{n_j\}$. A series $\sum_j x_j$ which is l^{∞} multiplier convergent is said to be *bounded multiplier convergent*.

We now establish the basic properties of Λ multiplier convergent series. We begin by considering boundedness properties of Λ multiplier convergent series.

Let $\sum_j x_j$ be a λ multiplier convergent series. The *summing operator* S (with respect to λ and $\sum_j x_j$) is the linear map $S : \lambda \to X$ defined by $St = S(\{t_j\}) = \sum_{j=1}^{\infty} t_j x_j$ for $t = \{t_j\} \in \lambda$. Recall that the β-dual of λ, λ^{β}, is defined to be

$$\lambda^{\beta} = \left\{ \{s_j\} : \sum_{j=1}^{\infty} s_j t_j = s \cdot t \text{ converges for every } t \in \lambda \right\}$$

and λ and λ^{β} form a dual pair under the bilinear pairing $s \cdot t$ (Appendix A).

If Y, Y' are a dual pair, a Hellinger-Toeplitz topology defined for dual pairs is a topology $w(Y, Y')$ with the property that whenever a linear map $T : Y \to Z$ is $\sigma(Y, Y') - \sigma(Z, Z')$ continuous, then T is $w(Y, Y') - w(Z, Z')$ continuous (Appendix A.1).

Theorem 2.2. *Let X be a Hausdorff locally convex TVS (LCTVS) and $\sum_j x_j$ a λ multiplier convergent series in X. The summing operator $S : \lambda \to X$ is $\sigma(\lambda, \lambda^\beta) - \sigma(X, X')$ continuous and, therefore, $w(\lambda, \lambda^\beta) - w(X, X')$ continuous with respect to any Hellinger-Toeplitz topology w.*

Proof: Let $x' \in X'$, $t \in \lambda$. Then

$$\langle x', St \rangle = \sum_{j=1}^{\infty} t_j \langle x', x_j \rangle = \{ \langle x', x_j \rangle \} \cdot t$$

since $\{ \langle x', x_j \rangle \} \in \lambda^\beta$ by the convergence of the series. This implies that S is $\sigma(\lambda, \lambda^\beta) - \sigma(X, X')$ continuous. The last statement follows from the definition of Hellinger-Toeplitz topologies.

Theorem 2.2 gives a boundedness result for multiplier convergent series

Corollary 2.3. *If B is $\sigma(\lambda, \lambda^\beta)$ bounded, then $SB = \{ \sum_{j=1}^{\infty} t_j x_j : t \in B \}$ is bounded in X.*

For topological sequence spaces we also have a boundedness result for the sums of multiplier convergent series.

Corollary 2.4. *Let λ be a K-space. If $\lambda^\beta \subset \lambda'$ and $B \subset \lambda$ is bounded, then $SB = \{ \sum_{j=1}^{\infty} t_j x_j : t \in B \}$ is bounded in X.*

Proof: Since B is bounded in X and $\lambda^\beta \subset \lambda'$, B is $\sigma(\lambda, \lambda^\beta)$ bounded so the result follows from Corollary 2.3.

For a general condition which guarantees that $\lambda^\beta \subset \lambda'$, we have

Proposition 2.5.

(i) If λ is a barrelled K-space, then $\lambda^\beta \subset \lambda'$.
(ii) If λ is an AK-space, then $\lambda' \subset \lambda^\beta$.
(iii) If λ is a barrelled AK-space, then $\lambda' = \lambda^\beta$.

Proof: (i): Let $s \in \lambda^{\beta}$. For each n define $f_n : \lambda \to \mathbb{R}$ by $f_n(t) = \sum_{j=1}^{n} s_j t_j$. Since λ is a K-space, each f_n is continuous. Now $f_n(t) \to \sum_{j=1}^{\infty} s_j t_j = s \cdot t$ so the linear functional $t \to s \cdot t$ is continuous since λ is barrelled. Therefore, $s \in \lambda'$.

(ii): Let $f \in \lambda'$. Set $s_j = \langle f, e^j \rangle$. If $t \in \lambda$, then $t = \sum_{j=1}^{\infty} t_j e^j$ (convergence in λ) so $\langle f, t \rangle = \sum_{j=1}^{\infty} t_j \langle f, e^j \rangle = \sum_{j=1}^{\infty} t_j s_j = s \cdot t$ so $f = s \in \lambda^{\beta}$.

(iii) follows from (i) and (ii).

Proposition 2.5 is applicable, in particular, if λ is a Banach or Frechet space.

Concerning the strong boundedness of partial sums of multiplier convergent series, we have

Corollary 2.6. *Suppose that λ is a barrelled AB space (Appendix B.3) and $\sum_j x_j$ is λ multiplier convergent. If $B \subset \lambda$ is bounded, then $\{\sum_{j=1}^{n} t_j x_j : n \in \mathbb{N}, t \in B\}$ is $\beta(X, X')$ bounded.*

Proof: Let $P_n : \lambda \to \lambda$ be the sectional operator $P_n(t) = \sum_{j=1}^{n} t_j e^j$. By the AB assumption $\{P_n : n\}$ is pointwise bounded on λ and, therefore, $\{P_n : n\}$ is equicontinuous by the barrelledness assumption. Since λ is barrelled, λ has the strong topology $\beta(\lambda, \lambda')$ so $\{P_n B : n\}$ is $\beta(\lambda, \lambda')$ bounded. By Proposition 2.5, $\lambda^{\beta} \subset \lambda'$ so $\{P_n B : n\}$ is $\beta(\lambda, \lambda^{\beta})$ bounded. The result now follows from Theorem 2.2 since the strong topology is a Hellinger-Toeplitz topology.

In particular, Corollary 2.6 is applicable to bounded multiplier convergent series.

Corollary 2.7. *Let $\sum_j x_j$ be bounded multiplier convergent (l^{∞} multiplier convergent). Then $\{\sum_{j=1}^{\infty} t_j x_j : \|\{t_j\}\|_{\infty} \leq 1\}$ is $\beta(X, X')$ bounded.*

Since m_0 is barrelled (Theorem 7.59 or [Sw1] 4.7.9), we also have

Corollary 2.8. *Let $\sum_j x_j$ be subseries convergent (m_0 multiplier convergent). Then $\{\sum_{j \in \sigma} x_j : \sigma \subset \mathbb{N}\}$ is $\beta(X, X')$ bounded.*

The condition that $\lambda^{\beta} \subset \lambda'$ in Corollaries 2.6, 2.7 and 2.8 is important.

Example 2.9. Let $\lambda = c_{00}$ with the sup-norm. For any $\{x_j\} \subset X$ the series $\sum_j x_j$ is c_{00} multiplier convergent. Take any unbounded sequence

$\{x_j\}$ in X. Then $\{e^k : k \in \mathbb{N}\}$ is bounded in c_{00}, but $\{\sum_{j=1}^k e_j^k x_j = x_k : k \in \mathbb{N}\}$ is not bounded in X. Note that $c'_{00} = l^1 \subset c_{00}^\beta = s$.

We have another result for boundedness in topological sequence spaces. This result requires a gliding hump assumption. An interval in $\mathbb{N}$ is a set of the form $I = \{k \in \mathbb{N} : m \le k \le n\}$, where $m \le n$. A sequence of intervals $\{I_j\}$ is increasing if $\max I_k < \min I_{k+1}$ for all k. If $x = \{x_k\}$ is any sequence (scalar or vector) and $\sigma \subset \mathbb{N}$, $\chi_\sigma x$ will denote the coordinatewise product of χ_σ and x.

Definition 2.10. Let λ be a K-space. Then λ has the zero gliding hump property (0-GHP) if whenever $\{I_j\}$ is an increasing sequence of intervals and $\{t^j\} \subset \lambda$ converges to 0 in λ, there is a subsequence $\{n_j\}$ such that the coordinatewise sum of the series $\sum_{j=1}^\infty \chi_{I_{n_j}} t^{n_j}$ belongs to λ.

For examples of spaces with 0-GHP, see Appendix B.

Theorem 2.11. *Let λ be a K-space with 0-GHP. If $\sum_j x_j$ is λ multiplier convergent in X, then the summing operator $S : \lambda \to X$ is sequentially continuous and, therefore, bounded.*

Proof: Suppose the conclusion fails. Then there exist a closed neighborhood of 0, U, in X and a null sequence $\{t^i\}$ in λ such that $\sum_{j=1}^\infty t_j^i x_j \notin U$ for every i. Set $m_1 = 1$. There exists $n_1 > m_1$ such that $\sum_{j=1}^{n_1} t_j^{m_1} x_j \notin U$. Pick a closed symmetric neighborhood of 0, V, such that $V + V \subset U$. Since $\lim_i t_j^i = 0$ for every j, there exists $m_2 > m_1$ such that $\sum_{j=1}^{n_1} t_j^{m_2} x_j \in V$. There exists $n_2 > n_1$ such that $\sum_{j=1}^{n_2} t_j^{m_2} x_j \notin U$. Hence, $\sum_{j=n_1+1}^{n_2} t_j^{m_2} x_j \notin V$. Continuing this construction produces increasing sequences $\{m_j\}, \{n_j\}$ such that

$$(*) \quad \sum_{j \in I_k} t_j^{m_k} x_j \notin V, \text{ where } I_k = [n_k + 1, n_{k+1}].$$

By 0-GHP, there is a subsequence $\{p_k\}$ such that $t = \sum_{k=1}^\infty \chi_{I_{p_k}} t^{p_k} \in \lambda$. Hence, $\sum_{j=1}^\infty t_j x_j = \sum_{k=1}^\infty \sum_{j \in I_{p_k}} t_j^{m_{p_k}} x_j$ should converge. But, this contradicts $(*)$.

Corollary 2.12. *Let the assumptions be as in Theorem 2.11. If $B \subset \lambda$ is bounded, the $SB \subset X$ is bounded.*

Remark 2.13. Note that $\lambda = c_{00}$ does not have 0-GHP so this assumption in Theorem 2.11 cannot be dropped. See Example 2.9.

We next consider uniform convergence for multiplier convergent series when the multipliers range over certain bounded subsets of the multiplier space. These results also require gliding hump assumptions. In the following definition a sign is a variable which assumes the values $\{\pm 1\}$.

Definition 2.14. Let λ be a K-space and $\Lambda \subset \lambda$. The set Λ has the signed strong gliding hump property (signed-SGHP) if for every bounded sequence $\{t^k\}$ in Λ and every increasing sequence of intervals $\{I_k\}$, there is a subsequence $\{n_k\}$ and a sequence of signs $\{s_k\}$ such that the coordinatewise sum $\sum_{k=1}^{\infty} s_k \chi_{I_{n_k}} t^{n_k}$ belongs to Λ. The set has the strong gliding hump property (SGHP) if the signs in the definition above can all be chosen to be equal to 1.

See Appendix B for examples. For example, l^{∞} has SGHP and bs has signed-SGHP but not SGHP. $\Lambda = \{\chi_{\sigma} : \sigma \subset \mathbb{N}\} = M_0$ has SGHP whereas $m_0 = spanM_0$ does not.

We first establish a lemma.

Lemma 2.15. Let $\sum_j x_j$ be Λ multiplier convergent where $\Lambda \subset \lambda$. If the series $\sum_{j=1}^{\infty} t_j x_j$ do not converge uniformly for $t \in B \subset \Lambda$, then there exist a symmetric neighborhood of $0, V$, in X, $t^k \in B$ and an increasing sequence of intervals $\{I_k\}$ such that $\sum_{j \in I_k} t_j^k x_j \notin V$.

Proof: If the series $\sum_{j=1}^{\infty} t_j x_j$ do not converge uniformly for $t \in B$, there exists a symmetric neighborhood, U, of 0 such that for every k there exist $t^k \in B, m_k \geq k$ such that $\sum_{j=m_k}^{\infty} t_j^k x_j \notin U$. For $k = 1$, let m_1 and $t^1 \in B$ satisfy this condition so $\sum_{j=m_1}^{\infty} t_j^1 x_j \notin U$. Pick a symmetric neighborhood of 0, V, such that $V + V \subset U$. There exists $n_1 > m_1$ such that $\sum_{j=n_1+1}^{\infty} t_j^1 x_j \in V$. Then

$$\sum_{j=m_1}^{n_1} t_j^1 x_j = \sum_{j=m_1}^{\infty} t_j^1 x_j - \sum_{j=n_1+1}^{\infty} t_j^1 x_j \notin V.$$

Put $I_1 = [m_1, n_1]$ and continue the construction.

Theorem 2.16. Let λ be a K-space and $\Lambda \subset \lambda$ have signed-SGHP. If the series $\sum_j x_j$ is Λ multiplier convergent, then the series $\sum_{j=1}^{\infty} t_j x_j$ converge uniformly for t belonging to bounded subsets of Λ.

Proof: Suppose that $B \subset \Lambda$ is bounded but the series $\sum_{j=1}^{\infty} t_j x_j$ do not converge uniformly for $t \in B$. Let the notation be as in Lemma 2.15. Let n_k and s_k be as in the definition of signed-SGHP above and let $t = \sum_{k=1}^{\infty} s_k \chi_{I_{n_k}} t^{n_k} \in \Lambda$. Then $\sum_{j=1}^{\infty} t_j x_j$ does not converge since $\sum_{j \in I_{n_k}} t_j x_j = s_k \sum_{j \in I_{n_k}} t_j^{n_k} x_j \notin V$, i.e., $\sum_{j=1}^{\infty} t_j x_j$ does not satisfy the Cauchy condition.

Theorem 2.16 implies two well known results for bounded multiplier and subseries convergent series which we now state.

Corollary 2.17. *Let $\sum_j x_j$ be l^{∞} multiplier convergent. Then the series $\sum_{j=1}^{\infty} t_j x_j$ converge uniformly for $\|\{t_j\}\|_{\infty} \leq 1$.*

Proof: l^{∞} has SGHP.

Corollary 2.18. *Let $\sum_j x_j$ be subseries convergent. Then the series $\sum_{j=1}^{\infty} t_j x_j$ converge uniformly for $t \in M_0 = \{\chi_{\sigma} : \sigma \subset \mathbb{N}\}$.*

Proof: M_0 has SGHP and is a bounded subset of m_0.

A sequence space λ is *normal (solid)* if $t \in \lambda$ and $|s_j| \leq |t_j|$ for all j implies that $s = \{s_j\} \in \lambda$. For example, c_0 and $l^p, 0 < p \leq \infty$, are normal whereas c and m_0 are not normal. From Corollary 2.17, we have

Corollary 2.19. *Let λ be a normal K-space with signed-SGHP and with the property that $\{s \in \lambda : |s| \leq |t|\}$ is bounded for every $t \in \lambda$. If $\sum_j x_j$ is λ multiplier convergent and $t \in \lambda$, then the series $\sum_{j=1}^{\infty} s_j x_j$ converge uniformly for $|s_j| \leq |t_j|$.*

Without some assumptions on the multiplier space, the conclusion of Theorem 2.16 may fail even when the multiplier space satisfies WGHP or 0-GHP (see Appendix B).

Example 2.20. The series $\sum_j e^j$ is l^p multiplier convergent in $(l^p, \|\cdot\|_p)$ for any $1 \leq p < \infty$, but the series $\sum_{j=1}^{\infty} t_j e^j$ do not converge uniformly for $\|\{t_j\}\|_p \leq 1$ [take $t^k = e^k$ so $\sum_{j=1}^{\infty} t_j^k e^j = e^k$]. Note that l^p has both WGHP and 0-GHP (Appendix B).

The SGHP assumption in Theorem 2.16 is only a sufficient condition for the uniform convergence conclusion.

Example 2.21. Let $x = \{x_j\} \in l^2$. Then the series $\sum_j x_j e^j$ is l^2 multiplier convergent in l^1. If $B \subset l^2$ is bounded, $M = \sup\{\|t\|_2 : t \in B\}$ and $t \in B$, then

$$\left\| \sum_{j=n}^{\infty} t_j x_j e^j \right\|_1^2 = \left(\sum_{j=n}^{\infty} |t_j x_j| \right)^2 \leq \sum_{j=n}^{\infty} |t_j|^2 \sum_{j=n}^{\infty} |x_j|^2 \leq M \sum_{j=n}^{\infty} |x_j|^2 \to 0$$

so the conclusion of Theorem 2.16 holds but l^2 does not have SGHP.

Another uniform convergence result holds for multiplier spaces satisfying 0-GHP.

Theorem 2.22. *Let λ be a K-space with 0-GHP and let $\sum_j x_j$ be λ multiplier convergent. If $t^i \to 0$ in λ, then the series $\sum_{j=1}^{\infty} t_j^i x_j$ converge uniformly for $i \in \mathbb{N}$.*

Proof: Suppose the conclusion fails to hold. Then there exists a closed neighborhood, U, of 0 such that for every k there exist $p_k, m_k > k$ such that $\sum_{j=m_k}^{\infty} t_j^{p_k} x_j \notin U$. For $k = 1$, let $p_1, m_1 > 1$ satisfy this condition so

$$\sum_{j=m_1}^{\infty} t_j^{p_1} x_j \notin U.$$

Pick a symmetric neighborhood of $0, V$, such that $V + V \subset U$. There exists $n_1 > m_1$ such that $\sum_{j=n_1+1}^{\infty} t_j^{p_1} x_j \in V$. Then

$$\sum_{j=m_1}^{n_1} t_j^{p_1} x_j = \sum_{j=m_1}^{\infty} t_j^{p_1} x_j - \sum_{j=n_1+1}^{\infty} t_j^{p_1} x_j \notin V.$$

There exists N_1 such that $\sum_{j=m}^{n} t_j^i x_j \in V$ for $1 \leq i \leq p_1, n \geq m \geq N_1$. Let $p_2, m_2 > N_1, n_2$ satisfy the conditions above for N_1. Note that we must have $p_2 > p_1$. Continuing this construction produces increasing sequences $\{p_j\}, \{m_j\}, \{n_j\}$ such that

$$(\#) \quad \sum_{j=m_k}^{n_k} t_j^{p_k} x_j \notin V.$$

Set $I_k = [m_k, n_k]$. By 0-GHP, since $\{t^{p_k}\} \to 0$, there is a subsequence $\{q_k\}$ of $\{p_k\}$ such that $t = \sum_{k=1}^{\infty} \chi_{I_{q_k}} t^{q_k} \in \lambda$. Since $\sum_{j=1}^{\infty} t_j x_j$ converges, we should have that $\sum_{j \in I_{q_k}} t_j^{q_k} x_j \to 0$ contradicting the condition $(\#)$.

Without the 0-GHP assumption the conclusion of Theorem 2.22 may fail.

Example 2.23. Let $\lambda = c_{00} = X$. Then $\sum_{j=1}^{\infty} j e^j$ is c_{00} multiplier convergent in c_{00}. Now $t^i = e^i/i \to 0$ but the series $\sum_{j=1}^{\infty} t_j^i j e^j$ do not converge uniformly.

We can obtain another uniform convergence result from the continuity of the summing operator.

Theorem 2.24. *Let λ be a complete, metrizable AK-space and let X be a Mackey space (i.e., X has the Mackey topology). If $\sum_j x_j$ is λ multiplier convergent in X and $K \subset \lambda$ is compact, then the series $\sum_{j=1}^{\infty} t_j x_j$ converge uniformly for $t \in K$.*

Proof: Since λ is metrizable, λ carries the Mackey toplology ([Sw2] 18.8) so the summing operator $S : \lambda \to X$ is continuous (Theorem 2.2). Since K is compact, $\lim_n \sum_{j=n}^{\infty} t_j e^j = 0$ uniformly for $t \in K$ ([Sw2] 10.15). Thus, $\lim_n \sum_{j=n}^{\infty} t_j S e^j = \lim_n \sum_{j=n}^{\infty} t_j x_j = 0$ uniformly for $t \in K$.

We next consider uniform convergence for families of multiplier convergent series. The β-dual of $\Lambda \subset \lambda$ with respect to X is defined to be

$$\Lambda^{\beta X} = \{\{x_j\} : \sum_j x_j \text{ is } \Lambda \text{ multiplier convergent}\}.$$

If $t \in \lambda$ and $x \in \lambda^{\beta X}$, we write $t \cdot x = \sum_{j=1}^{\infty} t_j x_j$ [see Appendix B]. We define the topology $\omega(\lambda^{\beta X}, \lambda)$ on $\lambda^{\beta X}$ to be the weakest topology on $\lambda^{\beta X}$ such that the mappings

$$x = \{x_j\} \to \sum_{j=1}^{\infty} t_j x_j = t \cdot x$$

from $\lambda^{\beta X}$ into X are continuous for every $t \in \lambda$. Thus, if X is the scalar field, then $\lambda^{\beta X} = \lambda^{\beta}$ and $\omega(\lambda^{\beta X}, \lambda) = \sigma(\lambda^{\beta}, \lambda)$.

We now give a definition for another gliding hump property which will be used.

Definition 2.25. Let $\Lambda \subset \lambda$. The space Λ has the signed weak gliding hump property (signed-WGHP) if whenever $t \in \Lambda$ and $\{I_j\}$ is an increasing sequence of intervals, there is a subsequence $\{n_j\}$ and a sequence of signs $\{s_j\}$ such that the coordinatewise sum of the series $\sum_{j=1}^{\infty} s_j \chi_{I_{n_j}} t \in \Lambda$. The space Λ has the weak gliding hump property (WGHP) if the signs above can all be chosen equal to 1 for every $t \in \lambda$.

Examples of spaces with signed-WGHP and WGHP are given in Appendix B. For example, any monotone space such as $c_{00}, c_0, l^p \ (0 < p \leq \infty)$, and m_0 has WGHP while the space bs has signed-WGHP but not WGHP.

Theorem 2.26. *Let $\Lambda \subset \lambda$ have signed-WGHP. If $\{x^k\} \subset \Lambda^{\beta X}$ is such that $\lim_k t \cdot x^k$ exists for every $t \in \Lambda$ and $\lim_k x_j^k$ exists for every j, then for every $t \in \Lambda$ the series $\sum_{j=1}^{\infty} t_j x_j^k$ converge uniformly for $k \in \mathbb{N}$.*

Proof: If the conclusion fails, then

> (∗) *there exists a neighborhood of $0, U$, in X such that for every n there exist $k_n, n_n > m_n > n$ such that $\sum_{j=m_n}^{n_n} t_j x_j^{k_n} \notin U$.*

By (∗) for $n = 1$, there exist $k_1, n_1 > m_1$ such that $\sum_{j=m_1}^{n_1} t_j x_j^{k_1} \notin U$. There exists $m' > n_1$ such that $\sum_{j=m}^{n} t_j x_j^k \in U$ for $n > m > m', 1 \le k \le k_1$. By (∗) there exist $k_2, n_2 > m_2 > m'$ such that $\sum_{j=m_2}^{n_2} t_j x_j^{k_2} \notin U$. Hence, $k_2 > k_1$. Continuing this construction produces increasing sequences $\{k_i\}, \{m_i\}, \{n_i\}$ with $m_i < n_i < m_{i+1}$ and

> (#) $\quad x^{k_i} \cdot \chi_{I_i} t \notin U$, where $I_i = [m_i, n_i]$.

Define the matrix M by

$$M = [m_{ij}] = [x^{k_i} \cdot \chi_{I_j} t].$$

We show that M is a signed $\mathcal{K}$-matrix (Appendix D.3). First, the columns of M converge by hypothesis. Second, given any increasing sequence of integers, there is a subsequence $\{p_k\}$ and a sequence of signs $\{s_k\}$ such that $z = \{z_j\} = \sum_{j=1}^{\infty} s_j \chi_{I_{p_j}} t \in \Lambda$. Then

$$\sum_{j=1}^{\infty} s_j m_{ip_j} = \sum_{j=1}^{\infty} s_j x^{k_i} \cdot \chi_{I_{p_j}} t = x^{k_i} \cdot z$$

and $\lim x^{k_i} \cdot z$ exists. Hence, M is a signed $\mathcal{K}$-matrix. By the signed version of the Antosik-Mikusinski Matrix Theorem (Appendix D.3), the diagonal of M converges to 0. But, this contradicts (#).

From Theorem 2.26, we can obtain an important weak sequential completeness result due to Stuart ([St1], [St2], [Sw1]). First, we establish a lemma.

Lemma 2.27. *Let $\Lambda \subset \lambda$. If $\{x^k\} \subset \Lambda^{\beta X}$ is such that $\lim_k t \cdot x^k$ exists for every $t \in \Lambda$, $\lim_k x_j^k = x_j$ exists for each j and for each $t \in \Lambda$ the series $\sum_{j=1}^{\infty} t_j x_j^k$ converge uniformly for $k \in \mathbb{N}$, there exists $x \in \Lambda^{\beta X}$ such that $t \cdot x^k \to t \cdot x$ for every $t \in \Lambda$.*

Proof: Set $x = \{x_j\}$. We claim that $x \in \Lambda^{\beta X}$ and $t \cdot x^k \to t \cdot x$ for every $t \in \Lambda$. Put $u = \lim t \cdot x^k$. It suffices to show that $u = \sum_{j=1}^{\infty} t_j x_j$. Let U be a balanced neighborhood of 0 in X and pick a balanced neighborhood V such that $V + V + V \subset U$. There exists p such that $\sum_{j=n}^{\infty} t_j x_j^k \in V$ for $n \geq p, k \in \mathbb{N}$. Fix $n \geq p$. Pick $k = k_n$ such that $\sum_{j=1}^{\infty} t_j x_j^k - u \in V$ and $\sum_{j=1}^{n} t_j (x_j^k - x_j) \in V$. Then

$$\sum_{j=1}^{n} t_j x_j - u = (\sum_{j=1}^{\infty} t_j x_j^k - u) + \sum_{j=1}^{n} t_j (x_j - x_j^k) - \sum_{j=n+1}^{\infty} t_j x_j^k \in V + V + V \subset U$$

and the result follows.

Stuart's result now follows from Lemma 2.27 and Theorem 2.26.

Corollary 2.28. *(Stuart) Let λ have signed-WGHP and let X be sequentially complete. Then $(\lambda^{\beta X}, \omega(\lambda^{\beta X}, \lambda))$ is sequentially complete.*

Proof: If $\{x^k\}$ is $\omega(\lambda^{\beta X}, \lambda)$ Cauchy and X is sequentially complete, then $\lim_k t \cdot x^k$ exists for every $t \in \lambda$ so Theorem 2.26 and Lemma 2.27 apply.

Since any monotone space has WGHP, Corollary 2.28 applies to monotone spaces, in particular to c_0, l^p $(0 < p \leq \infty)$ and m_0. Corollary 2.28 also applies to the space of bounded series bs which has signed-WGHP but not WGHP as originally noted by Stuart (Appendix B).

A subset F of $\lambda^{\beta X}$ is said to be conditionally $\omega(\lambda^{\beta X}, \lambda)$ sequentially compact if every sequence $\{x^k\} \subset F$ has a subsequence which is such that $\lim t \cdot x^k$ exists for every $t \in \lambda$ ([Din]). From Theorem 2.26 we have

Corollary 2.29. *Let λ have signed-WGHP. If $F \subset \lambda^{\beta X}$ is conditionally $\omega(\lambda^{\beta X}, \lambda)$ sequentially compact and $t \in \lambda$, then the series $\sum_{j=1}^{\infty} t_j x_j$ converge uniformly for $x \in F$.*

Without the gliding hump assumptions, the conclusions in Theorem 2.26 and Corollary 2.28 may fail.

Example 2.30. Let $\lambda = c$ so $\lambda^{\beta} = l^1$. Then $\{e^k\}$ is $\omega(l^1, c) = \sigma(l^1, c)$ Cauchy, but if e is the constant sequence $\{1\}$, the series $\sum_{j=1}^{\infty} e_j^k e_j$ do not converge uniformly and the sequence $\{e^k\}$ is not $\sigma(l^1, c)$ convergent.

We next consider another uniform convergence result with another gliding hump assumption.

Definition 2.31. The space λ has the infinite gliding hump property (∞-GHP) if whenever $t \in \lambda$ and $\{I_j\}$ is an increasing sequence of intervals, there exist a subsequence $\{n_j\}$ and $a_{n_j} > 0, a_{n_j} \to \infty$ such that every subsequence of $\{n_j\}$ has a further subsequence $\{p_j\}$ such that the coordinate sum $\sum_{j=1}^{\infty} a_{p_j} \chi_{I_{p_j}} t \in \lambda$.

The term "infinite gliding hump" is used to suggest that the "humps", $\chi_{I_{p_j}} t$, are multiplied by a sequence of scalars which tend to ∞; there are other gliding hump properties where the humps are multiplied by elements of classical sequence spaces (Appendix B).

Examples of spaces with ∞-GHP are given in Appendix B. For example, $\lambda = l^p$ ($0 < p < \infty$) and $\lambda = cs$ have ∞-GHP. The spaces l^{∞}, m_0, bs and bv do not have ∞-GHP.

Theorem 2.32. *Assume that λ has ∞-GHP. If $B \subset \lambda^{\beta X}$ is pointwise bounded on λ, then for every $t \in \lambda$ the series $\sum_{j=1}^{\infty} t_j x_j$ converge uniformly for $x \in B$.*

Proof: If the conclusion fails, there exist $\epsilon > 0$, a continuous semi-norm p on X, $\{x^k\} \subset B$ and subsequences $\{m_k\}, \{n_k\}$ with $m_1 < n_1 < m_2 < \ldots$ and

$$(*) \quad p(\sum_{l=m_k}^{n_k} t_l x_l^k) > \epsilon.$$

Put $I_k = [m_k, n_k]$. By ∞-GHP there exist $\{p_k\}, a_{p_k} > 0, a_{p_k} \to \infty$ such that any subsequence of $\{p_k\}$ has a further subsequence $\{q_k\}$ such that

$$\sum_{k=1}^{\infty} a_{q_k} \chi_{I_{q_k}} t \in \lambda.$$

Define a matrix

$$M = [m_{ij}] \text{ by } m_{ij} = \sum_{l \in I_{p_j}} a_{p_j} t_l x_l^{p_i} / a_{p_i}.$$

We claim that M is a $\mathcal{K}$ matrix (Appendix D.2). First, the columns of M converge to 0 since B is pointwise bounded on λ and $1/a_{p_i} \to 0$. Next, given any subsequence of $\{p_j\}$ there is a further subsequence $\{q_j\}$ such that $u = \sum_{k=1}^{\infty} a_{q_k} \chi_{I_{q_k}} t \in \lambda$. Then

$$\sum_{j=1}^{\infty} m_{iq_j} = (1/a_{p_i}) \sum_{l=1}^{\infty} u_l x_l^{p_i} = (1/a_{p_i}) x^{p_i} \cdot u \to 0.$$

Hence, M is a $\mathcal{K}$ matrix so the diagonal of M converges to 0 by the Antosik-Mikusinski Matrix Theorem (Appendix D.2). But, this contradicts $(*)$.

From Lemma 2.27 and Theorem 2.32 we obtain another weak sequential completeness result.

Corollary 2.33. *Assume that λ has ∞-GHP and X is sequentially complete. Then $(\lambda^{\beta X}, \omega(\lambda^{\beta X}, \lambda))$ is sequentially complete.*

Remark 2.34. The signed-WGHP and ∞-GHP are independent so Corollaries 2.28 and 2.33 cover different spaces. For example, the space bs has signed-WGHP but not ∞-GHP while the space bv_0 has ∞-GHP but not signed-WGHP.

We next consider uniform convergence results when the elements range over both subsets of $\lambda^{\beta X}$ and λ. These results have stronger conclusions but require stronger assumptions. First, we consider an improvement of Theorem 2.26.

Theorem 2.35. *Assume that $\Lambda \subset \lambda$ has signed-SGHP and $\lim_k x_j^k = x_j$ exists for each j. If $\{x^k\} \subset \lambda^{\beta X}$ is such that $\lim t \cdot x^k$ exists for every $t \in \Lambda$ and $B \subset \Lambda$ is bounded, then the series $\sum_{j=1}^{\infty} t_j x_j^k$ converge uniformly for $k \in \mathbb{N}, t \in B$.*

Proof: If the conclusion fails,

$(*)$ there is a neighborhood U of 0 such that for every n there exist $k_n, t^n \in B, n_n > m_n > n$ such that $\sum_{j=m_n}^{n_n} t_j^n x_j^{k_n} \notin U$.

By $(*)$ for $n = 1$ there exist $k_1, t^1 \in B, n_1 > m_1 > 1$ such that $\sum_{j=m_1}^{n_1} t_j^1 x_j^{k_1} \notin U$. By Theorem 2.16 there exists $m' > n_1$ such that $\sum_{j=p}^{q} t_j x_j^k \in U$ for $1 \le k \le k_1, t \in B, q \ge p \ge m'$. By $(*)$ there exist $k_2, t^2 \in B, n_2 > m_2 > m'$ such that $\sum_{j=m_2}^{n_2} t_j^2 x_j^{k_2} \notin U$. Hence, $k_2 > k_1$. Continuing this construction produces increasing sequences $\{k_i\}, \{m_i\}, \{n_i\}, m_{i+1} > n_i > m_i, t^i \in B$ such that

$$(**) \quad \sum_{j=m_i}^{n_i} t_j^i x_j^{k_i} \notin U.$$

Set $I_i = [m_i, n_i]$ and define a matrix

$$M = [m_{ij}] = [x^{k_i} \cdot \chi_{I_j} t^j].$$

We claim that M is a signed $\mathcal{K}$-matrix (Appendix D.3). First, the columns of M converge by hypothesis. Next, for any increasing sequence of positive integers, there is a subsequence $\{p_j\}$ and a sequence of signs $\{s_j\}$ such that $z = \sum_{j=1}^{\infty} s_j \chi_{I p_j} t^{p_j} \in \Lambda$. Then

$$\sum_{j=1}^{\infty} s_j m_{i p_j} = \sum_{j=1}^{\infty} x^{k_i} \cdot s_j \chi_{I p_j} t^{p_j} = x^{k_i} \cdot z$$

and $\lim x^{k_i} \cdot z$ exists. Thus, M is a signed $\mathcal{K}$-matrix so by the signed version of the Antosik-Mikusinski Matrix Theorem the diagonal of M converges to 0 (Appendix D.3). But, this contradicts $(**)$.

Since $M_0 \subset m_0$ has SGHP and l^{∞} has SGHP, we have the following corollaries.

Corollary 2.36. *Let $\sum_j x_{ij}$ be subseries convergent for every $i \in \mathbb{N}$ and suppose that $\lim_i \sum_{j \in \sigma} x_{ij}$ exists for every $\sigma \subset \mathbb{N}$. Then the series $\sum_{j \in \sigma} x_{ij}$ converge uniformly for $i \in \mathbb{N}, \sigma \subset \mathbb{N}$.*

Corollary 2.37. *Let $\sum_j x_{ij}$ be bounded multiplier convergent for every $i \in \mathbb{N}$ and suppose $\lim_i \sum_{j=1}^{\infty} t_j x_{ij}$ exists for every $\{t_j\} \in l^{\infty}$. Then the series $\sum_{j=1}^{\infty} t_j x_{ij}$ converge uniformly for $i \in \mathbb{N}, \|\{t_j\}\|_{\infty} \leq 1$.*

Corollary 2.38. *Assume that λ has signed-SGHP. If $F \subset {}^{\beta X}$ is conditionally $\omega(\lambda^{\beta X}, \lambda)$ sequentially compact and $B \subset \lambda$ is bounded, then the series $\sum_{j=1}^{\infty} t_j x_j$ converge uniformly for $x \in F, t \in B$.*

Example 2.30 shows that the gliding hump property in Theorem 2.35 is important.

We have another uniform convergence result generalizing Theorem 2.22.

Theorem 2.39. *Assume that λ has 0-GHP. If $\{x^k\} \subset \lambda^{\beta X}$ is such that $\lim t \cdot x^k$ exists for every $t \in \lambda$ and $t^k \to 0$ in λ, then the series $\sum_{j=1}^{\infty} t_j^l x_j^k$ converge uniformly for $k, l \in \mathbb{N}$.*

Proof: If the conclusion fails,

$(*)$ there is a neighborhood U of 0 such that for every n there exist $k_n, l_n, n_n > m_n > n$ such that

$$\sum_{j=m_n}^{n_n} t_j^{l_n} x_j^{k_n} \notin U.$$

By $(*)$ for $n = 1$, there exist $k_1, l_1, m_1 < n_1$ such that $\sum_{j=m_1}^{n_1} t_j^{l_1} x_j^{k_1} \notin U$. By Theorem 2.22 there exists $m' > n_1$ such that $\sum_{j=p}^{q} t_j^l x_j^k \in U$ for

$i \in \mathbb{N}, 1 \leq k \leq k_1, q > p > m'$. By $(*)$ there exist $k_2, l_2, n_2 > m_2 > m'$ such that $\sum_{j=m_2}^{n_2} t_j^{l_2} x_j^{k_2} \notin U$. Hence, $k_2 > k_1$ and $l_2 > l_1$. Continuing this construction produces increasing sequences $\{k_i\}, \{l_i\}, \{m_i\}, \{n_i\}$ with $m_{i+1} > n_i > m_i$ and

$$(**) \quad \sum_{j=m_i}^{n_i} t_j^{l_i} x_j^{k_i} \notin U.$$

Set $I_i = [m_i, n_i]$. Define a matrix

$$M = [m_{ij}] = [x^{k_i} \cdot \chi_{I_j} t^{l_j}].$$

We claim that M is a $\mathcal{K}$-matrix (Appendix D.2). First the columns of M converge by hypothesis. Next, given any increasing sequence of positive integers there is a subsequence $\{p_j\}$ such that $z = \sum_{j=1}^{\infty} \chi_{I_{p_j}} t^{p_j} \in \lambda$. Then $\sum_{j=1}^{\infty} m_{ip_j} = x^{k_i} \cdot z$ and $\lim_i x^{k_i} \cdot z$ exists. By the Antosik-Mikusinski Matrix Theorem (Appendix D.2), the diagonal of M converges to 0. But, this contradicts $(**)$.

The 0-GHP hypothesis in Theorem 2.39 is important.

Example 2.40. Let $\lambda = c_{00}$ so $\lambda^{\beta} = s$. Let $x^k = \sum_{j=1}^{k} e^j \in s$ and $t^l = \sum_{j=1}^{l} e^j / l$ so $\{x^k\}$ is $\sigma(s, c_{00})$ Cauchy and $t^l \to 0$ in $(c_{00}, \|\cdot\|_{\infty})$. Then $\sum_{j=N}^{\infty} t_j^l x_j^k = (k-l)/l$ if $k \geq l \geq N$ so the series do not converge uniformly for $k, l \in \mathbb{N}$ although the series do converge uniformly for fixed k or l. This shows that the 0-GHP cannot be dropped in Theorem 2.39.

Corollary 2.41. *Assume that λ has 0-GHP. If $F \subset \lambda^{\beta X}$ is conditionally $\omega(\lambda^{\beta X}, \lambda)$ sequentially compact and $t^l \to 0$ in λ, then the series $\sum_{j=1}^{\infty} t_j^l x_j$ converge uniformly for $l \in \mathbb{N}, x \in F$.*

We next consider compactness in the range of the summing operator associated with a multiplier convergent series. We first establish a basic lemma.

Lemma 2.42. *Let $\Lambda \subset \lambda$. If $\sum_j x_j$ is Λ multiplier convergent and the series $\sum_{j=1}^{\infty} t_j x_j$ converge uniformly for $t \in \Lambda$, then the summing operator $S : \Lambda \to X, St = \sum_{j=1}^{\infty} t_j x_j$, is continuous with respect to the topology p of coordinatewise convergence on Λ and the topology of X.*

Proof: Let $t^{\delta} = \{t_j^{\delta}\}$ be a net in Λ which converges to $t \in \Lambda$ with respect to p. Let U be a neighborhood of 0 in X and pick a symmetric neighborhood V such that $V + V + V \subset U$. There exists n such that $\sum_{j=n}^{\infty} t_j x_j \in V$ for every $t \in \Lambda$. There exists δ such that $\alpha \geq \delta$ implies $\sum_{j<n} (t_j^{\alpha} - t_j) x_j \in V$. Thus, for $\alpha \geq \delta$,

$$S(t^{\alpha}) - S(t) = \sum_{j<n} (t_j^{\alpha} - t_j) x_j + \sum_{j=n}^{\infty} t_j^{\alpha} x_j - \sum_{j=n}^{\infty} t_j x_j \in V + V + V \subset U$$

so S is continuous with respect to p.

From Theorem 2.16 and Lemma 2.42, we have

Corollary 2.43. *Let λ be a K-space and $\Lambda \subset \lambda$ have signed-SGHP and let $\sum_j x_j$ be Λ multiplier convergent. If $\Lambda \subset \lambda$ is bounded, then the summing operator $S : \Lambda \to X$ is continuous with respect to p.*

Without some assumption on the multiplier space, the conclusion of Corollary 2.43 may fail.

Example 2.44. Let $\Lambda = \{ t \in l^p : \|t\|_p \leq 1 \}$ for $1 \leq p < \infty$. The series $\sum e^j$ is l^p multiplier convergent in l^p but the summing operator $S = I : \Lambda \to l^p$ is not continuous with respect to p [$e^j \to 0$ in p but $e^j \not\to 0$ in l^p].

From Corollary 2.43 we have the following compactness result.

Theorem 2.45. *Let λ be a K-space with signed-SGHP and let $\sum_j x_j$ be λ multiplier convergent. If $\Lambda \subset \lambda$ is bounded in λ and compact with respect to p, then $S\Lambda = \{ \sum_{j=1}^{\infty} t_j x_j : t \in \Lambda \}$ is compact in X.*

From Theorem 2.45, we obtain a well known result for bounded multiplier convergent series.

Corollary 2.46. *Let $\sum_j x_j$ be l^{∞} multiplier convergent. Then $\{ \sum_{j=1}^{\infty} t_j x_j : \|t\|_{\infty} \leq 1 \}$ is compact in X.*

Proof: The unit ball of l^{∞} is compact with respect to p so Theorem 2.45 applies.

Similarly, from Theorem 2.45, we obtain a well known result for subseries convergent series.

Corollary 2.47. *Let $\sum_j x_j$ be subseries convergent. Then $\{ \sum_{j \in \sigma} x_j : \sigma \subset \mathbb{N} \}$ is compact in X.*

Proof: The set $\{\sigma : \sigma \subset \mathbb{N}\}$ is compact with respect to p so Theorem 2.45 applies.

It is an interesting result that the "converse" to Corollary 2.47 holds.

Theorem 2.48. *Let X be a TVS and $\{x_j\} \subset X$. If $F = \{ \sum_{j \in \sigma} x_j : \sigma$ finite$\}$ is relatively compact in X, then $\sum_j x_j$ is subseries convergent.*

Proof: Since the closure of F is complete, it suffices to show that the series $\sum_j x_j$ is Cauchy. If this fails to hold, there exist a closed neighborhood of 0, U, and an increasing sequence of intervals $\{I_j\}$ such that $z_k = \sum_{j \in I_k} x_j \notin U$. Pick a symmetric neighborhood of 0, V, such that $V + V \subset U$. Since F is bounded, there exists k such that $F \subset kV$. Pick a symmetric neighborhood of 0, W, such that $W + ... + W(k$ terms$) \subset V$. F is relatively compact so there exist $z_1 + W, ..., z_n + W$ covering $Z = \{z_k : k \in \mathbb{N}\}$. At least one set $z_1 + W, ..., z_n + W$ contains infinitely many elements of Z, say, $\{z_{n_j}\} \subset z_1 + W$. Then

$$\sum_{j=1}^{k} z_{n_j} \in (z_1 + W) + ... + (z_1 + W)[k \text{ terms}] \subset kz_1 + V$$

so

$$kz_1 \in \sum_{j=1}^{k} z_{n_j} + V \subset F + V \subset kV + V \subset kU$$

which implies that $z_1 \in U$. This is a contradiction. The same argument can be applied to any subseries so the result follows.

Remark 2.49. Robertson has established a number of results connecting convergence of series and compactness of finite partial sums in [Ro1], [Ro2].

From Theorem 2.2 we also obtain the following compactness result.

Theorem 2.50. *Let $\sum_j x_j$ be λ multiplier convergent in the LCTVS X. If $\Lambda \subset \lambda$ is compact in the Hellinger-Toeplitz topology $w(\lambda, \lambda^\beta)$, then $S\Lambda = \{\sum_{j=1}^{\infty} t_j x_j : t \in \Lambda\}$ is $w(X, X')$ compact.*

Corollary 2.51. *If λ is a reflexive Banach space with $\lambda' = \lambda^\beta$ and $\sum_j x_j$ is λ multiplier convergent in the LCTVS X, then $\{\sum_{j=1}^{\infty} t_j x_j : \|t\| \leq 1\}$ is $\sigma(X, X')$ compact.*

Proof: The unit ball in a reflexive Banach space is weakly compact so Theorem 2.50 applies.

For conditions which guarantee that $\lambda' = \lambda^\beta$, see Proposition 2.5. In particular, Corollary 2.51 applies to $\lambda = l^p$ for $1 < p < \infty$.

Remark 2.52. In Chapter 4 we show that certain series which are λ multiplier convergent with respect to the weak topology are actually λ multiplier convergent with respect to stronger topologies such as the Mackey topology. This implies that the conclusions of some of the results above can be strengthened.

We will now consider the relationship between subseries convergent series and bounded multiplier convergent series. It is also useful to discuss several other notions of convergence for series which we do at this point. To discuss the relationship between subseries convergent series and bounded multiplier convergent series, we employ an interesting inequality of Rutherford and McArthur ([MR]).

Lemma 2.53. *Let p be a semi-norm on the vector space V, let $\sigma \subset \mathbb{N}$ be finite, $x_j \in V$, $t_j \in \mathbb{R}$ for $j \in \sigma$. Then*

$$p(\sum_{j \in \sigma} t_j x_j) \le 2 \sup_{j \in \sigma} |t_j| \, sup_{\sigma' \subset \sigma} p(\sum_{j \in \sigma'} x_j).$$

Proof: First assume that all t_j are non-negative with $t_1 \ge t_2 \ge ... \ge t_n \ge 0$. Then

$$p(\sum_{j=1}^{n} t_j x_j) = p(\sum_{j=1}^{n-1} (t_j - t_{j+1})(x_1 + ... + x_j) + t_n(x_1 + ... + x_n))$$

$$\le \sum_{j=1}^{n-1} (t_j - t_{j+1})p(x_1 + ... + x_j) + t_n p(x_1 + ... + x_n)$$

$$\le (\sum_{j=1}^{n-1} (t_j - t_{j+1}) + t_n) \sup_{\sigma' \subset \sigma} p(\sum_{j \in \sigma'} x_j)$$

$$= \sup_{j \in \sigma} |t_j| \, sup_{\sigma' \subset \sigma} p(\sum_{j \in \sigma'} x_j).$$

For the general case, apply the inequality above to the positive and negative scalars.

Theorem 2.54. *Let $\sum_j x_j$ be subseries convergent in the LCTVS X. Then the series $\sum_j x_j$ is l^∞ multiplier Cauchy. If X is sequentially complete, then the series $\sum_j x_j$ is l^∞ multiplier convergent; moreover the series $\sum_{j=1}^\infty t_j x_j$ converge uniformly for $\|t\|_\infty \leq 1$.*

Proof: Let p be a continuous semi-norm on X and let $\epsilon > 0$. From Corollary 2.18 there exists N such that $p(\sum_{j\in\sigma} x_j) \leq \epsilon$ when $\min \sigma \geq N$. The result now follows from the inequality in Lemma 2.53.

Without the sequential completeness assumption, the last statement in Theorem 2.54 may fail to hold.

Example 2.55. Let $X = m_0$ and let $s = \{s_j\} \in l^1$ with $s_j > 0$ for every j. Equip m_0 with the norm $\|t\|_s = \sum_{j=1}^\infty s_j |t_j|$. Then the series $\sum e^j$ is subseries convergent in $(m_0, \|\cdot\|_s)$ but is not bounded multiplier convergent since, for example, the series $\sum_j e^j / j$ does not converge to an element of m_0.

From Theorem 2.54 and Corollary 2.46, we have

Corollary 2.56. *Let X be a sequentially complete LCTVS. If $\sum_j x_j$ is subseries convergent, then $\{\sum_{j=1}^\infty t_j x_j : \|t\|_\infty \leq 1\}$ is compact in X.*

We now define two additional notions of convergence for series in TVS.

Definition 2.57. The series $\sum_j x_j$ is unconditionally convergent (rearrangement convergent) if the series $\sum_{j=1}^\infty x_{\pi(j)}$ converges for every permutation $\pi : \mathbb{N} \to \mathbb{N}$. The series $\sum_j x_j$ is unconditionally Cauchy if the series $\sum_{j=1}^\infty x_{\pi(j)}$ is Cauchy for every permutation.

Definition 2.58. The series $\sum_j x_j$ is unordered convergent if the net $\{\sum_{j\in\sigma} x_j : \sigma \in \mathcal{F}\}$ converges, where $\mathcal{F}$ is the family of all finite subsets of $\mathbb{N}$ ordered by inclusion. We write $\lim_{\mathcal{F}} \sum_{j\in\sigma} x_j$ for the limit of this net when the net converges.

We have the following relationships.

Theorem 2.59. *Let X be a TVS and $\{x_j\} \subset X$. Consider the following conditions:*
(i) the series $\sum_j x_j$ is unconditionally convergent,
(ii) the series $\sum_j x_j$ is unordered convergent,
(iii) the series $\sum_j x_j$ is subseries Cauchy,

(iv) the series $\sum_j x_j$ is unconditionally Cauchy.
Then $(i) \Rightarrow (ii) \Rightarrow (iii) \Rightarrow (iv)$.

Proof: Assume (i). Let $x = \sum_{j=1}^{\infty} x_j$. Assume the net $\{\sum_{j \in \sigma} x_j : \sigma \in \mathcal{F}\}$ does not converge to x. Then there exists a symmetric neighborhood, U, of 0 in X such that for every $\sigma \in \mathcal{F}$ there exists $\sigma' \in \mathcal{F}$, $\sigma' \supset \sigma$ with $x - \sum_{j \in \sigma'} x_j \notin U$. Pick a symmetric neighborhood of 0, V, such that $V + V \subset U$. There exists N such that $x - \sum_{j=1}^{n} x_j \in V$ for $n \geq N$. Let $d_1 = \{1, ..., N\}$ and let d_1' be as above. Set $d_2 = \{1, ..., \max d_1'\}$ and let d_2' be as above. Continue in this way to obtain a sequence $d_1, d_1', d_2, d_2', ...\,$. Define a permutation π of $\mathbb{N}$ by enumerating the elements of $d_1, d_1' \backslash d_2, d_2 \backslash d_1', d_2' \backslash d_2, ...\,$. The series $\sum_{j=1}^{\infty} x_{\pi(j)}$ is not convergent since

$$\sum_{j \in d_n' \backslash d_n} x_j = (\sum_{j \in d_n'} x_j - x) + (x - \sum_{j \in d_n} x_j) \notin V.$$

Hence, (i) implies (ii).

Assume (ii). Note that since $\lim_{\mathcal{F}} \sum_{j \in \sigma} x_j$ is unique, every rearrangement of $\sum_j x_j$ converges to the same limit, namely, $\lim_{\mathcal{F}} \sum_{j \in \sigma} x_j$. Let U be a symmetric neighborhood of 0 in X. There exists $\sigma_0 \in \mathcal{F}$ such that $\sum_{j \in \sigma} x_j - x \in U$ for every $\sigma \supset \sigma_0$. Let $\sum_j x_{n_j}$ be a subseries of $\sum_j x_j$. Pick $N > \max \sigma_0$. If $k > j \geq N$, then $\sum_{i=j}^{k} x_i \in U$ so $\sum x_{n_j}$ is Cauchy and (iii) holds.

Assume (iii). If (iv) fails, there exist a symmetric neighborhood of 0, U, in X and a permutation π of $\mathbb{N}$ and an increasing sequence $\{m_n\}$ such that $\sum_{i=m_n+1}^{m_{n+1}} x_{\pi(i)} \notin U$. Choose a subsequence $\{m_{n_j}\}$ of $\{m_n\}$ such that

$$\min\{\pi(i) : m_{n_j} + 1 \leq i \leq m_{n_{j+1}+1}\} > \max\{\pi(i) : m_{n_j} \leq i \leq m_{n_j+1}\}.$$

Arrange the integers $\pi(i), m_{n_j} + 1 \leq i \leq m_{n_{j+1}}, j \in \mathbb{N}$ into an increasing sequence $\{i_j\}$. Then $\sum_{j=1}^{\infty} x_{i_j}$ does not satisfy the Cauchy condition and (iii) fails. Hence, (iii) implies (iv).

Corollary 2.60. *Let X be a sequentially complete TVS. Then (i) and (ii) of Theorem 2.59 are equivalent to:*

(iii)' the series $\sum_j x_j$ is subseries convergent.

Finally, for LCTVS there is the notion of absolute convergence.

Definition 2.61. *Let X be an LCTVS. The series $\sum_j x_j$ is absolutely convergent if $\sum_{j=1}^{\infty} p(x_j) < \infty$ for every continuous semi-norm p on X.*

The notion of absolute convergence is a very strong condition. For example, in a Banach space every convergent series is absolutely convergent iff the space is finite dimensional (Dvoretsky-Rogers Theorem ([Sw2] 30.1)); in a Frechet space every convergent series is absolutely convergent iff the space is nuclear ([Sch] 10.7.2)). However, we do have

Proposition 2.62. *If the series $\sum_j x_j$ is absolutely convergent in the LCTVS X, then the series is subseries Cauchy; if X is sequentially complete, then the series is subseries convergent.*

The sequential completeness statement in the last part of Theorem 2.62 is important.

Example 2.63. Let $X = c_{00}$. Set $x_j = e^j/j^2$. Then the series $\sum_j x_j$ is absolutely convergent in X but not convergent.

Chapter 3

Applications of Multiplier Convergent Series

In this chapter we will give several applications of multiplier convergent series to various topics in locally convex spaces and vector valued measures. As before, throughout this chapter λ will denote a sequence space containing c_{00}, the space of sequences which are eventually 0 and X will denote a (Hausdorff) LCTVS. We begin by establishing a generalization of a result of G. Bennett ([Be]).

Proposition 3.1. If $\sum_j x_j$ is λ multiplier Cauchy in X, then $\{\langle x', x_j \rangle\} \in \lambda^\beta$ for every $x' \in X'$.

Proof: Let $t \in \lambda, x' \in X'$. Then $\left\langle x', \sum_{j=1}^\infty t_j x_j \right\rangle = \sum_{j=1}^\infty t_j \langle x', x_j \rangle$ converges so $\{\langle x', x_j \rangle\} \in \lambda^\beta$.

We consider the converse of Proposition 3.1 under additional assumptions.

Theorem 3.2. Let (λ, τ) be a metrizable AK-space such that $\lambda' = \lambda^\beta$. Then $\sum_j x_j$ is λ multiplier Cauchy in $(X, \tau(X, X'))$ iff $\{\langle x', x_j \rangle\} \in \lambda^\beta$ for every $x' \in X'$.

Proof: Suppose that $\{\langle x', x_j \rangle\} \in \lambda^\beta$ for every $x' \in X'$. Define a linear map $T : c_{00} \to X$ by $Tt = \sum_{j=1}^\infty t_j x_j$. If $x' \in X', t \in c_{00}$, then $\langle x', Tt \rangle = \sum_{j=1}^\infty t_j \langle x', x_j \rangle = t \cdot \{\langle x', x_j \rangle\}$ which implies by hypothesis that T is $\sigma(c_{00}, \lambda^\beta) - \sigma(X, X')$ continuous and ,therefore, $\tau(c_{00}, \lambda^\beta) - \tau(X, X')$ continuous. Since $\lambda' = (c_{00}, \tau \mid_{c_{00}})'$ and $\tau \mid_{c_{00}} = \tau(\lambda, \lambda') \mid_{c_{00}}$ (any metrizable space carries the Mackey topology ([Sw2] 18.8)), $\tau \mid_{c_{00}} = \tau(c_{00}, \lambda') = \tau(c_{00}, \lambda^\beta)$. Now, if $s \in \lambda, s = \sum_{j=1}^\infty s_j e^j$, where the convergence is in $\tau = \tau(\lambda, \lambda^\beta)$

25

by the AK assumption, so $\{\sum_{j=1}^{n} s_j e^j\}_n$ is $\tau(c_{00}, \lambda^\beta)$ Cauchy. Therefore, $\{\sum_{j=1}^{n} T(s_j e^j)\}_n = \{\sum_{j=1}^{n} s_j T e^j\}_n = \{\sum_{j=1}^{n} s_j x_j\}_n$ is $\tau(X, X')$ Cauchy. The converse is given in Proposition 3.1.

Corollary 3.3. *If λ satisfies the assumptions of Theorem 3.2 and $(X, \tau(X, X'))$ is sequentially complete, then $\sum_j x_j$ is λ multiplier convergent in $(X, \tau(X, X'))$ iff $\{\langle x', x_j \rangle\} \in \lambda^\beta$ for every $x' \in X'$.*

Corollary 3.4. *Let λ be as in Theorem 3.2 and let X be an FK-space. Then X contains λ iff $c_{00} \subset X$ and $\{\langle x', x_j \rangle\} \in \lambda^\beta$ for every $x' \in X'$.*

Remark 3.5. Bennett's result corresponds to the case where $\lambda = l^p$, $1 \le p < \infty$ or $\lambda = c_0$ in Corollaries 3.3 and 3.4. Note that Corollaries 3.3 and 3.4 also apply to the spaces cs and bv_0. For conditions which guarantee that $\lambda' = \lambda^\beta$, see Proposition 2.5.

Without some additional assumptions on the multiplier space λ, the converse of Proposition 3.1 may fail.

Example 3.6. Let $\lambda = l^\infty$ so $\lambda^\beta = l^1$. Let $X = c_0$ and consider the series $\sum e^j$ in c_0. If $s \in X' = l^1$, then $\{\langle s, e^j \rangle\} = \{s_j\} \in l^1 = \lambda^\beta$, but $\sum e^j$ is not l^∞ multiplier convergent in c_0.

We next consider results which involve series which are c_0 multiplier Cauchy and c_0 multiplier convergent. These series are often described in a different way which we now consider.

Definition 3.7. A series $\sum_j x_j$ in X is said to be weakly unconditionally Cauchy (wuc) if $\sum_{j=1}^{\infty} |\langle x', x_j \rangle| < \infty$ for every $x' \in X'$.

Note that a series $\sum_j x_j$ is wuc iff the series $\sum_j x_j$ is subseries Cauchy in the weak topology $\sigma(X, X')$. A series which is subseries convergent in the weak topology $\sigma(X, X')$ is wuc, but a wuc series may not be subseries convergent in the weak topology (consider the series $\sum_j e^j$ in c_0). We give several characterizations of wuc series.

Proposition 3.8. *Let $\{x_j\} \subset X$. The following are equivalent:*

(i) The series $\sum_j x_j$ is wuc.
(ii) $\{\langle x', x_j \rangle\} \in l^1$ for every $x' \in X'$.
(iii) The series $\sum_j x_j$ is c_0 multiplier Cauchy.
(iv) $\{\sum_{j \in \sigma} x_j : \sigma \text{ finite}\}$ is bounded in X.

(v) *For every continuous semi-norm p on X, there exists $M > 0$ such that $p(\sum_{j\in\sigma} t_j x_j) \leq M \|t\|_\infty$ for every $t \in l^\infty$ and σ finite.*

(vi) *The map $T : c_{00} \to X$, $Tt = \sum_{j=1}^\infty t_j x_j$, is linear and continuous.*

(vii) *The series $\sum_j x_j$ is c_0 multiplier Cauchy in $\sigma(X, X')$.*

Proof: Clearly (i) and (ii) are equivalent, and (ii) and (iii) are equivalent by Bennett's result in Theorem 3.2.

Assume that (i) holds. If $x' \in X'$ and σ is finite, then

$$\left| \left\langle x', \sum_{j\in\sigma} x_j \right\rangle \right| \leq \sum_{j=1}^\infty |\langle x', x_j \rangle| < \infty$$

so $\{\sum_{j\in\sigma} x_j : \sigma \text{ finite}\}$ is $\sigma(X, X')$ bounded and, therefore, bounded in X so (iv) holds.

Assume that (iv) holds. Let p be a continuous semi-norm on X. Set $M = 2\sup\{p(\sum_{j\in\sigma} x_j) : \sigma \text{ finite}\}$. By the McArthur/Rutherford inequality (Lemma 2.53), $p(\sum_{j\in\sigma} t_j x_j) \leq M \|t\|_\infty$ for every $t \in l^\infty$ so (v) holds.

That (v) implies (vi) is immediate.

Suppose that (vi) holds. Then the adjoint operator $T' : X' \to c'_{00} = l^1$ so $T'x' = \{\langle x', x_j \rangle\} \in l^1$. Therefore, $\sum_{j=1}^\infty s_j \langle x', x_j \rangle$ converges for every $s \in c_0$ and $\sum_{j=1}^\infty s_j x_j$ is $\sigma(X, X')$ Cauchy or $\sum_j x_j$ is c_0 multiplier Cauchy in $\sigma(X, X')$. Thus, (vii) holds.

Assume that (vii) holds. Then $\sum_{j=1}^\infty s_j \langle x', x_j \rangle$ converges for every $x' \in X'$ and for every $s \in c_0$. Hence, $\sum_{j=1}^\infty |\langle x', x_j \rangle| < \infty$ for every $x' \in X'$ and (i) holds.

Note that it follows from Proposition 3.8 that a continuous linear operator between LCTVS carries wuc series into wuc series (condition (iv)).

Corollary 3.9. *Let $\sum_j x_j$ be c_0 multiplier convergent in X. Then*

(i) *$\sum_j x_j$ is wuc,*

(ii) *for every continuous semi-norm p on X there exists $M > 0$ such that $p(\sum_{j=1}^\infty t_j x_j) \leq M \|t\|_\infty$ for every $t \in c_0$,*

(iii) *the linear map $T : c_0 \to X$, $Tt = \sum_{j=1}^\infty t_j x_j$, is continuous.*

Proof: (i) follows from Proposition 3.8 (iii); (ii) follows from Proposition 3.8 (v); (iii) follows directly from (ii).

We can now use the notions of wuc series and c_0 multiplier convergent series to give a characterization of a locally complete LCTVS due to Madrigal and Arrese ([MA]). Recall that an LCTVS X is *locally complete* if for every closed, bounded, absolutely convex set $B \subset X$, the space $X_B = spanB$ equipped with the Minkowski functional p_B of B in X_B is complete ([K2]).

Theorem 3.10. *The LCTVS X is locally complete iff every wuc series in X is c_0 multiplier convergent.*

Proof: Suppose that X is locally complete and let $\sum_j x_j$ be a wuc series in X. Then $S = \{\sum_{j \in \sigma} x_j : \sigma$ finite$\}$ is bounded in X by Proposition 3.8. Let B be the closed, absolutely convex hull of S so (X_B, p_B) is complete. Since S is bounded in (X_B, p_B), $\sum_j x_j$ is wuc in (X_B, p_B) by Proposition 3.8. By the completeness of (X_B, p_B) and condition (iii) of Proposition 3.8, $\sum_j x_j$ is c_0 multiplier convergent in (X_B, p_B). Since the inclusion of (X_B, p_B) into X is continuous, $\sum_j x_j$ is c_0 multiplier convergent in X.

Let B be a closed, bounded, absolutely convex subset of X and suppose that $\{x_j\}$ is Cauchy in (X_B, p_B). Pick an increasing sequence $\{n_j\}$ such that

$$p_B \left(x_{n_{j+1}} - x_{n_j} \right) < 1/j2^j$$

for every j and set $y_j = x_{n_{j+1}} - x_{n_j}$. Then $\sum_{j=1}^{\infty} j y_j$ is p_B absolutely convergent ($\sum_{j=1}^{\infty} p_B(jy_j) \leq \sum_{j=1}^{\infty} 1/2^j < \infty$) so by Proposition 3.8 $\sum_{j=1}^{\infty} j y_j$ is wuc in (X_B, p_B) and, therefore, $\sum_{j=1}^{\infty} j y_j$ is wuc in X. By hypothesis $\sum_{j=1}^{\infty} j y_j$ is c_0 multiplier convergent in X so the series $\sum_j y_j$ is convergent to, say, $y \in X$. Thus, $\sum_{j=1}^{k} y_j = x_{n_{k+1}} - x_{n_1} \to y$ or $x_{n_{j+1}} \to y + x_{n_1} = z$ in X. Now, $\{x_{n_j}\}$ is Cauchy in (X_B, p_B), $\{x_{n_j}\}$ converges in X to z and the topology p_B is linked to the relative topology of X_B from X so $\{x_{n_j}\}$ converges to z in X_B (Appendix A.4). Thus, X_B is complete with respect to p_B.

Theorem 3.10 has an interesting corollary due to Madrigal and Arrese ([MA]).

Corollary 3.11. *Let X be a locally complete LCTVS. The following are equivalent:*

(i) every wuc series in X is subseries convergent,
(ii) every wuc series in X is l^{∞} multiplier convergent,
(iii) every continuous linear operator $T : c_0 \to X$ has a compact extension $T : l^{\infty} \to X$.

Proof: Suppose that (i) holds. Let $\sum_j x_j$ be wuc and let $t \in l^{\infty}$. By Theorem 3.10, $\sum_j x_j$ is c_0 multiplier convergent. By Proposition 3.8 and

Corollary 3.9, the series $\sum_j t_j x_j$ is wuc. Hence, $\sum_j t_j x_j$ converges by (i) and (ii) holds.

Suppose that (ii) holds. Let $T : c_0 \to X$ be linear and continuous. Since $\sum e^j$ is wuc in c_0, $\sum_j Te^j$ is wuc in X. By (ii), $\sum_j Te^j$ is l^∞ multiplier convergent. By Corollary 2.46, $\{\sum_{j=1}^\infty t_j Te^j : \|t\|_\infty \le 1\}$ is compact. Therefore, by Theorem 2.2, $Tt = \sum_j t_j Te^j$, defines a compact operator from l^∞ into X which extends T. Hence, (iii) holds.

Suppose that (iii) holds. Let $\sum_j x_j$ be wuc in X. By Theorem 3.10, $\sum_j x_j$ is c_0 multiplier convergent so $Tt = \sum_{j=1}^\infty t_j x_j$ defines a continuous linear operator from c_0 into X by Corollary 3.9. By (iii) T is compact so $S = \{\sum_{j \in \sigma} x_j : \sigma \text{ finite}\}$ is relatively compact. By Theorem 2.48, $\sum_j x_j$ is subseries convergent.

Bessaga and Pelczynski have shown that a Banach space X contains no subspace isomorphic to c_0 iff every wuc series in X is subseries convergent ([BP]). We now extend this characterization to LCTVS. For this we require several preliminary lemmas.

Lemma 3.12. *Let* $x_{ij} \in \mathbb{R}$, $\varepsilon_{ij} > 0$ *for every* $i, j \in \mathbb{N}$. *If* $\lim_i x_{ij} = 0$ *for every* j *and* $\lim_j x_{ij} = 0$ *for every* i, *then there exists an increasing sequence* $\{m_j\}$ *such that* $|x_{m_i m_j}| \le \varepsilon_{ij}$ *for* $i \ne j$.

Proof: Set $m_1 = 1$. There exists $m_2 > m_1$ such that $|x_{m_1 j}| < \varepsilon_{12}$ and $|x_{i m_1}| < \varepsilon_{21}$ for all $i, j \ge m_2$. There exists $m_3 > m_2$ such that $|x_{m_1 j}| < \varepsilon_{13}, |x_{m_2 j}| < \varepsilon_{23}, |x_{i m_1}| < \varepsilon_{31}, |x_{i m_2}| < \varepsilon_{32}$ for all $i, j \ge m_3$. Now just continue.

Lemma 3.13. *Let* X *be a semi-normed space and* $x_{ij} \in X$ *for* $i, j \in \mathbb{N}$. *If* $\lim_i x_{ij} = 0$ *for every* j *and* $\lim_j x_{ij} = 0$ *for every* i, *then given* $\epsilon > 0$ *there exists a subsequence* $\{m_j\}$ *such that*

$$\sum_{i=1}^\infty \sum_{j \ne i} \|x_{m_i m_j}\| < \epsilon.$$

Proof: Pick $\epsilon_{ij} > 0$ such that $\sum_{i=1}^\infty \sum_{j=1}^\infty \epsilon_{ij} < \epsilon$. Let $\{m_j\}$ be the subsequence from Lemma 3.12 applied to the double sequence $\|x_{ij}\|$. Then $\|x_{m_i m_j}\| \le \epsilon_{ij}$ for $i \ne j$ so the result follows.

Lemma 3.14. *Let* X *be a semi-normed space that contains a* c_0 *multiplier convergent series* $\sum_j x_j$ *with* $\|x_j\| \ge \delta > 0$ *for every* j. *Then there exists a subsequence* $\{m_j\}$ *such that for any subsequence* $\{n_j\}$ *of* $\{m_j\}$, $T\{t_j\} = Tt = \sum_{j=1}^\infty t_j x_{n_j}$ *defines a topological isomorphism of* c_0 *into* X.

Proof: By replacing X by the linear subspace spanned by $\{x_j\}$, we may assume that X is separable. For each j pick $x'_j \in X'$, $\|x'_j\| \le 1$, such that $\langle x'_j, x_j \rangle = \|x_j\|$. By the Banach-Alaoglu Theorem, $\{x'_j\}$ has a subsequence which is weak* convergent to an element $x' \in X'$; to avoid cumbersome notation later, assume that $\{x'_j\}$ is weak* convergent to x'. Then

$$|\langle x'_j - x', x_j \rangle| \ge \delta - |\langle x', x_j \rangle| > \delta/2$$

for large j since $\langle x', x_j \rangle \to 0$; again to avoid cumbersome notation assume that $|\langle x'_j - x', x_j \rangle| \ge \delta/2$ for all j. The matrix

$$M = [\langle x'_i - x', x_j \rangle]$$

satisfies the assumption of Lemma 3.13 so let $\{m_j\}$ be the subsequence from Lemma 3.13 with $\epsilon = \delta/4$.

Now define a continuous linear operator $T : c_0 \to X$ by $Tt = \sum_{j=1}^{\infty} t_j x_{m_j}$ (Corollary 3.9). If $z'_i = x'_{m_i} - x'$, then by the conclusion of Lemma 3.13, we have

$$2\|T\{t_j\}\| \ge |\langle z'_i, T\{t_j\} \rangle| \ge |t_i \langle z'_i, x_{m_i} \rangle| - \sum_{j \ne i} |t_j \langle z'_i, x_{m_j} \rangle|$$

$$\ge |t_i| \delta/2 - \|\{t_j\}\|_{\infty} \delta/4.$$

Taking the supremum over all i in the inequality above gives

$$\|T\{t_j\}\| \ge (\delta/8) \|\{t_j\}\|_{\infty}$$

so T has a bounded inverse.

The same computation applies to any subsequence $\{n_j\}$ of $\{m_j\}$ so the result follows.

We now give a characterization of sequentially complete LCTVS which have the property that any wuc series is subseries convergent. In the statement below, if X is a semi-normed space, $B(X)$ denotes the closed unit ball of X.

Theorem 3.15. *Let X be a sequentially complete LCTVS. The following are equivalent:*

(i) *X contains no subspace (topologically) isomorphic to c_0.*
(ii) *If $\sum_j x_j$ is c_0 multiplier convergent in X, then $x_j \to 0$.*
(iii) *If $\sum_j x_j$ is c_0 multiplier convergent in X, then $\sum_j x_j$ is subseries convergent in X.*

(iv) If $\sum_j x_j$ is c_0 multiplier convergent in X, then $\sum_j x_j$ is bounded multiplier convergent in X.

(v) If $\sum_j x_j$ is c_0 multiplier convergent in X, then $\sum_{j=1}^{\infty} t_j x_j$ converges uniformly for $\{t_j\} \in B(l^{\infty})$.

(vi) If $\sum_j x_j$ is c_0 multiplier convergent in X, then $\sum_{j=1}^{\infty} t_j x_j$ converges uniformly for $\{t_j\} \in B(c_0)$.

(vii) If $\sum_j x_j$ is c_0 multiplier convergent in X, then $\sum_{j=1}^{\infty} t_j x_j$ converges uniformly for $\{t_j\} \in B(l^1)$.

(viii) Every continuous linear operator $T : c_0 \to X$ is compact and has a compact extension to l^{∞}.

Proof: (i) implies (ii): Suppose there exists a c_0 multiplier convergent series $\sum_j x_j$ with $x_j \nrightarrow 0$. Then we may assume there exists a continuous semi-norm p on X and $\delta > 0$ such that $p(x_j) \geq \delta$ for all j. By Lemma 3.14 there is a subsequence $\{m_i\}$ such that $H\{t_j\} = \sum_{j=1}^{\infty} t_j x_{m_j}$ defines a topological isomorphism from c_0 onto (Hc_0, p). Let I be the continuous inclusion operator from X onto (X, p). By Corollary 3.9, $T\{t_j\} = \sum_{j=1}^{\infty} t_j x_{m_j}$ defines a continuous linear operator from c_0 into X, and $T^{-1} = H^{-1}I$ is continuous so T defines a linear homeomorphism from c_0 into X.

(ii) implies (iii): Suppose there exists a c_0 multiplier convergent series $\sum_j x_j$ in X such that $\sum_j x_j$ diverges. Since X is sequentially complete, $\{s_n\} = \{\sum_{j=1}^{n} x_j\}$ is not Cauchy. Hence, there exist a neighborhood of 0, V, in X and an increasing sequence $\{n_j\}$ such that $y_j = s_{n_{j+1}} - s_{n_j} \notin V$ for all j. Since $\sum_j x_j$ is c_0 multiplier convergent, the series $\sum_{j=1}^{\infty} t_j y_j$ converges for every $\{t_j\} \in c_0$. By (ii), $y_j \to 0$. This contradiction shows that (ii) implies (iii).

That (iii) implies (iv) is given in Theorem 2.54.

That (iv) implies (v) is given in Theorem 2.54.

That (v) implies (vi) and (vi) implies (vii) is clear.

(vii) implies (ii): Suppose there is a c_0 multiplier convergent series $\sum_j x_j$ in X such that the series $\sum_{j=1}^{\infty} t_j x_j$ converges uniformly for $\{t_j\} \in B(l^1)$ but $x_j \nrightarrow 0$. There exists a neighborhood of 0, V, and a subsequence $\{x_{n_j}\}$ such that $x_{n_j} \notin V$ for every j. Let $t^k = \{t_j^k\} = e^{n_k} \in B(l^1)$. Then $\sum_{j=1}^{\infty} t_j^k x_j = x_{n_k} \notin V$ so the series $\sum_{j=1}^{\infty} t_j x_j$ fail to converge uniformly for $\{t_j\} \in B(l^1)$.

(viii) implies (i) since no continuous, linear, 1-1 map from c_0 into X can have a continuous inverse by the compactness of the map.

Finally, (iv) implies (viii): Let $T : c_0 \to X$ be linear and continuous and set $Te^j = x_j$. Then $\sum_j x_j$ is c_0 multiplier convergent and, hence, bounded multiplier convergent by (iv). By Corollary 2.46,

$$\{T\{t_j\} : \|\{t_j\}\|_\infty \leq 1\} = \{\sum_{j=1}^\infty t_j x_j : \|\{t_j\}\|_\infty \leq 1\}$$

is compact so (viii) holds.

Remark 3.16. The equivalence of (i) and (iii) for the case when X is a Banach space is a well known result of Bessaga and Pelczynski ([BP]). Bessaga and Pelczynski derive their result from results on basic sequences in B-spaces; Diestel and Uhl give a proof based on Rosenthal's Lemma ([DU] I.4.5). The equivalence of (i) and (viii) was noted by Li. The conditions (v), (vi) and (vii) are contained in [LB].

Without the sequential completeness assumption, the conclusions in Theorem 3.15 may fail.

Example 3.17. The series $\sum e^j$ is wuc in c_{00} with the sup-norm but is not subseries convergent. However, c_{00} being of countable algebraic dimension does not contain a subspace isomorphic to c_0.

We next derive a result of Pelczynski on unconditionally converging operators. A continuous linear operator T from a Banach space X into a Banach space Y is said to be *unconditionally converging* if T carries wuc series into subseries convergent series ([Pl]). A weakly compact operator is unconditionally converging [we give a proof of this fact in Chapter 4 after we establish the Orlicz-Pettis Theorem; recall an operator is weakly compact if it carries bounded sets into relatively weakly compact sets]. The identity on l^1 gives a example of an unconditionally converging operator which is not weakly compact [recall that a sequence in l^1 is weakly convergent iff the sequence is norm convergent; this result will be established in Chapter 7 when Hahn-Schur Theorems are derived; see also, [Sw2] 16.14].

Theorem 3.18. *Let X, Y be Banach spaces and $T : X \to Y$ a continuous linear operator which is not unconditionally converging. Then there exist topological isomorphisms $I_1 : c_0 \to X$ and $I_2 : c_0 \to Y$ such that $TI_1 = I_2$ [i.e., T has a bounded inverse on a subspace isomorphic to c_0].*

Proof: By hypothesis there exists a wuc series $\sum_j x_j$ in X such that $\sum_j T x_j$ is not subseries convergent. Since $\sum_j T x_j$ contains a subseries which is not convergent, we may as well assume that the series $\sum_j T x_j$ diverges. Thus, there exist $\delta > 0$ and a subsequence $\{n_j\}$ such that $\|z_j\| \geq$

δ, where $z_j = Tu_j$ and $u_j = \sum_{i=n_j+1}^{n_{j+1}} x_i$. By Proposition 3.8, the series $\sum_j u_j$ and $\sum_j Tu_j$ are both wuc. Since $\|x\| \geq \|Tx\| / \|x\|$ for $x \in X$, $\|u_j\| \geq \delta / \|T\|$. Applying Lemma 3.14 to the series $\sum_j u_j$ and $\sum_j Tu_j$, there is a subsequence $\{m_j\}$ such that $I_1\{t_j\} = \sum_{j=1}^{\infty} t_j u_{m_j}$ and $I_2 = \sum_{j=1}^{\infty} t_j Tu_{m_j}$ define isomorphisms from c_0 into X and Y, respectively. Obviously, $TI_1 = I_2$.

Remark 3.19. The converse of Theorem 3.18 holds and gives an interesting characterization of unconditionally converging operators (see [Ho]).

We next consider wuc series in the strong dual of an LCTVS.

Theorem 3.20. *Let X be a barrelled LCTVS. The following are equivalent:*

(i) $(X', \beta(X', X))$ *contains no subspace isomorphic to c_0,*
(ii) every wuc series $\sum_j x_j'$ in X' is $\beta(X', X)$ subseries convergent,
(iii) every series $\sum_j x_j'$ in X' which satisfies $\sum_{j=1}^{\infty} |\langle x_j', x \rangle| < \infty$ for every $x \in X$ is $\beta(X', X)$ subseries convergent,
(iv) every continuous linear operator $T : X \to l^1$ is compact [an operator T is compact if T carries bounded sets into relatively compact sets].

Proof: Conditions (i) and (ii) are equivalent by Theorem 3.15 since $\beta(X', X)$ is sequentially complete by the barrelledness of X ([Wi] 6.1.16 and 9.3.8).

Assume that (ii) holds. Let $\sum_j x_j'$ be such that $\sum_{j=1}^{\infty} |\langle x_j', x \rangle| < \infty$ for every $x \in X$. Then $\{\sum_{j \in \sigma} x_j' : \sigma \text{ finite}\}$ is weak* bounded and, therefore, $\beta(X', X)$ bounded since X is barrelled. Therefore, $\sum_j x_j'$ is wuc in $(X', \beta(X', X))$ by Proposition 3.8. Hence, $\sum_j x_j'$ is $\beta(X', X)$ subseries convergent by (ii) and (iii) holds.

Assume that (iii) holds. Let $T : X \to l^1$ be linear and continuous. Set $x_j' = T'e^j$. Now T' is $\beta(l^\infty, l^1) - \beta(X', X)$ continuous so $\{x_j'\}$ is $\beta(X', X)$ bounded. For $x \in X$, $Tx \in l^1$ we have

$$\sum_{j=1}^{\infty} |\langle x_j', x \rangle| = \sum_{j=1}^{\infty} |\langle T'e^j, x \rangle| = \sum_{j=1}^{\infty} |\langle e^j, Tx \rangle| < \infty.$$

By (iii), $\sum_j x_j'$ is $\beta(X', X)$ subseries convergent and, therefore, l^∞ multiplier convergent since $\beta(X', X)$ is sequentially complete as noted above (Theorem 2.54). Therefore, if $B \subset X$ is bounded, then

$$\limsup_n \sum_{x \in B}^{\infty} |\langle x_j', x \rangle| = \limsup_n \sum_{x \in B}^{\infty} |\langle e^j, Tx \rangle| = 0.$$

Hence, TB is relatively compact in l^1 ([Sw2] 10.15) and (iv) holds.

Assume that (iv) holds. Let $\sum_j x'_j$ be wuc in $(X', \beta(X', X))$. Define $T :$ $X \to l^1$ by $Tx = \{\langle x'_j, x \rangle\}$. T is obviously linear and is $\sigma(X, X') - \sigma(l^1, l^\infty)$ continuous since if $t \in l^\infty, x \in X$,

$$t \cdot Tx = \sum_{j=1}^{\infty} t_j \langle x'_j, x \rangle = \left\langle \sum_{j=1}^{\infty} t_j x'_j, x \right\rangle$$

[the series $\sum_j t_j x'_j$ is $\sigma(X', X)$ Cauchy and ,therefore, $\sigma(X', X)$ convergent since X is barrelled ([Wi] 9.3.8)]. Thus, T is $\beta(X, X') - \beta(l^1, l^\infty)$ continuous. By (iv), T is compact. If $B \subset X$ is bounded, TB is relatively compact in l^1 so

$$\lim_n \sup_{x \in B} \sum_{j=n}^{\infty} |\langle e^j, Tx \rangle| = \lim_n \sup_{x \in B} \sum_{j=n}^{\infty} |\langle x'_j, x \rangle| = 0$$

([Sw2] 10.15) and $\sum_j x'_j$ is $\beta(X', X)$ convergent. The same argument can be applied to every subseries of $\sum_j x'_j$ so (ii) holds.

Remark 3.21. If X is barrelled, then X' is weak* sequentially complete so condition (iii) is equivalent to the statement that every series $\sum_j x'_j$ in X' which is $\sigma(X', X)$ subseries convergent is $\beta(X', X)$ subseries convergent. This is the statement of an Orlicz-Pettis type Theorem which we will consider in Chapter 4.

Without the barrelledness assumption, the conclusion of Theorem 3.20 may fail.

Example 3.22. Let $X = c_{00}$ with the sup-norm. The series $\sum_j e^j$ in $l^1 = X'$ satisfies the condition (iii) in Theorem 3.20 but is not strongly subseries convergent in l^1 and l^1 contains no subspace isomorphic to c_0.

We next give a characterization of Banach-Mackey spaces in terms of multiplier convergent series. Recall that an LCTVS X is a *Banach-Mackey space* if every $\sigma(X, X')$ bounded subset of X is $\beta(X, X')$ bounded; i.e., if $B \subset X$ is pointwise bounded on X', then B is uniformly bounded on $\sigma(X', X)$ bounded subsets of X' ([Wi] 10.4.3). The Banach-Mackey Theorem states that any sequentially complete LCTVS is a Banach-Mackey space ([Wi] 10.4.8).

Let X be an LCTVS. Let X^b (X^s) be the space of all bounded (sequentially continuous) linear functionals on X. Since $X' \subset X^s \subset X^b$,

(X, X^s) and (X, X^b) both form dual pairs. We now give a characterization of Banach-Mackey spaces in terms of l^1 multiplier convergent series and the spaces X^s and X^b.

Theorem 3.23. *Let X be an LCTVS. The following are equivalent:*

(i) X is a Banach-Mackey space.
(ii) If $\{x_j'\}$ is $\sigma(X', X)$ bounded and $\{t_j\} \in l^1$, then $\sum_{j=1}^{\infty} t_j x_j' \in X^s$.
(iii) If $\{x_j'\}$ is $\sigma(X', X)$ bounded and $\{t_j\} \in l^1$, then $\sum_{j=1}^{\infty} t_j x_j' \in X^b$.
(iv) If $\{x_j'\}$ is $\sigma(X', X)$ Cauchy and $\langle x', x \rangle = \lim \langle x_j', x \rangle$ for $x \in X$, then $x' \in X^b$.

Proof: Suppose that (i) holds. Let $x_j \to 0$ in X. Then $\{x_j\}$ is bounded in X and, therefore, $\beta(X, X')$ bounded by (i). Hence,

$$M = \sup\{|\langle x_i', x_j \rangle| : i, j \in \mathbb{N}\} < \infty$$

and

$$\left| \sum_{j=n}^{\infty} t_j \langle x_j', x_i \rangle \right| \leq M \sum_{j=n}^{\infty} |t_j|$$

for $\{t_j\} \in l^1$. Therefore, the series $\sum_{j=1}^{\infty} t_j \langle x_j', x_i \rangle$ converge uniformly for $i \in \mathbb{N}$. Hence,

$$\lim_i \sum_{j=1}^{\infty} t_j \langle x_j', x_i \rangle = \sum_{j=1}^{\infty} t_j \lim_i \langle x_j', x_i \rangle = 0$$

so $\sum_{j=1}^{\infty} t_j x_j' \in X^s$ and (ii) holds.

That (ii) implies (iii) is immediate.

Assume that (iii) holds. We show that (i) holds. Let $A \subset X$ be $\sigma(X, X')$ bounded and $B \subset X'$ be $\sigma(X', X)$ bounded. We show that $\sup\{|\langle x', x \rangle| : x' \in B, x \in A\} < \infty$. If this fails to hold, there exist $\{x_j'\} \subset B$ and $\{x_j\} \subset A$ such that

$$(\#) \quad |\langle x_i', x_i \rangle| > i^2 \text{ for every } i.$$

Consider the matrix

$$M = [m_{ij}] = [(1/j) \langle x_j', (1/i)x_i \rangle].$$

We claim that M is a $\mathcal{K}$-matrix (Appendix D.2). First, the columns of M converge to 0 since $\{x_i\}$ is $\sigma(X, X')$ bounded. Given any subsequence $\{m_j\}$ pick a further subsequence $\{n_j\}$ such that $\sum_{j=1}^{\infty} 1/n_j < \infty$. By (iii)

$$\left\langle \sum_{j=1}^{\infty} (1/n_j)x_{n_j}', (1/i)x_i \right\rangle = \sum_{j=1}^{\infty} (1/n_j) \left\langle x_{n_j}', (1/i)x_i \right\rangle \to 0.$$

Hence, M is a $\mathcal{K}$-matrix so by the Antosik-Mikusinski Matrix Theorem (Appendix D.2) the diagonal of M converges to 0. But, this contradicts (#).

We next show that (i) implies (iv). Let $A \subset X$ be bounded. Then A is $\sigma(X, X')$ bounded by (i). Since $\{x'_j\}$ is $\beta(X', X)$ bounded, $\{\langle x'_j, x \rangle : x \in A, j \in \mathbb{N}\}$ is bounded. Therefore, $\{\langle x', x \rangle : x \in A\}$ is bounded. Therefore, $x' \in X^b$ and (iv) holds.

Suppose that (iv) holds. We show that (iii) holds and this will complete the proof. If $x \in X$ and $\{t_j\} \in l^1$, then $\lim_n \sum_{j=1}^n t_j \langle x'_j, x \rangle = \sum_{j=1}^\infty t_j \langle x'_j, x \rangle$. By (iv), $\sum_{j=1}^\infty t_j x'_j \in X^b$ and (iii) holds.

Theorem 3.23 is contained in [LS], Theorem 7, where other characterizations of Banach-Mackey spaces are given.

We make an interesting observation concerning Banach spaces with an unconditional Schauder basis. Let X be a Banach space. A sequence $\{b_j\} \subset X$ is a *Schauder basis* for X if every $x \in X$ has a unique series representation $x = \sum_{j=1}^\infty t_j b_j$; the linear functionals $f_j : X \to \mathbb{R}$ defined by $\langle f_j, x \rangle = t_j$ are called the coordinate functionals associated with the basis $\{b_j\}$. It is known that the coordinate functionals are equicontinuous ([Sw2] 10.10). If the series $x = \sum_{j=1}^\infty \langle f_j, x \rangle b_j$ is unconditionally convergent (subseries convergent) for every x, the basis $\{b_j\}$ is said to be unconditional.

Theorem 3.24. *Let $\{b_j\}$ be an unconditional basis for the Banach space X. If $x \in X$, the series $\sum_{j=1}^\infty t_j b_j$ converge uniformly for $|t_j| \leq |\langle f_j, x \rangle|$.*

Proof: Since X is complete, the series $\sum_{j=1}^\infty \langle f_j, x \rangle b_j$ is also bounded multiplier convergent (Theorem 2.54) so the result follows from Theorem 2.54.

In the last part of this chapter we present several applications of convergent series to topics in vector valued measures. Let $\mathcal{A}$ (Σ) be an algebra (σ-algebra) of subsets of a set S and let X be a TVS. A set function $\mu : \mathcal{A} \to X$ is *finitely additive (countably additive)* if $\mu(\emptyset) = 0$ and $\mu(A \cup B) = \mu(A) + \mu(B)$ when $A, B \in \mathcal{A}$ with $A \cap B = \emptyset$ ($\mu(A) = \sum_{j=1}^\infty \mu(A_j)$ when $\{A_j\} \subset \mathcal{A}$, is pairwise disjoint and $A = \cup_{j=1}^\infty A_j \in \mathcal{A}$). Note that if μ is countably additive, then the series $\sum_{j=1}^\infty \mu(A_j)$ is unconditionally convergent since the union $\cup_{j=1}^\infty A_j$ is independent of the ordering of the $\{A_j\}$.

Finitely additive set functions, even scalar valued functions, defined on algebras or σ-algebras are not necessarily bounded as the following examples show.

First, we give a simple example of an unbounded, finitely additive set function defined on an algebra.

Example 3.25. Let $\mathcal{A}$ be the algebra of finite/co-finite subsets of $\mathbb{N}$; i.e., $A \in \mathcal{A}$ iff either A or the complement of A, A^c, is finite. Define $\mu : \mathcal{A} \to \mathbb{R}$ by $\mu(A)$ equals the number of elements in A when A is finite and $\mu(A)$ equals minus the number of elements in A^c when A^c is finite. Then μ is finitely additive but not bounded.

To present an example of a finitely additive set function defined on a σ-algebra which is unbounded is more complicated. We present an example due to Giesy ([Gi]).

Lemma 3.26. *Let $\mathcal{A}$, $\mathcal{B}$ be algebras of subsets of a set S with $\mathcal{A} \subset \mathcal{B}$ and let $\alpha : \mathcal{A} \to \mathbb{R}$ be finitely additive. If $B \in \mathcal{B} \backslash \mathcal{A}$ and $b \in \mathbb{R}$, there exists $\beta : \mathcal{B} \to \mathbb{R}$ finitely additive such that β is an extension of α with $\beta(B) = b$.*

Proof: Let $\mathcal{S}(\mathcal{A})$ $(\mathcal{S}(\mathcal{B}))$ be the vector space of all $\mathcal{A}$ $(\mathcal{B})$ simple functions. Then α induces a linear functional $\alpha' : \mathcal{S}(\mathcal{A}) \to \mathbb{R}$ via integration, i.e., $\alpha'(f) = \int f d\alpha$. The linear functional α' has a linear extension, β', to $\mathcal{S}(\mathcal{B})$ such that $\beta'(\chi_B) = b$. Then $\beta(E) = \beta'(\chi_E)$ defines the desired finitely additive extension of α.

We now give an example of a real valued, finitely additive set function defined on the σ-algebra of Lebesgue measurable subsets of $\mathbb{R}$ which is not bounded.

Example 3.27. Let $\{E_j\}_{j=0}^\infty$ be a pairwise disjoint sequence of bounded intervals whose union is $\mathbb{R}$. Let $\mathcal{A}_k$ be the algebra generated by $\{\mathbb{R}, E_0, E_1, ..., E_k\}$ so $\mathcal{A}_0 \subset \mathcal{A}_1 \subset ... \subset \mathcal{M}$, where $\mathcal{M}$ is the σ-algebra of Lebesgue measurable subsets of $\mathbb{R}$. Set $\alpha_0 = 0$ on $\mathcal{A}_0$; let α_1 be a finitely additive extension of α_0 to $\mathcal{A}_1$ such that $\alpha_1(E_1) = 1$ (Lemma 3.26). Inductively, there is a sequence $\{\alpha_k\}$ of finitely additive set functions such that $\alpha_k : \mathcal{A}_k \to \mathbb{R}$, α_{k+1} extends α_k and $\alpha_k(E_k) = k$. Now $\mathcal{A} = \cup_{k=0}^\infty \mathcal{A}_k$ is an algebra and $\alpha = \cup_{k=0}^\infty \alpha_k$ is finitely additive on $\mathcal{A}$. By Lemma 3.26 there is a finitely additive extension of α, μ, to $\mathcal{M}$ and we have that $\mu(E_k) = k$ for every k so μ is not bounded.

We now give several conditions which characterize bounded, finitely additive set functions with values in LCTVS.

Theorem 3.28. *Let X be an LCTVS and let $\mu : \mathcal{A} \to \mathbb{R}$ be finitely additive. The following are equivalent:*

(i) μ is bounded,
(ii) for every pairwise disjoint sequence $\{A_j\}$ from $\mathcal{A}$, $\{\mu(A_j)\}$ is bounded,
(iii) for every pairwise disjoint sequence $\{A_j\}$ from $\mathcal{A}$, the series $\sum_j \mu(A_j)$ is c_0 multiplier Cauchy.

Proof: Clearly, (i) implies (ii).

Suppose that (ii) holds and μ is not bounded. If $E \in \mathcal{A}$, set $\mathcal{A}_E = \{A \cap E : A \in \mathcal{A}\}$. Suppose that $\mu(\mathcal{A}_E)$ is not absorbed by the absolutely convex neighborhood of 0, U, in X. Pick an absolutely convex neighborhood of 0, V, such that $V + V \subset U$. We claim that for every k there exist $n_k > k$ and a partition (A_k, B_k) of E with $A_k, B_k \in \mathcal{A}$ and $\mu(A_k) \notin n_k V, \mu(B_k) \notin n_k V$. For, there exists $n_k > k$ such that $\mu(E) \in n_k V$. But, $\mu(\mathcal{A}_E) \not\subseteq n_k(V + V)$ since $V + V \subset U$. Therefore, there exists $A_k \in \mathcal{A}_E$ such that $\mu(A_k) \notin n_k(V + V)$. Note that $\mu(A_k) \notin n_k V$. Put $B_k = E \backslash A_k$. Then $\mu(B_k) \notin n_k V$ since otherwise $\mu(A_k) = \mu(E) - \mu(B_k) \in n_k(V + V)$.

Since $\mu(\mathcal{A})$ is assumed to be unbounded, there exists an absolutely convex neighborhood of 0, U, in X such that $\mu(\mathcal{A})$ is not absorbed by U. Pick V as above. By the observation above there exist $n_1 > 1$ and a partition (A_1, B_1) of S such that $\mu(A_1) \notin n_1 V$ and $\mu(B_1) \notin n_1 V$. Either $\mu(\mathcal{A}_{A_1})$ or $\mu(\mathcal{A}_{B_1})$ is not absorbed by U since otherwise there exists m such that $\mu(\mathcal{A}_{A_1}) \subset mU$ and $\mu(\mathcal{A}_{B_1}) \subset mU$ and $\mu(\mathcal{A}_S) = \mu(\mathcal{A}) \subset m(U+U) \subset m(2U)$ since U is convex. Pick whichever of A_1 or B_1 satisfies this condition, label it F_1 and set $E_1 = S \backslash F_1$. Now treat F_1 as above to obtain a partition (E_2, F_2) of F_1 and $n_2 > n_1$ such that $\mu(E_2) \notin n_2 V, \mu(F_2) \notin n_2 V$ and $\mu(\mathcal{A}_{F_2})$ is not absorbed by U. Continuing this construction produces a pairwise disjoint sequence $\{E_k\}$ such that $\{\mu(E_k)\}$ is not absorbed by U. Thus, (ii) fails to hold so (ii) implies (i).

Suppose that (iii) holds, $\{A_j\} \subset \mathcal{A}$ is pairwise disjoint and $t \in c_0$. Then $\sum_j t_j \mu(A_j)$ is Cauchy so $\lim t_j \mu(A_j) = 0$. Since $t \in c_0$ is arbitrary, $\{\mu(A_j)\}$ is bounded. Thus, (ii) holds.

Suppose that (i) holds and let $\{A_j\} \subset \mathcal{A}$ be pairwise disjoint. Then

$$\left\{ \sum_{j \in \sigma} \mu(A_j) : \sigma \text{ finite} \right\} = \left\{ \mu(\cup_{j \in \sigma} A_j) : \sigma \text{ finite} \right\}$$

is bounded. Therefore, $\sum_j \mu(A_j)$ is c_0 multiplier Cauchy by Proposition 3.8.

If the LCTVS X is sequentially complete, the condition (iii) in Theorem 3.28 can be strengthened. To establish this we employ a very intersting and simple result of Li Ronglu ([LS] Corollary 10) which will be used several times later.

Lemma 3.29. *Let $\{E_j\}$ be a sequence of sets. Let G be an Abelian (Hausdorff) topological group and $f_j : E_j \to G$. If the series $\sum_{j=1}^{\infty} f_j(t_j)$ converges for every sequence $\{t_j\}$ with $t_j \in E_j$, then the series $\sum_{j=1}^{\infty} f_j(t_j)$ converge uniformly for all sequences $\{t_j\}$ with $t_j \in E_j$.*

Proof: If the conclusion fails to hold, there exist a neighborhood, U, of 0 in G and sequences $\{t_j^i\}_j, t_j^i \in E_j$, and an increasing sequence $\{n_i\}$ such that $\sum_{j=n_i}^{\infty} f_j(t_j^i) \notin U$. Pick a symmetric neighborhood of $0, V$, such that $V + V \subset U$. Since $\lim_k \sum_{j=k}^{\infty} f_j(t_j^1) = 0$ and $\sum_{j=n_1}^{\infty} f_j(t_j^1) \notin U$, there exists $m_1 > n_1$ such that $\sum_{j=n_1}^{m_1} f(t_j^1) \notin V$. Put $N_1 = 1$ and pick $n_{i_2} = N_2 > m_1$ such that $\sum_{j=N_2}^{\infty} f_j(t_j^{i_2}) \notin U$. As before pick $m_2 > N_2$ such that $\sum_{j=N_2}^{m_2} f_j(t_j^{i_2}) \notin V$. Continuing this construction produces increasing sequences $\{N_k\}, \{m_k\}$ and $\{i_k\}$ such that $N_k < m_k < N_{k+1}$ and $\sum_{j=N_k}^{m_k} f_j(t_j^{i_k}) \notin V$. Pick an arbitrary sequence $\{u_j\}$ with $u_j \in E_j$ for every j. Define a sequence $\{s_j\}$ with $s_j \in E_j$ by $s_j = t_j^{i_k}$ if $N_k \leq j \leq m_k$ and $s_j = u_j$ otherwise. If the series $\sum_{j=1}^{\infty} f_j(s_j)$ converges, there exists N such that $\sum_{j=m}^{n} f_j(s_j) \in V$ for $n > m \geq N$. But, $\sum_{j=N_k}^{m_k} f_j(s_j) = \sum_{j=N_k}^{m_k} f_j(t_j^{i_k}) \notin V$ for large k so the series $\sum_{j=1}^{\infty} f_j(s_j)$ does not satisfy the Cauchy condition and, therefore, does not converge. This contradicts the hypothesis.

To illustate the utility of Lemma 3.29, we derive a couple of previous results for series which were established by other means. First, we consider a version of Corollary 2.18.

Corollary 3.30. *Let X be a TVS and $\sum_j x_j$ a series in X which is subseries convergent. Then the series $\sum_{j \in \sigma} x_j$ converge uniformly for $\sigma \subset \mathbb{N}$.*

Proof: Let $E_j = \{0, 1\}$ for every j and define $f_j : E_j \to X$ by $f_j(0) = 0$ and $f_j(1) = x_j$. Then the conclusion follows directly from Lemma 3.29.

Next, we consider an improvement of Corollary 2.19.

Corollary 3.31. *Let X be a TVS and let λ be a normal space and let $\sum_j x_j$ be λ multipier convergent. If $t \in \lambda$, then the series $\sum_{j=1}^{\infty} s_j x_j$ converge uniformly for $|s_j| \leq |t_j|$.*

Proof: Let $E_j = \{t \in \mathbb{R} : |t| \le |t_j|\}$ and define $f_j : E_j \to X$ by $f_j(t) = tx_j$. Then the conclusion follows directly from Lemma 3.29.

Note that the space λ in Corollary 3.31 is not assume to be a K-space with signed-SGHP as in Corollary 2.19. From Corollary 3.31 we can obtain immediately Corollary 2.17.

Corollary 3.32. *Let X be a TVS and let $\sum_j x_j$ be l^∞ multiplier convergent. Then the series $\sum_{j=1}^{\infty} t_j x_j$ converge uniformly for $\|\{t_j\}\|_\infty \le 1$.*

Proof: Let t be the constant sequence with 1 in each coordinate. Then the result follows immediately from Corollary 3.31.

Notice that Corollaries 3.31 and 3.32 were proven in reverse order in Chapter 2.

We now show that Lemma 3.29 can be used to obtain an improvement in Theorem 3.28 when the space X is a sequentially complete LCTVS.

Corollary 3.33. *Let X be a sequentially complete LCTVS and $\mu : \mathcal{A} \to X$ finitely additive. The following are equivalent:*

(i) μ is bounded,
(ii) for every pairwise disjoint sequence $\{A_j\}$ from $\mathcal{A}$, $\{\mu(A_j)\}$ is bounded,
(iii)' for every pairwise disjoint sequence $\{A_j\}$ from $\mathcal{A}$, the series $\sum_j \mu(A_j)$ is c_0 multiplier convergent,
(iv) for every pairwise disjoint sequence $\{A_j\}$ from $\mathcal{A}$ and $t \in c_0$, the series $\sum_{j=1}^{\infty} s_j \mu(B_j)$ converge uniformly for $B_j \subset A_j$, $B_j \in \mathcal{A}$ and $|s_j| \le |t_j|$.

Proof: (i), (ii) and (iii)' are equivalent by Theorem 3.28.
Obviously, (iv) implies (iii)'.
Assume (iii)'. We apply Lemma 3.29. Set

$$E_j = \{(B, s) : B \in \mathcal{A}, B \subset A_j, |s| \le |t_j|\}$$

and define $f_j : E_j \to X$ by $f_j(B, s) = s\mu(B)$. Lemma 3.29 now gives the result.

We also have the following boundedness result for countably additive set functions defined on σ-algebras.

Corollary 3.34. *Let X be an LCTVS and Σ a σ-algebra. If $\mu : \Sigma \to X$ is countably additive, then μ is bounded.*

Proof: μ satisfies condition (ii) of Theorem 3.28.

Remark 3.35. The locally convex assumption in Corollary 3.34 is important. Turpin has given an example of a countably additive set function defined on a σ-algebra with values in a (non-locally convex) TVS which is unbounded ([Rol] 3.6.4).

We next consider an important property for vector valued set functions which was introduced by Rickart which lies between finite additivity and countable additivity.

Definition 3.36. Let X be a TVS and $\mu : \mathcal{A} \to X$ be finitely additive. Then μ is strongly bounded (strongly additive, exhaustive) if $\mu(A_j) \to 0$ for every pairwise disjoint sequence $\{A_j\}$ from $\mathcal{A}$.

A countably additive set function defined on a σ-algebra is obviously strongly bounded. We show below that bounded, finitely additive scalar valued set functions are strongly bounded and give an example of a bounded, finitely additive set function defined on a σ-algebra with values in a Banach space which is not strongly bounded.

Lemma 3.37. *Let $\{t_j\} \subset \mathbb{R}$ and assume that there exists $M \geq 0$ such that $\left| \sum_{j \in \sigma} t_j \right| \leq M$ for every finite $\sigma \subset \mathbb{N}$. Then $\sum_{j=1}^{\infty} |t_j| \leq 2M$.*

Proof: Let σ be finite. Set $\sigma_+ = \{j \in \sigma : t_j \geq 0\}$ and $\sigma_- = \{j \in \sigma : t_j < 0\}$. Then

$$\sum_{j \in \sigma_+} |t_j| = \sum_{j \in \sigma_+} t_j \leq M$$

and

$$\sum_{j \in \sigma_-} |t_j| = - \sum_{j \in \sigma_-} t_j \leq M$$

so $\sum_{j \in \sigma} |t_j| \leq 2M$. Since σ is arbitrary, $\sum_{j=1}^{\infty} |t_j| < 2M$.

Corollary 3.38. *Let $\mu : \mathcal{A} \to \mathbb{R}$ be finitely additive. Then μ is bounded iff μ is strongly bounded.*

Proof: If μ is strongly bounded, then μ is bounded since condition (ii) of Theorem 3.28 is satisfied.

Suppose that μ is bounded with $\sup\{|\mu(A)| : A \in \mathcal{A}\} = M < \infty$. Let $\{A_j\} \subset \mathcal{A}$ be pairwise disjoint. If $\sigma \subset \mathbb{N}$ is finite, then

$$\left|\sum_{j \in \sigma} \mu(A_j)\right| = |\mu(\cup_{j \in \sigma} A_j)| \leq M$$

so by Lemma 3.37, $\sum_{j=1}^{\infty} |\mu(A_j)| \leq 2M$. In particular, $\mu(A_j) \to 0$ so μ is strongly bounded.

Remark 3.39. The proof of Corollary 3.38 shows that if $\mu : \mathcal{A} \to \mathbb{R}$ is bounded and finitely additive and $\{A_j\}$ is pairwise disjoint, then the series $\sum_{j=1}^{\infty} \mu(A_j)$ is absolutely convergent. Thus, if μ fails to be countably additive, the series $\sum_{j=1}^{\infty} \mu(A_j)$ converges but may fail to converge to the "proper value", namely, $\mu(\cup_{j=1}^{\infty} A_j)$.

For vector valued set functions we have the following boundedness result.

Corollary 3.40. *Let X be an LCTVS. If $\mu : \mathcal{A} \to X$ is strongly bounded, then μ is bounded.*

Proof: For each $x' \in X'$, $x' \circ \mu = x'\mu : \mathcal{A} \to \mathbb{R}$ is strongly bounded so $\{\langle x', \mu(A)\rangle : A \in \mathcal{A}\}$ is bounded by Corollary 3.38. Thus, $\{\mu(A) : A \in \mathcal{A}\}$ is weakly bounded in X and, therefore, bounded in X.

The example indicated in Remark 3.35 shows that the local convex assumption in Corollary 3.40 is important. The converse of Corollary 3.38 is false, in general.

Example 3.41. Let $\mathcal{M}$ be the σ-algebra of Lebesgue measurable subsets of $[0, 1]$. Define $\mu : \mathcal{M} \to L^{\infty}[0, 1]$ by $\mu(E) = \chi_E$. Then μ is bounded, finitely additive but not strongly bounded [take any pairwise disjoint sequence from $\mathcal{M}$ with positive Lebesgue measure].

We have a series characterization of strongly additive set functions.

Theorem 3.42. *Let X be a TVS and $\mu : \mathcal{A} \to X$ be finitely additive. The following are equivalent:*

(i) μ is strongly bounded,

(ii) for every pairwise disjoint sequence $\{A_j\} \subset \mathcal{A}$, the series $\sum_j \mu(A_j)$ is Cauchy.

Proof: That (ii) implies (i) is clear.

Assume that (ii) fails to hold. Then there exist a neighborhood U of 0 in X and an increasing sequence of intervals $\{I_j\}$ such that $\sum_{j \in I_k} \mu(A_j) \notin U$ for all k. If $B_k = \cup_{j \in I_k} A_j$, then $\{B_k\}$ is pairwise disjoint and $\mu(B_k) \not\to 0$ so (i) fails.

If X is sequentially complete, using Li's Lemma 3.29 we can strengthen condition (ii).

Corollary 3.43. *Let X be a sequentially complete TVS and $\mu : \mathcal{A} \to X$ be finitely additive. The following are equivalent:*

(i) μ is strongly bounded,
(ii)' for every pairwise disjoint sequence $\{A_j\} \subset \mathcal{A}$, the series $\sum_j \mu(A_j)$ converges,
(iii) for every pairwise disjoint sequence $\{A_j\} \subset \mathcal{A}$, the series $\sum_j \mu(B_j)$ converge uniformly for $B_j \subset A_j, B_j \in \mathcal{A}$.

Proof: That (i) and (ii)' are equivalent follows from Theorem 3.42. Obviously (iii) implies (ii)'.

Assume that (ii)' holds. We establish (iii) by using Lemma 3.29. Set $E_j = \{B \in \mathcal{A} : B \subset A_j\}$ and define $f_j : E_j \to X$ by $f_j(B) = \mu(B)$. Lemma 3.29 now gives the result.

We consider the semi-variation of set functions $\mu : \mathcal{A} \to X$ with values in a normed space X. The semi-variation is useful in discussing topics in vector measures and vector integration ([DS], [DU] I.1).

Definition 3.44. For $A \in \mathcal{A}$ the semi-variation of μ on A is defined to be

$$\|\mu\|(A) = \sup \left\{ \left\| \sum_{j=1}^{n} t_j \mu(A_j) \right\| : \{A_j\}_{j=1}^{n} \text{ a partition of } A \text{ and } |t_j| \le 1 \right\}.$$

We have the following properties of the semi-variation. In the proposition below, the variation of a real valued set function ν is denoted by $|\nu|$ ([Sw3] 2.2.1.7).

Proposition 3.45. *Let $\mu : \mathcal{A} \to X$.*

(i) $\|\mu\|(A) = \sup\{|x'\mu|(A) : \|x'\| \le 1\}$,
(ii) $\sup\{\|\mu(B)\| : B \subset A, B \in \mathcal{A}\} \le \|\mu\|(A) \le 2\sup\{\|\mu(B)\| : B \subset A, B \in \mathcal{A}\}$.

Proof: (i): Let $\{A_1, ..., A_n\}$ be a partition of A and let $|t_j| \leq 1$ for $j = 1, ..., n$. Then

$$\left\|\sum_{j=1}^n t_j \mu(A_j)\right\| = \sup\{\left|\left\langle x', \sum_{j=1}^n t_j \mu(A_j)\right\rangle\right| : \|x'\| \leq 1\}$$
$$\leq \sup\{\sum_{j=1}^n |\langle x', t_j \mu(A_j)\rangle| : \|x'\| \leq 1\}$$
$$\leq \sup\{\sum_{j=1}^n |\langle x', \mu(A_j)\rangle| : \|x'\| \leq 1\}$$
$$\leq \sup\{\sum_{j=1}^n |x'\mu|(A_j) : \|x'\| \leq 1\} = \sup\{|x'\mu|(A) : \|x'\| \leq 1\}.$$

Therefore, $\|\mu\|(A) \leq \sup\{|x'\mu|(A) : \|x'\| \leq 1\}$.

For the reverse inequality, let $x' \in X'$, $\|x'\| \leq 1$ and $\{A_1, ..., A_n\}$ be a partition of A. Then

$$\sum_{j=1}^n |\langle x', \mu(A_j)\rangle| = \sum_{j=1}^n (sign x'\mu(A_j)) x'\mu(A_j)$$
$$= \left|\left\langle x', \sum_{j=1}^n (sign x'\mu(A_j))\mu(A_j)\right\rangle\right|$$
$$\leq \left\|\sum_{j=1}^n (sign x'\mu(A_j))\mu(A_j)\right\| \leq \|\mu\|(A).$$

Therefore, $|x'\mu|(A) \leq \|\mu\|(A)$ and $\|\mu\|(A) \geq \sup\{|x'\mu(A)| : \|x'\| \leq 1\}$. Thus, (i) holds.

For (ii), recall that for scalar set functions ν we have that

$$\sup\{|\nu(B)| : B \subset A, B \in \mathcal{A}\} \leq |\nu|(A) \leq 2\sup\{|\nu(B)| : B \subset A, B \in \mathcal{A}\}$$

([Sw3] 2.2.7). Let $\|x'\| \leq 1$. Then

$$\sup\{\|\mu(B)\| : B \subset A, B \in \mathcal{A}\}$$
$$= \sup\{|x'\mu(B)| : B \subset A, B \in \mathcal{A}, \|x'\| \leq 1\}$$
$$\leq \sup\{|x'\mu|(B) : B \subset A, B \in \mathcal{A}, \|x'\| \leq 1\} = \sup\{|x'\mu|(A) : \|x'\| \leq 1\}$$
$$\leq 2\sup\{|x'\mu(B)| : B \subset A, B \in \mathcal{A}, \|x'\| \leq 1\}$$
$$= 2\sup\{\|\mu(B)\| : B \subset A, B \in \mathcal{A}\}.$$

Thus. (ii) follows from (i).

Thus, from Proposition 3.45 μ has finite semi-variation iff μ is bounded. Conditions for μ to be bounded are given in Theorem 3.28 and Corollary 3.33.

We next show that the stronger conclusion of Corollary 3.43 can be used to establish a strong boundedness property for strongly bounded set functions.

Proposition 3.46. *Let X be a Banach space and $\mu : \mathcal{A} \to X$ be strongly bounded. If $\{A_j\} \subset \mathcal{A}$ is pairwise disjoint, then for every $\varepsilon > 0$ there exists an N such that $\|\mu\|(\cup_{j=m}^n A_j) < \varepsilon$ for $n > m \geq N$. In particular, the semi-variation is strongly bounded in the sense that $\|\mu\|(A_j) \to 0$.*

Proof: By Corollary 3.43 there exists N such that $\left\|\sum_{j=m}^\infty \mu(B_j)\right\| < \varepsilon$ for $B_j \subset A_j, B_j \in \mathcal{A}, m \geq N$. Therefore, $\left\|\sum_{j=m}^n \mu(B_j)\right\| \leq 2\varepsilon$ for $B_j \subset$

$A_j, B_j \in \mathcal{A}$, $n > m \geq N$. Suppose $n > m \geq N$ and $B \in \mathcal{A}$ with $B \subset \cup_{j=m}^n A_j$. Then $B = \cup_{j=m}^n B \cap A_j$ so $\|\mu(B)\| = \left\| \sum_{j=m}^n \mu(B \cap A_j) \right\| \leq 2\varepsilon$. By Proposition 3.45, $\|\mu\| \left(\cup_{j=m}^n A_j \right) \leq 4\varepsilon$.

Using Theorem 3.15 we can derive a result of Diestel connecting bounded, finitely additive set functions and strongly bounded set functions ([DU] I.4.2). For this we first present an example.

Example 3.47. Let $\mathcal{A}$ be the algebra of finite/co-finite subsets of $\mathbb{N}$; $\mathcal{A} = \{A : \text{either } A \text{ or } A^c \text{ is finite}\}$. Define $\mu : \mathcal{A} \to c_0$ by $\mu(A) = \chi_A$ if A is finite and $\mu(A) = -\chi_{A^c}$ if A^c is finite. Then μ is bounded and finitely additive but not strongly bounded since $\mu(\{j\}) = e^j \not\to 0$.

Theorem 3.48. *Let X be a sequentially complete LCTVS. Then X contains no subspace isomorphic to c_0 iff every bounded, finitely additive X valued set function defined on an algebra of sets is strongly bounded.*

Proof: Example 3.47 shows that if X contains a subspace isomorphic to c_0, then there is a bounded, finitely additive X valued set function defined on an algebra which is not strongly bounded.

Suppose that X contains a subspace isomorphic to c_0 and $\mu : \mathcal{A} \to X$ is a bounded, finitely additive set function defined on an algebra $\mathcal{A}$. Let $\{A_j\}$ be a pairwise disjoint sequence from $\mathcal{A}$. By Corollary 3.33 $\sum_j \mu(A_j)$ is c_0 multiplier convergent and by Theorem 3.15 the series $\sum_j \mu(A_j)$ is subseries convergent. Hence, $\mu(A_j) \to 0$ and μ is strongly bounded.

Finally, in this section we consider the class of vector valued measures of bounded variation. For simplicities sake, we consider only the case of set functions with values in a normed space.

Recall that a series $\sum_j x_j$ in a normed space is absolutely convergent iff $\sum_{j=1}^\infty \|x_j\| < \infty$. An absolutely convergent series is obviously subseries Cauchy so if X is a Banach space an absolutely convergent series is subseries convergent. The converse holds in a finite dimensional space but not in infinite dimensional spaces [consider $\sum_j (1/j)e^j$ in c_0 or recall the Dvoretsky-Rogers Theorem ([Day], [Sw2] 30.1.1)].

Definition 3.49. Let X be a normed space and $\mu : \mathcal{A} \to X$ be finitely additive. If $E \in \mathcal{A}$, the variation of μ on E is defined to be

$$|\mu|(E) = \sup \left\{ \sum_{j=1}^n \|\mu(A_j)\| : \{A_j\}_{j=1}^n \text{ is a partition of } E \text{ with } A_j \in \mathcal{A} \right\}.$$

If $|\mu|\,(S) < \infty$, μ is said to have bounded variation.

It is routine to show that the variation $|\mu| : \mathcal{A} \to [0,\infty]$ is finitely additive. Since $\|\mu(A)\| \le |\mu|\,(A)$ for $A \in \mathcal{A}$, if μ has bounded variation, then μ is bounded. If $\mu : \Sigma \to \mathbb{R}$ is countably additive, then μ has bounded variation ([Sw3] 2.2.1). This statement is false for vector valued set functions as the following example shows.

Example 3.50. Let $\mathcal{P}$ be power set of $\mathbb{N}$. Let X be a normed space and $\sum_j x_j$ subseries convergent in X. Define $\mu : \mathcal{P} \to X$ by $\mu(\sigma) = \sum_{j\in\sigma} x_j$. Then it is easily seen that μ is countably additive and bounded [indeed $\{\mu(\sigma) : \sigma \subset \mathbb{N}\}$ is relatively compact by Theorem 2.47]. However, μ has bounded variation iff $\sum_{j=1}^{\infty} \|\mu(\{j\})\| = \sum_{j=1}^{\infty} \|x_j\| < \infty$, i.e., iff $\sum_j x_j$ is absolutely convergent. Thus, if X is infinite dimensional, by the Dvoretsky-Rogers Theorem ([Day], [Sw2] 30.1.1), there is a countably additive X valued set function defined on a σ-algebra which is of infinite variation.

We have a characterization of set functions having bounded variation in terms of absolutely converging series.

Theorem 3.51. *Let* $\mu : \mathcal{A} \to X$ *be finitely additive. The following are equivalent:*

(i) *μ has bounded variation,*
(ii) *for every pairwise disjoint sequence $\{A_j\} \subset \mathcal{A}$, $\sum_j \mu(A_j)$ is absolutely convergent,*
(iii) *for every pairwise disjoint sequence $\{A_j\} \subset \mathcal{A}$, the series $\sum_{j=1}^{\infty} \|\mu(B_j)\|$ converge uniformly for $B_j \subset A_j$ with $B_j \in \mathcal{A}$.*

Proof: That (i) implies (ii) is clear.

Suppose that (ii) holds. We establish (iii) by using Li's Lemma 3.29. Set $E_j = \{B \in \mathcal{A} : B \subset A_j\}$ and define $f_j : E_j \to \mathbb{R}$ by $f_j(B) = \|\mu(B)\|$. Then Lemma 3.29 gives (iii) immediately.

Clearly (iii) implies (ii).

Suppose that (ii) holds but (i) fails. Note that μ is bounded by Theorem 3.28. Set $M = \sup\{\|\mu(A)\| : A \in \mathcal{A}\}$. There exists a partition $\{A_1^1, ..., A_n^1, A_{n+1}^1\}$ of S such that

$$\sum_{j=1}^{n+1} \|\mu(A_j^1)\| > M + 1,$$

where some $\{A_j^1\}$, say, A_{n+1}^1 satisfies $|\mu|\,(A_{n+1}^1) = \infty$ since $|\mu|$ is finitely additive. Then

$$\sum_{j=1}^{n} \left\| \mu(A_j^1) \right\| \geq 1 + M - \left\| \mu(A_{n+1}^1) \right\| \geq 1.$$

Now treat A_{n+1}^1 as S above to obtain a partition of A_{n+1}^1, $\{A_1^2, ..., A_m^2, A_{m+1}^2\}$ with $\sum_{j=1}^{m} \left\| \mu(A_j^2) \right\| \geq 2$ and $|\mu|\,(A_{m+1}^2) = \infty$. Continuing this construction produces a pairwise disjoint sequence $\{A_1^1, ..., A_n^1, A_1^2, ..., A_m^2, ...\}$ which violates condition (ii).

The equivalence of (i) and (ii) was established by Thorpe ([Thr]).

Chapter 4

The Orlicz-Pettis Theorem

As noted earlier the classical version of the Orlicz-Pettis Theorem for normed spaces asserts that a series in a normed space which is subseries convergent in the weak topology of the space is subseries convergent in the norm topology of the space ([Or], [Pe]). The theorem was originally established by Orlicz for weakly sequentially complete spaces but was evidently known in full generality by the Polish mathematicians as it appears as a statement in Banach's book ([Ba]). The first version available in English was established by Pettis in [Pe] where it was used to treat topics in vector valued integration — the Pettis integral. The theorem was extended to locally convex spaces by McArthur ([Mc]). For historical discussions of the theorem see [Ka3], [DU], or [FL]. Since a series is subseries convergent iff the series is m_0 multiplier convergent, it is natural to ask what sequence spaces λ have the property that series which are λ multiplier convergent in the weak topology are λ multiplier convergent in some stronger topology such as the Mackey topology. We will refer to such results as Orlicz-Pettis Theorems.

The locally convex topologies which we utilize will all be polar topologies which are described briefly in Appendix A. We record the polar topologies which we will encounter. Let X, X' be a pair of spaces in duality with the duality pairing $\langle , \rangle$. The weak topology $\sigma(X, X')$ (strong topology $\beta(X, X')$) is the polar topology generated by the finite subsets ($\sigma(X', X)$ bounded subsets) of X'. The Mackey topology is the polar topology $\tau(X, X')$ generated by the absolutely convex, $\sigma(X', X)$ compact subsets of X'. We will also use two other polar topologies. The polar topology $\lambda(X, X')(\gamma(X, X'))$ is the polar topology on X generated by the family of all $\sigma(X', X)$ compact subsets of X' [conditionally $\sigma(X', X)$ sequentially

49

compact subsets of X'; a subset $B \subset X'$ is conditionally $\sigma(X', X)$ sequentially compact if every sequence $\{x'_j\} \subset B$ has a subsequence $\{x'_{n_j}\}$ such that $\lim \langle x'_{n_j}, x \rangle$ exists for every $x \in X$ ([Din])]. Obviously, $\lambda(X, X')$ is stronger than the Mackey topology $\tau(X, X')$ and can be strictly stronger ([K1] 21.4). The topologies $\lambda(X, X')$ and $\gamma(X, X')$ are not comparable.

We recall some basic results from Appendix A.3-6.

Definition 4.1. Let X be a vector space and σ and τ two vector topologies on X. We say that τ is linked to σ if τ has a neighborhood base at 0 consisting of σ closed sets. [The terminology is that of Wilansky ([Wi] 6.1.9).]

For example, the polar topologies $\beta(X, X'), \tau(X, X'), \gamma(X, X')$ and $\lambda(X, X')$ are linked to the weak topology $\sigma(X, X')$.

Lemma 4.2. *Let X be a vector space and σ and τ two vector topologies on X such that τ is linked to σ.*

(i) If $\{x_j\} \subset X$ is τ Cauchy and if $\sigma - \lim x_j = x$, then $\tau - \lim x_j = x$.
(ii) If (X, σ) is sequentially complete and $\sigma \subset \tau$, then (X, τ) is sequentially complete.

Remark 4.3. It is important that the topologies σ and τ are linked in Lemma 4.2. For example, consider the space c with its weak topology $\sigma(c, l^1)$ and the topology of pointwise convergence p. The series $\sum_j e^j$ is p convergent, the partial sums of the series are $\sigma(c, l^1)$ Cauchy, but the series is not $\sigma(c, l^1)$ convergent.

Lemma 4.4. *Let X be a vector space and σ and τ two vector topologies on X such that τ is linked to σ. If every series $\sum_j x_j$ which is σ subseries convergent satisfies $\tau - \lim x_j = 0$, then every series in X which is σ subseries convergent is τ subseries convergent.*

The proofs of the lemmas can be found in Appendix A.3-6.

Throughout this chapter λ will denote a scalar sequence space which contains c_{00}, the space of sequences which are eventually 0. If

$$\lambda^\beta = \left\{ \{s_j\} : \sum_{j=1}^\infty s_j t_j \text{ converges for every } \{t_j\} \in \lambda \right\}$$

is the β-dual of λ, we write $s \cdot t = \sum_{j=1}^\infty s_j t_j$ for $\{s_j\} \in \lambda^\beta$ and $\{t_j\} \in \lambda$. Note that λ and λ^β are in duality with respect to the bilinear pairing $s \cdot t$.

Recall that if λ has a vector topology τ, then (λ, τ) is an AK-space if $t = \tau - \lim \sum_{j=1}^{n} t_j e^j = \sum_{j=1}^{\infty} t_j e^j$ for each $t \in \lambda$ [Appendix B.2]. We show that the conclusion of any Orlicz-Pettis Theorem for a Hellinger-Toeplitz topology is characterized by the AK-property. Recall that a locally convex topology $w(X, X')$ defined for dual pairs X, X' is said to be a Hellinger-Toeplitz topology if whenever a linear map $T : X \to Y$ is $\sigma(X, X')-\sigma(Y, Y')$ continuous, then T is also $w(X, X')-w(Y, Y')$ continuous ([Wi] 11.1.5 or see Appendix A.1; note that Hellinger-Toeplitz topologies must be defined for dual pairs). For example, the polar topologies $\beta(X, X'), \tau(X, X'), \gamma(X, X')$ and $\lambda(X, X')$ are Hellinger-Toeplitz topologies [Appendix A.2].

Theorem 4.5. *Let w be a Hellinger-Toeplitz topology for dual pairs. The following are equivalent:*

(i) *For every dual pair X, X' a series which is λ multiplier convergent for the weak topology $\sigma(X, X')$ is λ multiplier convergent with respect to $w(X, X')$.*

(ii) *$(\lambda, w(\lambda, \lambda^\beta))$ is an AK-space.*

Proof: Assume (i). Then $\sum_j e^j$ is λ multiplier convergent with respect to $\sigma(\lambda, \lambda^\beta)$ so by (i), $\sum_j e^j$ is λ multiplier convergent with respect to $w(\lambda, \lambda^\beta)$. But, this means that if $t \in \lambda$, then $t = \sum_{j=1}^{\infty} t_j e^j$, where the series is $w(\lambda, \lambda^\beta)$ convergent so (ii) holds.

Assume (ii). Let $\sum_j x_j$ be λ multiplier convergent with respect to $\sigma(X, X')$. Consider the summing operator $S : \lambda \to X$, $St = \sum_{j=1}^{\infty} t_j x_j$ $[\sigma(X, X')$ limit $]$. By Theorem 2.2, S is $\sigma(\lambda, \lambda^\beta) - \sigma(X, X')$ continuous and, therefore, $w(\lambda, \lambda^\beta) - w(X, X')$ continuous. If $t = \{t_j\} \in \lambda$, then $t = w(\lambda, \lambda^\beta) - \lim \sum_{j=1}^{n} t_j e^j$ so $Tt = w(X, X') - \lim \sum_{j=1}^{n} t_j x_j = \sum_{j=1}^{\infty} t_j x_j$. Hence, (i) holds.

Condition (i) is, of course, just the conclusion of the Orlicz-Pettis Theorem for the Hellinger Toeplitz topology $w(X, X')$. Thus, in order to establish an Orlicz-Pettis Theorem for a Hellinger-Toeplitz topology, it suffices to check the AK-property for the topology $w(\lambda, \lambda^\beta)$. We now give several examples where this is the case.

Corollary 4.6. *Let λ be a barrelled AK-space. If $\sum_j x_j$ is λ multiplier convergent with respect to the weak topology $\sigma(X, X')$, then $\sum_j x_j$ is λ multiplier with respect to the strong topology $\beta(X, X')$.*

Proof: A barrelled space always carries the strong topology so the result follows from Theorem 4.5 since the strong topology is a Hellinger-Toeplitz topology.

Remark 4.7. Corollary 4.6 applies to any Banach [Frechet] AK-space. In particular, Corollary 4.6 applies to the spaces $\lambda = c_0, l^p \ (1 \le p < \infty), cs$ or bv_0 [Appendix B].

In general, Orlicz-Pettis Theorems do not hold for the strong topology even in the case of subseries convergent series as the following example shows.

Example 4.8. The series $\sum_j e^j$ is subseries convergent in l^∞ with respect to the weak topology $\sigma(l^\infty, l^1) = \sigma(l^\infty, (l^\infty)^\beta)$ but is not subseries convergent with respect to the strong topology $\beta(l^\infty, l^1) = \|\cdot\|_\infty$.

We consider Orlicz-Pettis Theorems for the strong topology in Chapter 5. These results require more stringent assumptions on the multiplier space λ.

We next establish an AK theorem for a general class of sequence spaces. Recall that a sequence space λ has the signed weak gliding hump property (signed-WGHP) if whenever $t \in \lambda$ and $\{I_j\}$ is an increasing sequence of intervals, then there exist a sequence of signs $\{s_j\}$ and a subsequence $\{n_j\}$ such that the coordinate sum $\sum_{j=1}^\infty s_j \chi_{I_{n_j}} t \in \lambda$; if the signs can all be chosen to be equal to 1, then λ is said to have the weak gliding hump property (WGHP). For examples, see Appendix B.

Theorem 4.9. *Assume that λ has signed-WGHP. Then*

(i) $(\lambda, \gamma(\lambda, \lambda^\beta))$ *is an AK-space.*
(ii) $(\lambda, \lambda(\lambda, \lambda^\beta))$ *is an AK-space.*

Proof: (i): Since $\gamma(\lambda, \lambda^\beta)$ is linked to $\sigma(\lambda, \lambda^\beta)$, it suffices to show that for every $t \in \lambda$ the series $\sum_{j=1}^\infty t_j e^j$ is $\gamma(\lambda, \lambda^\beta)$ Cauchy (Lemma 4.2). Suppose that there exist $\epsilon > 0, K \subset \lambda^\beta$ which is conditionally $\sigma(\lambda^\beta, \lambda)$ sequentially compact and increasing intervals $\{I_j\}$ such that

$$\sup_{u \in K} \left| u \cdot \sum_{j \in I_k} t_j e^j \right| > \epsilon.$$

For each k pick $u^k \in K$ such that

$$(*) \quad \left| u^k \cdot \sum_{j \in I_k} t_j e^j \right| > \epsilon.$$

There exists an increasing sequence $\{n_k\}$ such that $\{u_{n_k}\}$ is $\sigma(\lambda^\beta, \lambda)$ Cauchy. Define the matrix

$$M = [m_{ij}] = [u^{n_i} \cdot \sum_{l \in I_{n_j}} t_l e^l].$$

We show that M is a signed $\mathcal{K}$-matrix (Appendix D.3). First the columns of M converge since $\{u^{n_i}\}$ is $\sigma(\lambda^\beta, \lambda)$ Cauchy. Next, if $\{p_j\}$ is an increasing sequence, there is a subsequence $\{q_j\}$ of $\{p_j\}$ and a sequence of signs $\{s_j\}$ such that $v = \sum_{j=1}^\infty s_j \sum_{l \in I_{n_{q_j}}} t_l e^l \in \lambda$. Then

$$\sum_{j=1}^\infty s_j m_{iq_j} = u^{n_i} \cdot v$$

so $\lim \sum_{j=1}^\infty s_j m_{iq_j}$ exists since $\{u^{n_i}\}$ is $\sigma(\lambda^\beta, \lambda)$ Cauchy. Hence, M is a signed $\mathcal{K}$-matrix and by the signed version of the Antosik-Mikusinski Matrix Theorem (Appendix D.3), the diagonal of M converges to 0. But, this contradicts $(*)$ and establishes (i).

(ii): Consider λ with the Mackey topology $\tau(\lambda, \lambda^\beta)$ so the dual of $(\lambda, \tau(\lambda, \lambda^\beta))$ is λ^β. We claim that $(\lambda, \tau(\lambda, \lambda^\beta))$ is $\tau(\lambda, \lambda^\beta)$ separable. This follows since $(\lambda, \sigma(\lambda, \lambda^\beta))$ is an AK-space so the $\sigma(\lambda, \lambda^\beta)$ closure of $S = span\{e^k : k \in \mathbb{N}\}$ is $\sigma(\lambda, \lambda^\beta)$ dense in λ. But, S has the same closure in $\sigma(\lambda, \lambda^\beta)$ and $\tau(\lambda, \lambda^\beta)$ so S is $\tau(\lambda, \lambda^\beta)$ dense in λ and $(\lambda, \tau(\lambda, \lambda^\beta))$ is $\tau(\lambda, \lambda^\beta)$ separable. This implies that $\sigma(\lambda^\beta, \lambda)$ compact sets are sequentially compact ([Wi] 9.5.3). Now, the proof of part (i) may be repeated using a $\sigma(\lambda^\beta, \lambda)$ compact (sequentially compact) set $K \subset \lambda^\beta$.

From Theorems 4.5 and 4.9, we obtain an Orlicz-Pettis Theorem for λ multiplier convergent series.

Corollary 4.10. *Assume that λ has signed-WGHP and let X be an LCTVS. If $\sum_j x_j$ is λ multiplier convergent with respect to the weak topology $\sigma(X, X')$, then $\sum_j x_j$ is λ multiplier convergent with respect to the topologies $\gamma(X, X')$ and $\lambda(X, X')$. In particular, if $\sum_j x_j$ is λ multiplier convergent with respect to $\sigma(X, X')$, then $\sum_j x_j$ is λ multiplier convergent with respect to the Mackey topology $\tau(X, X')$.*

Corollary 4.10 contains an Orlicz-Pettis Theorem with respect to the topology $\gamma(X, X')$ for a multiplier space λ which has the signed-WGHP. We observe here that any multiplier space for which the Orlicz-Pettis Theorem holds with respect to the topology $\gamma(X, X')$ has the property that the topology $\sigma(\lambda^\beta, \lambda)$ is sequentially complete [thus, Corollary 4.10 implies Stuart's sequential completeness result in 2.28]. For this observation, let $\{y^k\}$ be $\sigma(\lambda^\beta, \lambda)$ Cauchy. Set $y_j = \lim_k y^k \cdot e^j = \lim_k y_j^k$ and $y = \{y_j\}$. We claim that $y \in \lambda^\beta$ and $y^k \to y$ in $\sigma(\lambda^\beta, \lambda)$. Then $\sum_j e^j$ is λ multiplier convergent with respect to $\sigma(\lambda, \lambda^\beta)$ and is, therefore, λ multiplier convergent with respect to $\gamma(\lambda, \lambda^\beta)$ by hypothesis. Let $t \in \lambda$ and let $\epsilon > 0$. There exists N such that $\left| \sum_{j=m}^n y_j^k t_j \right| < \epsilon$ for $n > m \geq N$ and for all $k \in \mathbb{N}$ by the $\gamma(\lambda, \lambda^\beta)$ convergence. Then $\left| \sum_{j=m}^n y_j t_j \right| \leq \epsilon$ for $n > m \geq N$ so $\sum_j y_j t_j$ converges and $y \in \lambda^\beta$. Pick M such that $k \geq M$ implies that $\left| \sum_{j=1}^{N-1} (y_j^k t_j - y_j t_j) \right| < \epsilon$. If $k \geq M$, then

$$\left| \sum_{j=1}^\infty y_j^k t_j - \sum_{j=1}^\infty y_j t_j \right| \leq \left| \sum_{j=1}^{N-1} (y_j^k t_j - y_j t_j) \right| + \left| \sum_{j=N}^\infty y_j^k t_j \right| + \left| \sum_{j=N}^\infty y_j t_j \right| < 3\epsilon$$

so $y^k \to y$ in $\sigma(\lambda^\beta, \lambda)$.

Appendix B gives a list of sequence spaces with signed-WGHP to which Corollary 4.10 applies. In particular, the space m_0 has signed-WGHP (being monotone and having WGHP) so Corollary 4.10 applies to subseries convergent series. We give a formal statement of the subseries result.

Corollary 4.11. *Let X be an LCTVS. If $\sum_j x_j$ is subseries convergent with respect to the weak topology $\sigma(X, X')$, then $\sum_j x_j$ is subseries convergent with respect to the topologies $\gamma(X, X')$ and $\lambda(X, X')$. In particular, then $\sum_j x_j$ is subseries convergent with respect to the Mackey topology $\tau(X, X')$.*

The usual statement of the Orlicz-Pettis Theorem for subseries convergent series and the Mackey topology was established by McArthur ([Mc]). The statement for the topology $\lambda(X, X')$ was established Bennett and Kalton in [BK]. The version for $\gamma(X, X')$ is given by Dierolf in [Die].

We give an example covered by Corollary 4.10 but not by Corollary 4.11.

Example 4.12. Consider the series $\sum_j (1/j) e^j$ in cs, the space of convergent series (Appendix B). This series is obviously not subseries convergent in cs with respect to the norm topology. However, if $\lambda = bs$, the space of bounded series (Appendix B), and if $t \in bs$, then the series $\sum_j (t_j/j)e^j$ converges in cs since $\{t_j/j\} \in cs$ because $\{1/j\} \in bv_0 = (bs)^\beta$, the space of null sequences with bounded variation (Appendix B). Thus, $\sum_j (1/j)e^j$ is bs multiplier convergent in cs but not subseries convergent in cs.

Without some assumption on the multiplier space λ, the conclusion of Corollary 4.10 may fail.

Example 4.13. Let $c_c = c_0 \oplus span\{1,1,1,...\}$, the space of sequences which are eventually constant (Appendix B). If X is a TVS, then a series $\sum_j x_j$ in X is c_c multiplier convergent in X iff the series $\sum_j x_j$ converges in X. The series $\sum_j (e^{j+1} - e^j)$ is $\sigma(c_0, l^1)$ convergent in c_0 (to $-e^1$) and, therefore, c_c multiplier convergent with respect to $\sigma(c_0, l^1)$ but is not c_c multiplier convergent with respect to the norm or Mackey topology of c_0.

The space l^∞ is monotone and, therefore, has WGHP so Corollary 4.10 applies to l^∞ or bounded multiplier convergent series. We give a formal statement of this version of the Orlicz-Pettis Theorem.

Corollary 4.14. *Let X be an LCTVS. If the series $\sum_j x_j$ is l^∞ multiplier convergent with respect to the weak topology $\sigma(X, X')$, then the series $\sum_j x_j$ is l^∞ multiplier convergent with respect to the topologies $\gamma(X, X')$ and $\lambda(X, X')$. In particular, if the series $\sum_j x_j$ is l^∞ multiplier convergent with respect to the weak topology, then the series is l^∞ multiplier convergent with respect to the Mackey topology $\tau(X, X')$.*

We can use Corollary 4.10 to give a generalization of an interesting and useful Orlicz-Pettis Theorem due to Kalton ([Ka3]).

Theorem 4.15. *Let λ have signed-WGHP. Let X, X' be a pair of vector spaces in duality and suppose that τ is a polar topology from this duality which is separable. If $\sum_j x_j$ is λ multiplier convergent with respect to $\sigma(X, X')$, then $\sum_j x_j$ is λ multiplier convergent with respect to τ.*

Proof: Let $D = \{d_k : k \in \mathbb{N}\}$ be τ dense in X and let $\mathcal{A}$ be a family of $\sigma(X', X)$ bounded sets which generate the polar topology $\tau, \tau = \tau_\mathcal{A}$ (Appendix A). If $A \in \mathcal{A}$, we show that A is conditionally $\sigma(X', X)$ sequentially compact and the result will follow from Corollary 4.10. Let $\{y_k\} \subset A$. Since $\{y_k\}$ is pointwise bounded on D, a diagonal procedure implies that there exists a subsequence $\{y_{n_k}\}$ of $\{y_k\}$ which converges pointwise on D ([DeS] 26.10). We claim that $\{y_{n_k}\}$ is $\sigma(X', X)$ Cauchy. Let $x \in X$. There exists a net $\{d^\alpha\}$ in D such that $\{d^\alpha\}$ converges to x with respect to τ. If $\epsilon > 0$, there exists β such that $\left|\langle y, x - d^\beta \rangle\right| < \epsilon/3$ for $y \in A$. There exists n such that $j, k \geq n$ implies that $\left|\langle y_{n_j} - y_{n_k}, d^\beta \rangle\right| < \epsilon/3$. If $j, k \geq n$, then

$$\left|\langle y_{n_j} - y_{n_k}, x \rangle\right| \leq \left|\langle y_{n_k}, x - d^\beta \rangle\right| + \left|\langle y_{n_j} - y_{n_k}, d^\beta \rangle\right| + \left|\langle y_{n_j}, d^\beta - x \rangle\right| < \epsilon.$$

Hence, $\lim \langle y_{n_k}, x \rangle$ exists so $\{y_{n_k}\}$ is $\sigma(X', X)$ Cauchy as claimed.

Theorem 4.15 has an application to the strong topology.

Corollary 4.16. *Let λ have signed-WGHP. If $(X, \beta(X, X'))$ is separable and $\sum_j x_j$ is λ multiplier convergent with respect to $\sigma(X, X')$, then $\sum_j x_j$ is λ multiplier convergent with respect to $\beta(X, X')$.*

Example 4.8 shows that the separability assumption in Theorem 4.15 and Corollary 4.16 is important. The subseries version of Theorem 4.15 is due to Kalton ([Ka3]).

Remark 4.17. If X is a barreled AK-space, Corollary 4.16 applies. In particular, Corollary 4.16 applies to c_0, l^p $(1 \leq p < \infty), cs$ or bv_0 with respect to their normed topologies.

In Corollary 4.11 we showed that a series which is subseries convergent with respect to the weak topology $\sigma(X, X')$ is actually subseries convergent with respect to two stronger polar topologies. Dierolf has shown that there is a strongest polar topology which has the same subseries convergent series as the weak topology ([Die]). He has established a similar result for bounded multiplier convergent series. We now establish both of these results of Dierolf and then give a generalization of his results to λ multiplier convergent series.

As established in Theorem 4.5 the conclusion of the Orlicz-Pettis Theorem with respect to a Hellinger-Toeplitz topology is associated with the AK property for the Hellinger-Toeplitz topology on the multiplier space. A series is subseries convergent iff the series is m_0 multiplier convergent, but m_0 is not an AK-space with respect to its "natural" topology, the $\|\cdot\|_\infty$ topology. We now define the Dierolf topology and show that m_o is an AK-space with respect to this topology. We also show that the Dierolf topology is a Hellinger-Toeplitz topology so Theorem 4.5 is applicable.

In treating the Dierolf topology we use some basic properties of l^1 which we now state for convenience.

Proposition 4.18. *For the space l^1, we have the following properties:*

(i) The topologies $\|\cdot\|_1$, $\sigma(l^1, l^\infty)$ *and* $\sigma(l^1, m_0)$ *have the same convergent sequences, Cauchy sequences, the same bounded sets and the same compact sets.*

(ii) A subset K *of* l^1 *is* $\|\cdot\|_1$ *relatively compact iff* $\lim_n \sum_{j=n}^{\infty} |t_j| = 0$ *uniformly for* $\{t_j\} \in K$.

For Proposition 4.18 see [K1] 22.4. For (ii) see [Sw2] 10.1.15. Part (i) will be established later in Chapter 7.

Let X, X' be vector spaces in duality. Let $\mathcal{M} = \mathcal{M}(X, X')$ be the family of all subsets $M \subset X'$ such that M is $\sigma(X', X)$ bounded and for every linear, continuous map

$$T : (X', \sigma(X', X)) \to (l^1, \sigma(l^1, m_0)),$$

TM is relatively compact in $(l^1, \|\cdot\|_1)$.

Definition 4.19. The Dierolf topology, $\delta_1(X, X')$, on X is the polar topology, $\tau_{\mathcal{M}}$, of uniform convergence on the elements of $\mathcal{M}$ (Appendix A).

From Theorem A.2 of Appendix A, we easily have

Theorem 4.20. $\delta_1(X, X')$ *is a Hellinger-Toeplitz topology.*

We now show that m_0 is an AK-space under the Dierolf topology $\delta_1(X, X')$. Actually, we have a stronger result.

Theorem 4.21. *For* $t \in m_0$, $\delta_1(m_0, l^1) - \lim_n \sum_{j=1}^{n} t_j e^j = t$, *uniformly for* $\|t\|_\infty \leq 1$. *In particular,* $(m_0, \delta_1(m_0, l^1))$ *is an AK-space.*

Proof: Let $M \in \mathcal{M}$ (relative to the duality between m_0 and l^1). Then M is relatively compact in $(l^1, \|\cdot\|_1)$ so by Proposition 4.18, $\lim_n \sum_{j=n}^{\infty} |s_j| = 0$ uniformly for $s \in M$. Thus, for $s \in M$ and $t \in m_0$ with $\|t\|_\infty \leq 1$, we have

$$\left| \sum_{j=n}^{\infty} s_j t_j \right| \leq \sum_{j=n}^{\infty} |s_j|.$$

Therefore, $\lim_n s \cdot \sum_{j=n}^{\infty} t_j e^j = 0$ uniformly for $s \in M, \|t\|_\infty \leq 1$, so $\delta_1(m_0, l^1) - \lim_n \sum_{j=1}^{n} t_j e^j = t$, uniformly for $\|t\|_\infty \leq 1$.

From Theorems 4.5 and 4.21, we obtain an Orlicz-Pettis Theorem for the Dierolf topology $\delta_1(X, X')$.

Theorem 4.22. *Let* X *be an LCTVS. If the series* $\sum_j x_j$ *is subseries convergent with respect to* $\sigma(X, X')$, *then* $\sum_j x_j$ *is subseries convergent with respect to* $\delta_1(X, X')$.

Actually, from Theorems 4.20, 4.21 and 4.22, we have an improvement of Corollary 4.11.

Theorem 4.23. *Let X be an LCTVS. If the series $\sum_j x_j$ is subseries convergent with respect to $\sigma(X, X')$, then $\delta_1(X, X') - \lim_n \sum_{j=1}^n t_j x_j = \sum_{j=1}^\infty t_j x_j$ uniformly for $t \in m_0, \|t\|_\infty \le 1$.*

If X is an LCTVS, it is clear that $\delta_1(X, X')$ is stronger than $\lambda(X, X')$, the topology of uniform convergence on $\sigma(X', X)$ compact sets, since any $\sigma(X', X)$ compact set belongs to the family $\mathcal{M}$ [relative to the duality between X and X'] by Proposition 4.18. Thus, Theorem 4.22 gives an improvement to the Orlicz-Pettis Theorem given in Corollary 4.11.

Remark 4.24. The topology $\delta_1(X, X')$ is stronger than the topology $\gamma(X, X')$, the topology of uniform convergence on the conditionally $\sigma(X', X)$ sequentially compact sets. For if $A \subset X'$ is conditionally $\sigma(X', X)$ sequentially compact and

$$T : (X', \sigma(X', X))) \to (l^1, \sigma(l^1, m_0))$$

is linear and continuous, then TA is conditionally $\sigma(l^1, m_0)$ sequentially compact. By Proposition 4.18, TA is relatively $\|\cdot\|_1$ compact since $\|\cdot\|_1$ and $\sigma(l^1, m_0)$ have the same Cauchy sequences. Thus, Theorem 4.22 gives an improvement of Corollary 4.11 for the topology $\gamma(X, X')$.

We now show that $\delta_1(X, X')$ is the strongest polar topology with the same subseries convergent series as $\sigma(X, X')$. For this we require a slight refinement of the statement in Proposition 4.18 (i).

Proposition 4.25. *Let $K \subset l^1$. The following are equivalent:*

(i) K is relatively $\|\cdot\|_1$ compact,
(ii) $\lim_n \sum_{j=n}^\infty |t_j| = 0$ uniformly for $t \in K$,
(iii) for each $s \in l^\infty$, $\lim_n \sum_{j=n}^\infty s_j t_j = 0$ uniformly for $t \in K$,
(iv) for each $s \in m_0$, $\lim_n \sum_{j=n}^\infty s_j t_j = 0$ uniformly for $t \in K$.

Proof: (i) and (ii) are equivalent by Proposition 4.18 and clearly (ii) implies (iii) implies (iv).

Suppose that (iv) holds but (ii) fails. Then there exists $\epsilon > 0$ such that for every k there exist $m_k > k$ and $t^k \in K$ such that $\sum_{i=m_k}^\infty |t_i^k| > 5\epsilon$. In particular, there exist $m_1 > 1, t^1 \in K$ such that $\sum_{i=m_1}^\infty |t_i^1| > 5\epsilon$. There exists $n_1 > m_1$ such that $\sum_{i=n_1+1}^\infty |t_i^1| < \epsilon$. Therefore, $\sum_{i=m_1}^{n_1} |t_i^1| < 4\epsilon$. Put $I_1 = [m_1, n_1], I_1^+ = \{i \in I_1 : t_i^1 \ge 0\}$ and $I_1^- = \{i \in I_1 : t_i^1 < 0\}$. Either

$\left|\sum_{i \in I_1^+} t_i^1\right| > 2\epsilon$ or $\left|\sum_{i \in I_1^-} t_i^1\right| > 2\epsilon$; pick one of these which satisfies this inequality and label it J_1 so $\left|\sum_{i \in J_1} t_i^1\right| > 2\epsilon$. Continuing this construction produces an increasing sequence of finite subsets J_k of $\mathbb{N}$, $t^k \in K$, $m_k < n_k < m_{k+1} < \ldots$ with

$$J_k \subset [m_k, n_k], \left|\sum_{i \in J_k} t_i^k\right| > 2\epsilon, \sum_{i=n_k+1}^{\infty} |t_i^k| < \epsilon.$$

Put $s = \sum_{i=1}^{\infty} \chi_{J_i}$ [coordinate sum] so $s \in m_0$. Then

$$\left|\sum_{i=m_k}^{\infty} s_i t_i^k\right| \geq \left|\sum_{i \in J_k} t_i^k\right| - \sum_{i=n_k+1}^{\infty} |t_i^k| > \epsilon$$

so (iv) fails to hold.

Theorem 4.26. $\delta_1(X, X')$ *is the strongest polar topology with the same subseries convergent series as* $\sigma(X, X')$.

Proof: From Theorem 4.22, $\delta_1(X, X')$ and $\sigma(X, X')$ have the same sub-series convergent series. Suppose α is a polar topology with the same sub-series convergent series as $\sigma(X, X')$ and let α be the topology of uniform convergence on the family $\mathcal{A}$ of $\sigma(X', X)$ bounded subsets of X' (Appendix A). Let $A \in \mathcal{A}$. We show that $A \in \mathcal{M}$ [with respect to the duality between X and X']. Let

$$T : (X', \sigma(X', X))) \to (l^1, \sigma(l^1, m_0))$$

be linear and continuous. Then

$$T' : (m_0, \sigma(m_0, l^1)) \to (X, \sigma(X, X'))$$

is linear and continuous. Now $\sum_j e^j$ is m_0 multiplier convergent with respect to $\sigma(m_0, l^1)$ so $\sum_j T'e^j$ is m_0 multiplier convergent with respect to $\sigma(X, X')$ and, therefore, with respect to α. Thus, if $s \in m_0$,

$$\lim_n \sum_{j=n}^{\infty} s_j \langle x', T'e^j \rangle = \lim_n \sum_{j=n}^{\infty} s_j \langle Tx', e^j \rangle = 0$$

uniformly for $x' \in A$. From Proposition 4.25, TA is relatively $\|\cdot\|_1$ compact so $A \in \mathcal{M}$. Thus, $\mathcal{A} \subset \mathcal{M}$ and α is weaker than $\delta_1(X, X')$.

As we have seen the Dierolf topology is the strongest polar topology which has the same subseries convergent series as the weak topology. Twed-dle has shown that there is a strongest locally convex topology which has the same subseries convergent series as the weak topology ([Tw]). We now give a description of the Tweddle topology. Let X be an LCTVS. Let $\mathcal{O}$ be the set of all X valued series $\sum_j x_j$ which are $\sigma(X, X')$ subseries convergent. If $\sum_j x_j \in \mathcal{O}$, we write $\sum_{j=1}^{\infty} x_j$ for the $\sigma(X, X')$ sum of this series. Let $X^\#$ be the space of all linear functionals x' on X such that

$$\sum_{j=1}^{\infty} \langle x', x_j \rangle = \left\langle x', \sum_{j=1}^{\infty} x_j \right\rangle \quad \text{for all} \quad \sum_j x_j \in \mathcal{O}.$$

The Mackey topology, $\tau(X, X^\#)$, is the Tweddle topology on X and is denoted by $t(X, X')$. We have the following important property of the Tweddle topology.

Theorem 4.27. *The Tweddle topology $t(X, X')$ is the strongest locally convex topology on X which has the same subseries convergent series as the weak topology $\sigma(X, X')$.*

Proof: Suppose that ν is a locally convex topology on X which has the same subseries convergent series as $\sigma(X, X')$. Let $H' = (X, \nu)'$. Then for $\sum_j x_j \in \mathcal{O}$ and $x' \in H'$, we have $\sum_{j=1}^{\infty} \langle x', x_j \rangle = \left\langle x', \sum_{j=1}^{\infty} x_j \right\rangle$ so $x' \in X^\#$ and $H' \subset X^\#$. Therefore, $\tau(X, H')$ is weaker than $\tau(X, X^\#) = t(X, X')$. But, $\nu \subset \tau(X, H')$ so ν is weaker than $t(X, X')$.

We give an example where the Tweddle topology is strictly stronger than the Mackey topology by computing the space $X^\#$ and comparing it to X'. The example uses the Nikodym Boundedness Theorem for countably additive set functions which we prove later in Theorem 4.60 (see also [Sw3] 2.8.8 for a statement).

Example 4.28. Let Σ be a σ-algebra of subsets of a set S. Let $B(S, \Sigma)$ be the space of all bounded, real valued Σ-measurable functions defined on S. Let $ca(\Sigma)$ be the space of all real valued, countably additive set functions defined on Σ and let $\Gamma = span\{\delta_t : t \in S\}$, where δ_t is the Dirac measure concentrated at t. The weak topology $\sigma(B(S, \Sigma), \Gamma)$ is just the topology of pointwise convergence, p, on $B(S, \Sigma)$ so $(B(S, \Sigma), p)' = \Gamma$. We show that $B(S, \Sigma)^\#$ (with respect to p) is $ca(\Sigma)$ so $t(B(S, \Sigma), \Gamma) = \tau(B(S, \Sigma), ca(\Sigma))$ is strictly stronger than $\tau(B(S, \Sigma), \Gamma)$.

First, suppose $f \in B(S, \Sigma)^\#$. Then f induces a set function on Σ, still denoted by f, defined by $f(E) = \langle f, \chi_E \rangle$ for $E \in \Sigma$. We claim that $f \in ca(\Sigma)$. Let $\{E_j\} \subset \Sigma$ be pairwise disjoint. Then $\sum_j \chi_{E_j}$ is subseries convergent with respect to p and the series $\sum_j \chi_{E_j}$ converges to $\chi_{\cup_{j=1}^{\infty} E_j}$ with respect to p. Therefore,

$$\left\langle f, \chi_{\cup_{j=1}^{\infty} E_j} \right\rangle = f(\cup_{j=1}^{\infty} E_j) = \sum_{j=1}^{\infty} \left\langle f, \chi_{E_j} \right\rangle = \sum_{j=1}^{\infty} f(E_j),$$

and f is countably additive. Thus, $B(S, \Sigma)^{\#} \subset ca(\Sigma)$.

Next, let $\nu \in ca(\Sigma)$ and let $\sum_j g_j$ be subseries convergent with respect to p. We claim that $\{\sum_{j \in \sigma} g_j : \sigma \text{ finite}\}$ is bounded with respect to the sup-norm, $\|\cdot\|_{\infty}$, on $B(S, \Sigma)$. If this is not the case, for every k there exist finite σ_k and $t_k \in S$ such that

$$(*) \qquad \left| \sum_{j \in \sigma_k} g_j(t_k) \right| > k.$$

Since $\sum_j g_j$ is subseries convergent in $B(S, \Sigma)$ with respect to p, the series $\sum_j \{g_j(t_k)\}_{k=1}^{\infty}$ is subseries convergent in l^{∞} with respect to the topology of coordinatewise convergence in l^{∞}. Define $\mu_k : 2^{\mathbb{N}} \to \mathbb{R}$ by $\mu_k(\sigma) = \sum_{j \in \sigma} g_j(t_k)$. Note that $\mu_k \in ca(2^{\mathbb{N}})$ and if $\sigma \subset \mathbb{N}$,

$$\sup_k |\mu_k(\sigma)| = \sup \left| (\sum_{j \in \sigma} g_j)(t_k) \right| < \infty.$$

By the Nikodym Boundedness Theorem (Theorem 4.60 and or [Sw3] 2.8.8),

$$\sup_k \sup_{\sigma} |\mu_k(\sigma)| = \sup_{\sigma} \sup_k \left| (\sum_{j \in \sigma} g_j)(t_k) \right| < \infty.$$

But, this contradicts $(*)$.

If $\{n_k\}$ is a subsequence, then $\{\sum_{j=1}^{k} g_{n_j}\}$ is uniformly bounded on S by the claim established above so by the Bounded Convergence Theorem,

$$\lim_k \sum_{j=1}^{k} \int_S g_{n_j} d\nu = \int_S \sum_{j=1}^{\infty} g_{n_j} d\nu.$$

Therefore, $\nu \in B(S, \Sigma)^{\#}$ and $B(S, \Sigma)^{\#} = ca(\Sigma)$.

There is an analogous Dierolf topology for bounded multiplier convergent series which we will now describe. The proofs of the various properties of the Dierolf topology for bounded multiplier convergent series are almost identical to the proofs for the Dierolf topology for the subseries case so we will give the appropriate statements but omit the proofs.

Let X, X' be in duality. Let $\mathcal{N} = \mathcal{N}(X, X')$ be the family of all $\sigma(X', X)$ bounded subsets $N \subset X'$ such that for every continuous linear operator

$$T : (X', \sigma(X', X)) \to (l^1, \sigma(l^1, l^\infty))$$

TN is relatively compact in $(l^1, \|\cdot\|_1)$ [again note Proposition 4.25].

Definition 4.29. The Dierolf topology for bounded multiplier convergent series, $\delta_2(X, X')$, on X is the polar topology, $\tau_\mathcal{N}$, of uniform convergence on the elements of $\mathcal{N} = \mathcal{N}(X, X')$ [Appendix A].

We now state the analogues of Theorems 4.20, 4.21, 4.22 and 4.23 for the Dierolf topology $\delta_2(X, X')$.

Theorem 4.30. *The topology $\delta_2(X, X')$ is a Hellinger-Toeplitz topology.*

Theorem 4.31. *For $t \in l^\infty$,*

$$\delta_2(l^\infty, l^1) - \lim_n \sum_{j=1}^n t_j e^j = t$$

uniformly for $\|t\|_\infty \leq 1$. In particular, $(l^\infty, \delta_2(l^\infty, l^1))$ is an AK-space.

Theorem 4.32. *Let X be an LCTVS. If the series $\sum_j x_j$ is bounded multiplier convergent in the weak topology $\sigma(X, X')$, then the series $\sum_j x_j$ is bounded multiplier convergent in the Dierolf topology $\delta_2(X, X')$.*

Theorem 4.33. *Let X be an LCTVS. If the series $\sum_j x_j$ is bounded multiplier convergent in the weak topology $\sigma(X, X')$, then $\delta_2(X, X') - \lim_n \sum_{j=1}^n t_j e^j = t$ uniformly for $\|t\|_\infty \leq 1$.*

The analogues of the statements in Remark 4.24 also hold for the Dierolf topology $\delta_2(X, X')$.

Remark 4.34. $\delta_2(X, X')$ is stronger than $\lambda(X, X')$, the topology of uniform convergence on the $\sigma(X', X)$ compact subsets of X'. Thus, Theorem 4.32 gives an improvement to the statement in Corollary 4.14 for the topology $\lambda(X, X')$.

Remark 4.35. $\delta_2(X, X')$ is stronger than $\gamma(X, X')$, the topology of uniform convergence on the $\sigma(X', X)$ bounded sets which are conditionally $\sigma(X', X)$ sequentially compact. Thus, Theorem 4.32 gives an improvement to the statement in Corollary 4.14 for the topology $\gamma(X, X')$.

Finally, the analogue of Theorem 4.26 holds for the Dierolf topology $\delta_2(X, X')$.

Theorem 4.36. *The Dierolf topology $\delta_2(X, X')$ is the strongest polar topology on an LCTVS X with the same bounded multiplier convergent series as the weak topology $\sigma(X, X')$.*

The proof of this theorem proceeds as the proof of Theorem 4.26 except that the statement in Proposition 4.25 (iii) is used in place of the statement in Proposition 4.25 (iv).

Using the Dierolf topologies for m_0 and l^∞ multiplier convergent series as models, we show that an analogous topology can be defined for arbitrary sequence spaces of multipliers, and when the multiplier space λ satisfies the signed-WGHP, we compare the general Dierolf topology to the topologies $\lambda(X, X')$ and $\gamma(X, X')$.

Definition 4.37. Let λ be a sequence space containing c_{00}. A subset $K \subset \lambda^\beta$ has uniform tails if for every $s \in \lambda$, $\lim_n \sum_{j=n}^{\infty} s_j t_j = 0$ uniformly for $t \in K$.

From Proposition 4.25, if $\lambda = m_0$ or $\lambda = l^\infty$, a subset $K \subset l^1 = \lambda^\beta$ has uniform tails iff K is relatively $\|\cdot\|_1$ compact.

Let X be an LCTVS. We say that a $\sigma(X', X)$ bounded subset $D \subset X'$ belongs to $\mathcal{D}_\lambda$ iff for every continuous linear operator

$$T : (X', \sigma(X', X)) \to (\lambda^\beta, \sigma(\lambda^\beta, \lambda))$$

the subset $TD \subset \lambda^\beta$ has uniform tails.

Definition 4.38. The general Dierolf topology on X, denoted by $\mathcal{D}_\lambda(X, X')$, is the polar topology $\tau_{\mathcal{D}_\lambda}$ of uniform convergence on the elements of $\mathcal{D}_\lambda$.

From Proposition 4.25 it follows that $\delta_1(X, X') = \mathcal{D}_{m_0}(X, X')$ and $\delta_2(X, X') = \mathcal{D}_{l^\infty}(X, X')$ so it is reasonable to refer to $\mathcal{D}_\lambda(X, X')$ as a Dierolf topology with respect to λ.

In order to establish the basic property of the general Dierolf topology we prove the following result.

Proposition 4.39. *There is a 1-1 correspondence between $\sigma(X, X')$ λ multiplier convergent series $\sum_j x_j$ and continuous linear operators*

$$T : (X', \sigma(X', X)) \to (\lambda^\beta, \sigma(\lambda^\beta, \lambda)).$$

The correspondence is given by $x_j = T'e^j$.

Proof: Suppose that $\sum_j x_j$ is λ multiplier convergent with respect to $\sigma(X, X')$. Define the summing operator $S : \lambda \to X$ by $St = \sum_{j=1}^{\infty} t_j x_j$ $[\sigma(X, X')$ sum]. Then S is $\sigma(\lambda, \lambda^\beta) - \sigma(X, X')$ continuous by Theorem 2.2. Therefore, $T = S' : X' \to \lambda^\beta$ is $\sigma(X', X) - \sigma(\lambda^\beta, \lambda)$ continuous and $T'x' = \{\langle x', x_j \rangle\}$.

If

$$T : (X', \sigma(X', X)) \to (\lambda^\beta, \sigma(\lambda^\beta, \lambda))$$

is linear and continuous, then $S = T' : \lambda \to X$ is $\sigma(\lambda, \lambda^\beta) - \sigma(X, X')$ continuous. Now $\sum_j e^j$ is $\sigma(\lambda, \lambda^\beta)$ λ multiplier convergent so $\sum_{j=1}^{\infty} T'e^j = \sum_{j=1}^{\infty} Se^j = \sum_{j=1}^{\infty} x_j$ is $\sigma(X, X')$ λ multiplier convergent and the correspondence follows.

Remark 4.40. Thus, it follows from Proposition 4.39, to check that a subset $D \subset X'$ belongs to $\mathcal{D}_\lambda$, it suffices to show that if $t \in \lambda$, then $\lim_n \sum_{j=n}^{\infty} t_j \langle x', x_j \rangle = 0$ uniformly for $x' \in D$ whenever $\sum_j x_j$ is λ multiplier convergent with respect to $\sigma(X, X')$.

Theorem 4.41. *The general Dierolf topology $\mathcal{D}_\lambda(X, X')$ is the strongest polar topology on X with the same λ multiplier convergent series as $\sigma(X, X')$.*

Proof: Let $\sum_j x_j$ be λ multiplier convergent with respect to $\sigma(X, X')$. Let $S : \lambda \to X$ be the summing operator with respect to $\sum_j x_j$, $St = \sum_{j=1}^{\infty} t_j x_j$ $[\sigma(X, X')$ sum]. By Theorem 2.2, S is $\sigma(\lambda, \lambda^\beta) - \sigma(X, X')$ continuous and $S' = T : (X', \sigma(X', X)) \to (\lambda^\beta, \sigma(\lambda^\beta, \lambda))$ is linear and continuous. Let $D \in \mathcal{D}_\lambda$. Then TD has uniform tails. Therefore, for every $s \in \lambda$, $\lim_n \sum_{j=n}^{\infty} s_j \langle x', x_j \rangle = 0$ uniformly for $x' \in D$ since $Tx' = \{\langle x', x_j \rangle\}$. That is, the series $\sum_j s_j x_j$ converges in $\mathcal{D}_\lambda(X, X')$.

Suppose that α is a polar topology with the same λ multiplier convergent series as $\sigma(X, X')$. Let α be the polar topology of uniform convergence on the family $\mathcal{A}$ of $\sigma(X', X)$ bounded sets (Appendix A). Let $A \in \mathcal{A}$. If $\sum_j x_j$ is λ multiplier convergent with respect to $\sigma(X, X')$, then $\sum_j x_j$ is λ multiplier convergent with respect to α so if $t \in \lambda$, then $\lim_n \sum_{j=n}^{\infty} t_j \langle x', x_j \rangle = 0$ uniformly for $x' \in A$. Then $A \in \mathcal{D}_\lambda$ by Remark 4.40 and α is weaker than $\mathcal{D}_\lambda(X, X')$.

As in Theorem 4.20 we can show that $\mathcal{D}_\lambda(X, X')$ is a Hellinger-Toeplitz topology.

Theorem 4.42. $\mathcal{D}_\lambda(X, X')$ *is a Hellinger-Toeplitz topology.*

Proof: From Theorem A.2 of Appendix A, we need to show that if $D \in \mathcal{D}_\lambda$ with respect to the dual pair Y, Y' and

$$T : (X, \sigma(X, X')) \to (Y, \sigma(Y, Y'))$$

is linear and continuous, where X, X' is another dual pair, then $T'D \in \mathcal{D}_\lambda$ with respect to the dual pair X, X'. Let

$$U : (X', \sigma(X', X)) \to (\lambda^\beta, \sigma(\lambda^\beta, \lambda))$$

be linear and continuous. Then

$$UT' : (Y', \sigma(Y', Y)) \to (\lambda^\beta, \sigma(\lambda^\beta, \lambda))$$

is linear and continuous so $UT'D$ has uniform tails in λ^β so $T'D \in \mathcal{D}_\lambda$ with respect to the dual pair X, X'.

From Theorems 4.5, 4.41 and 4.42, we have

Theorem 4.43. $(\lambda, \mathcal{D}_\lambda(\lambda, \lambda^\beta))$ *is an AK-space.*

We next compare the general Dierolf topology to the topologies $\lambda(X, X')$ and $\gamma(X, X')$. This will require an additional assumption on the multiplier space λ. We first establish a lemma.

Lemma 4.44. *Suppose $M \subset X'$ is such that there exist a series $\sum_j x_j$ which is λ multiplier convergent with respect to $\sigma(X, X')$ and a $t \in \lambda$ such that the series $\sum_{j=1}^\infty t_j \langle x', x_j \rangle$ do not converge uniformly for $x' \in M$. Then there exist $\epsilon > 0, \{x'_k\} \subset M$ and an increasing sequence of intervals $\{I_k\}$ with*

$$\left| \sum_{j \in I_k} t_j \langle x'_k, x_j \rangle \right| > \epsilon$$

for every k.

Proof: If the series do not converge uniformly for $x' \in M$, then there exists $\epsilon > 0$ such that for every k there exist $m_k > k, x' = x'(k) \in M$ such that $\left| \sum_{j=m_k}^\infty t_j \langle x', x_j \rangle \right| > 2\epsilon$. In particular, there exist $m_1, x'_1 \in M$ with $\left| \sum_{j=m_1}^\infty t_j \langle x'_1, x_j \rangle \right| > 2\epsilon$. Since the series $\sum_j t_j x_j$ is $\sigma(X, X')$ convergent, there exists $n_1 > m_1$ such that $\left| \sum_{j=n_1}^\infty t_j \langle x'_1, x_j \rangle \right| < \epsilon$. Thus, if $I_1 = [m_1, n_1]$, then $\left| \sum_{j \in I_1} t_j \langle x'_1, x_j \rangle \right| > \epsilon$. Now just continue the construction.

We first compare the general Dierolf topology with the topology $\gamma(X, X')$ of uniform convergence on $\sigma(X', X)$ bounded sets which are conditionally $\sigma(X', X)$ sequentially compact.

Theorem 4.45. *Let λ have signed-WGHP. Then $\gamma(X, X') \subset \mathcal{D}_\lambda(X, X')$.*

Proof: Suppose that $K \subset X'$ is $\sigma(X', X)$ bounded and conditionally $\sigma(X', X)$ sequentially compact. If K does not belong to $\mathcal{D}_\lambda$, then by Remark 4.40 and Lemma 4.44 there exist a multiplier convergent series $\sum_j x_j$ with respect to $\sigma(X, X')$, $t \in \lambda$, $\epsilon > 0$, $\{x'_k\} \subset K$ and an increasing sequence of intervals $\{I_k\}$ such that

$$(*) \quad \left| \sum_{j \in I_k} t_j \langle x'_k, x_j \rangle \right| > \epsilon$$

for every k. By the conditional $\sigma(X', X)$ sequential compactness of K, we may assume that $\lim \langle x'_k, x \rangle$ exists for every $x \in X$. Define a matrix

$$M = [m_{ij}] = [\sum_{l \in I_j} t_l \langle x'_i, x_l \rangle].$$

We show that M is a signed $\mathcal{K}$-matrix (Appendix D.3). First, the columns of M converge by the compactness condition. Next, if $\{p_j\}$ is an increasing sequence there is a further subsequence $\{q_j\}$ of $\{p_j\}$ and a sequence of signs $\{s_j\}$ such that $u = \{u_j\} = \sum_{j=1}^\infty s_j \chi_{I_{q_j}} t \in \lambda$. Then

$$\sum_{j=1}^\infty s_j m_{iq_j} = \left\langle x'_i, \sum_{j=1}^\infty u_j x_j \right\rangle$$

so $\lim_i \sum_{j=1}^\infty s_j m_{iq_j}$ exists. Thus, M is a signed $\mathcal{K}$-matrix. By the signed version of the Antosik-Mikusinski Matrix Theorem the diagonal of M converges to 0 (Appendix D.3). But, this contradicts $(*)$.

We next consider the topology $\lambda(X, X')$ of uniform convergence on $\sigma(X', X)$ compact subsets of X'.

Theorem 4.46. *Let λ have signed-WGHP. Then $\lambda(X, X') \subset \mathcal{D}_\lambda(X, X')$.*

Proof: Let $K \subset X'$ be $\sigma(X', X)$ compact. Assume that K does not belong to $\mathcal{D}_\lambda$ and let the notation be as in the proof of Theorem 4.45. Let $X_0 = span\{x_k : k \in \mathbb{N}\}$. The set $\{x'_k : k \in \mathbb{N}\}$ is relatively $\sigma(X'_0, X_0)$ compact and, therefore, relatively $\sigma(X'_0, X_0)$ sequentially compact since X_0 is separable ([Wi] 9.5.3). Therefore, we may assume that $\lim \langle x'_k, x \rangle$

exist for every $x \in X$. The proof can now be completed as in the proof of Theorem 4.45.

Without the assumption on the multiplier space λ, the inclusions in Theorems 4.45 and 4.46 may fail to hold.

Example 4.47. Let $\lambda = c_c$, the space of sequences which are eventually constant (Appendix B). Then a series in a TVS is λ multiplier convergent iff the series is convergent. If one has a series which is $\sigma(X, X')$ convergent but not $\tau(X, X')$ convergent, then the series is λ multiplier convergent with respect to $\sigma(X, X')$ but not λ multiplier convergent with respect to $\tau(X, X')$. [For example, take $\sum_j (e^{j+1} - e^j)$ in c_0.] Thus, $\tau(X, X')$ is not contained in $\mathcal{D}_\lambda(X, X')$ by Theorem 4.41. Since both topologies $\lambda(X, X')$ and $\gamma(X, X')$ contain $\tau(X, X')$, this shows that the containments in Theorems 4.45 and 4.46 do not hold.

Remark 4.48. Theorems 4.45 and 4.46 contain the results in Corollaries 4.11 and 4.14 as special cases.

We now compare the general Dierolf topology with the strong topology when the multiplier space has the ∞-GHP. Recall that λ has ∞-GHP if whenever $t \in \lambda$ and $\{I_j\}$ is an increasing sequence of intervals, there exist a subsequence $\{n_j\}$ and $a_{n_j} > 0, a_{n_j} \to \infty$ such that every subsequence of $\{n_j\}$ has a further subsequence $\{p_j\}$ such that the coordinate sum of the series $\sum_{j=1}^\infty a_{p_j} \chi_{I_{p_j}} t \in \lambda$ (Appendix B; examples are given in Appendix B).

Theorem 4.49. *Let λ have ∞-GHP. Then $\beta(X, X') \subset \mathcal{D}_\lambda(X, X')$.*

Proof: Let $B \subset X'$ be $\sigma(X', X)$ bounded. Assume that B does not belong to $\mathcal{D}_\lambda$. Let the notation be as in Theorem 4.45 so

$$(*) \quad \left| \sum_{j \in I_k} t_j \langle x'_k, x_j \rangle \right| > \epsilon$$

for every k with $x'_k \in B$. By the ∞-GHP there exist $\{p_k\}, a_{p_k} > 0, a_{p_k} \to \infty$ such that every subsequence of $\{p_k\}$ has a further subsequence $\{q_k\}$ such that $s = \{s_j\} = \sum_{k=1}^\infty a_{q_k} \chi_{I_{q_k}} t \in \lambda$. Define a matrix

$$M = [m_{ij}] = \left[\sum_{l \in I_j} t_l a_{p_j} \langle x'_i / a_{p_i}, x_l \rangle \right].$$

We claim that M is a $\mathcal{K}$-matrix (Appendix D.2). First, the columns of M converge to 0 since $x_i'/a_{p_i} \to 0$ in $\sigma(X', X)$. Next, given a subsequence of $\{p_j\}$ let $\{q_j\}$ be a subsequence as above. Then

$$\sum_{j=1}^{\infty} m_{iq_j} = \sum_{j=1}^{\infty} \sum_{l \in I_{q_j}} s_l \langle x_i'/a_{p_i}, x_l \rangle = \left\langle x_i'/a_{p_i}, \sum_{l=1}^{\infty} s_l x_l \right\rangle \to 0,$$

where $\sum_{l=1}^{\infty} s_l x_l$ is the $\sigma(X, X')$ sum of the series. Thus, M is a $\mathcal{K}$-matrix so the diagonal of M converges to 0 by the Antosik-Mikusinski Matrix Theorem (Appendix D.2). But, this contradicts $(*)$.

There is also an analogue of the Tweddle topology for λ multiplier convergent series when the multiplier space has signed-WGHP. If X is an LCTVS, let $X^{\#}(= X_{\lambda}^{\#})$ be the space of all linear functionals x' on X satisfying

$$\sum_{j=1}^{\infty} t_j \langle x', x_j \rangle = \left\langle x', \sum_{j=1}^{\infty} t_j x_j \right\rangle$$

for every $t \in \lambda$ and every $\sigma(X, X')$ λ multiplier convergent series $\sum_j x_j$, where $\sum_{j=1}^{\infty} t_j x_j$ is the $\sigma(X, X')$ sum of the series. We define the Tweddle topology on X to be $\mathcal{D}_{\lambda}(X, X^{\#})$ and denote the topology by $t_{\lambda}(X, X')$. Thus, $t_{m_0}(X, X') = t(X, X')$. We show that $t_{\lambda}(X, X')$ is the strongest locally convex topology on X with the same λ multiplier convergent series as $\sigma(X, X')$.

Theorem 4.50. *Let λ have signed-WGHP. Then $t_{\lambda}(X, X')$ is the strongest locally convex topology on X with the same λ multiplier convergent series as $\sigma(X, X')$.*

Proof: If $\sum_j x_j$ is λ multiplier convergent with respect to $\sigma(X, X')$, then $\sum_j x_j$ is λ multiplier convergent with respect to $\sigma(X, X^{\#})$ by the definition of $X^{\#}$. By Theorem 4.41, $\sum_j x_j$ is λ multiplier convergent with respect to $t_{\lambda}(X, X') = \mathcal{D}_{\lambda}(X, X^{\#})$.

Suppose α is a locally convex topology with the same λ multiplier convergent series as $\sigma(X, X')$. Put $H' = (X, \alpha)'$. Then $H' \subset X^{\#}$. By Theorem 4.41,

$$\alpha \subset \tau(X, H') \subset \mathcal{D}_{\lambda}(X, H') \subset \mathcal{D}_{\lambda}(X, X^{\#}) = t_{\lambda}(X, X').$$

We now present several applications of the Orlicz-Pettis theorems to various topics in functional analysis and measure theory. As a first application we present the original application by Pettis to vector valued measures.

Theorem 4.51. *Let Σ be a σ-algebra of subsets of a set S and let X be an LCTVS. If $\mu : \Sigma \to X$ is such that $x' \circ \mu = x'\mu : \Sigma \to \mathbb{R}$ is countably additive for every $x' \in X'$, then μ is countably additive with respect to the original topology of X. That is, if μ is countably additive with respect to the weak topology, then μ is countably additive with respect to the original topology.*

Proof: Let $\{A_j\} \subset \Sigma$ be pairwise disjoint. If $\{A_{n_j}\}$ is a subsequence, then $A = \cup_{j=1}^{\infty} A_{n_j} \in \Sigma$ and $\langle x', \mu(A) \rangle = \sum_{j=1}^{\infty} \langle x', \mu(A_{n_j}) \rangle$ for every $x' \in X'$. Thus, the series $\sum_j \mu(A_j)$ is subseries convergent with respect to the weak topology $\sigma(X, X')$. By the Orlicz-Pettis Theorem in Corollary 4.11, the series is subseries convergent with respect to the original topology of X. That is, μ is countably additive with respect to the original topology of X.

For vector valued set functions defined on algebras, we have

Theorem 4.52. *Let $\mathcal{A}$ be an algebra of subsets of a set S and let X be a weakly sequentially complete LCTVS. If $\mu : \mathcal{A} \to X$ is such that $x'\mu$ is countably additive for every $x' \in X'$, then μ is countably additive with respect to the original topology of X.*

Proof: Let $\{A_j\} \subset \mathcal{A}$ be pairwise disjoint with $A = \cup_{j=1}^{\infty} A_j \in \mathcal{A}$. Then $\langle x', \mu(A) \rangle = \sum_{j=1}^{\infty} \langle x', \mu(A_j) \rangle$ for every $x' \in X'$ and the series $\sum_j \mu(A_j)$ is unconditionally convergent with respect to $\sigma(X, X')$ since the union $\cup_{j=1}^{\infty} A_j$ is independent of the ordering of the $\{A_j\}$ [note that we cannot assert that the series is subseries convergent since $\cup_{j=1}^{\infty} A_{n_j}$ may not belong to $\mathcal{A}$ for arbitrary subsequences]. By Corollary 2.60 the series $\sum_j \mu(A_j)$ is subseries convergent with respect to $\sigma(X, X')$. By the Orlicz-Pettis Theorem in Corollary 4.11, the series $\sum_j \mu(A_j)$ is subseries convergent with respect to the original topology of X. Thus, μ is countably additive with respect to the original topology of X.

As an application of Theorem 4.52 and the Orlicz-Pettis Theorem, we have

Theorem 4.53. *Let X be a weakly sequentially complete LCTVS. Then X contains no subspace isomorphic to c_0.*

Proof: Let $\sum_j x_j$ be c_0 multiplier convergent in X. By Proposition 3.8, $\sum_{j=1}^{\infty} t_j \langle x', x_j \rangle$ converges for every $t \in l^{\infty}, x' \in X'$. Hence, the partial sums of the series $\sum_j t_j x_j$ are $\sigma(X, X')$ Cauchy and, therefore, $\sigma(X, X')$ convergent for every $t \in l^\infty$. That is, the series $\sum_j x_j$ is l^∞ multiplier convergent in the weak topology. By the Orlicz-Pettis Theorem in Corollary 4.14 the series $\sum_j x_j$ is l^∞ multiplier convergent in the original topology of X. Thus, condition (iv) of Theorem 3.15 holds. The proofs that (iv) implies (viii) implies (i) in Theorem 3.15 do not employ the sequential completeness of X so the result follows from these proofs.

Corollary 4.54. *If X is a semi-reflexive space, then X contains no subspace isomorphic to c_0.*

Proof: $(X, \sigma(X, X'))$ is boundedly complete and, therefore, sequentially complete ([Wi] 10.2.4) so the result follows from Theorem 4.53.

Corollary 4.55. *If X is a barrelled LCTVS, then $(X', \sigma(X', X))$ contains no subspace isomorphic to c_0.*

Proof: $(X', \sigma(X', X))$ is sequentially complete ([Wi] 9.3.8). But,

$$(X', \sigma(X', X))' = X$$

so $(X', \sigma(X', X))$ is weakly sequentially complete.

As another application of the Orlicz-Pettis Theorem, we derive a result similar in spirit to Theorem 3.2 but with different hypothesis ([KG] 3.10.5).

Theorem 4.56. *Let X be weakly sequentially complete and let λ be monotone. Then a series $\sum_j x_j$ in X is λ multiplier convergent in X iff $\{\langle x', x_j \rangle\} \in \lambda^\beta$ for every $x' \in X'$.*

Proof: If $\sum_j x_j$ is λ multiplier convergent, then $\{\langle x', x_j \rangle\} \in \lambda^\beta$ for every $x' \in X'$ by Proposition 3.1.

Conversely, suppose that $\{\langle x', x_j \rangle\} \in \lambda^\beta$ for every $x' \in X'$. Let $t = \{t_j\} \in \lambda$. Since λ is monotone, the partial sums of any subseries of $\sum_j t_j x_j$ is $\sigma(X, X')$ Cauchy and, therefore, $\sigma(X, X')$ convergent by hypothesis. That is, the series $\sum_j t_j x_j$ is $\sigma(X, X')$ subseries convergent and, therefore, convergent in X by the Orlicz-Pettis Theorem in Corollary 4.11.

Note that the assumption that λ is monotone in Theorem 4.56 is an algebraic condition whereas the assumptions in Theorem 3.2 are topological. Theorem 4.56, however, has the weak sequential completeness assumption.

We can also use the Orlicz-Pettis Theorem to derive a result of Pelczynski on weakly compact and unconditionally converging operators (recall

Theorem 3.18). A continuous linear operator T between normed spaces X and Y is *weakly compact* if T carries bounded sets into relatively weakly compact sets and T is *unconditionally converging* if T carries wuc series into subseries convergent series.

Theorem 4.57. *([Pl]) If $T : X \to Y$ is a weakly compact operator from the normed space X into the normed space Y, then T is unconditionally converging.*

Proof: Let $\sum_j x_j$ be wuc. Then $\{\sum_{j \in \sigma} x_j : \sigma \ finite\}$ is bounded by Proposition 3.8. Thus, $\{\sum_{j \in \sigma} T x_j : \sigma \ finite\}$ is relatively $\sigma(Y, Y')$ compact. By Theorem 2.48, the series $\sum_j T x_j$ is $\sigma(Y, Y')$ subseries convergent. By the Orlicz-Pettis Theorem in Corollary 4.11, the series $\sum_j T x_j$ is subseries convergent in Y.

The identity operator on l^1 shows that the converse of Theorem 4.57 is false, in general; however, there are spaces for which the converse does hold (see [Pl]).

As another application of the Orlicz-Pettis Theorem, we derive a version of the Nikodym Boundedness Theorem for countably additive set functions. Dunford and Schwartz refer to the Nikodym Boundedness Theorem as a "striking improvement of the principle of uniform boundedness" ([DS] p. 309). The theorem states that a family of countably additive signed measures defined on a σ-algebra which is pointwise bounded on the σ-algebra is uniformly bounded on the entire σ-algebra. For the proof of the theorem, we first establish a result which is central to most of the proofs of the theorem (see, however, Dunford and Schwartz where there is a proof based on the Baire Category Theorem ([DS] IV.9.8)).

Let $\mathcal{A}$ be an algebra of subsets of a set S, and let $ba(\mathcal{A})$ be the space of all real valued, bounded, finitely additive set functions defined on $\mathcal{A}$.

Lemma 4.58. *Let $M \subset ba(\mathcal{A})$ be pointwise bounded on $\mathcal{A}$. Then $\sup\{|\mu(A)| : \mu \in M, A \in \mathcal{A}\} < \infty$ iff $\sup\{|\mu_i(A_i)| : i \in \mathbb{N}\} < \infty$ for every pairwise disjoint sequence $\{A_i\}$ from $\mathcal{A}$ and every sequence $\{\mu_i\} \subset M$.*

Proof: Suppose $\sup\{|\mu(A)| : \mu \in M, A \in \mathcal{A}\} = \infty$. Note that for each $r > 0$ there exist a partition (E, F) of S with $E, F \in \mathcal{A}$ and $\mu \in M$ such that $\min\{|\mu(E)|, |\mu(F)|\} > r$. [This follows since $|\mu(E)| > r + \sup\{|\nu(S)| : \nu \in M\}$ implies $|\mu(S \setminus E)| \geq |\mu(E)| - |\mu(S)| > r$.] Hence, there exist $\mu_1 \in M$ and a partition (E_1, F_1) of S such that $\min\{\{|\mu_1(E_1)|, |\mu_1(F_1)|\} > 1$.

Now either

$$\sup\{|\mu(E_1 \cap A)| : \mu \in M, A \in \mathcal{A}\} = \infty$$

or

$$\sup\{|\mu(F_1 \cap A)| : \mu \in M, A \in \mathcal{A}\} = \infty.$$

Pick whichever of E_1 or F_1 satisfies this condition and label it B_1 and set $A_1 = S \setminus B_1$. Now treat B_1 as S above to obtain a partition (A_2, B_2) of B_1 and $\mu_2 \in M$ such that $|\mu_2(A_2)| > 2$ and $\sup\{|\mu(B_2 \cap A)| : \mu \in M, A \in \mathcal{A}\} = \infty$. Continuing this construction produces a sequence $\{\mu_i\} \subset M$ and a pairwise disjoint sequence $\{A_i\}$ from $\mathcal{A}$ such that $|\mu_i(A_i)| > i$. This establishes the sufficiency; the necessity is clear.

Let Σ be a σ-algebra of subsets of a set S. Let $\mathcal{S}(\Sigma)$ be the vector space of all real valued Σ-simple functions and let $ca(\Sigma)$ be the space of all countably additive signed measures $\mu : \Sigma \to \mathbb{R}$. Then $\mathcal{S}(\Sigma)$, $ca(\Sigma)$ form a dual pair via the integration pairing $\langle \mu, f \rangle = \int f d\mu, f \in \mathcal{S}(\Sigma), \mu \in ca(\Sigma)$. We now give our proof of the Nikodym Boundedness Theorem.

Theorem 4.59. *Let $M \subset ca(\Sigma)$ be such that $\sup\{|\mu(E)| : \mu \in M\} < \infty$ for every $E \in \Sigma$. Then $\sup\{|\mu(E)| : \mu \in M, E \in \Sigma\} < \infty$.*

Proof: By Lemma 4.58 it suffices to show that $\sup\{|\mu_i(A_i)| : i \in \mathbb{N}\} < \infty$ for every $\{\mu_i\} \subset M$ and pairwise disjoint sequence $\{A_i\}$ from Σ or that $(1/i)\mu_i(A_i) \to 0$. The series $\sum_i \chi_{A_i}$ is $\sigma(\mathcal{S}(\Sigma), ca(\Sigma))$ subseries convergent by the countable additivity of the members of $ca(\Sigma)$. By the version of the Orlicz-Pettis Theorem for the topology $\lambda(\mathcal{S}(\Sigma), ca(\Sigma))$ the series $\sum_i \chi_{A_i}$ converges in $\lambda(\mathcal{S}(\Sigma), ca(\Sigma))$. In particular, $\lim_n \sum_{j=n}^{\infty} (1/k)\mu_k(A_j) = 0$ uniformly for $k \in \mathbb{N}$ since $\{(1/k)\mu_k\}$ is $\sigma(ca(\Sigma), \mathcal{S}(\Sigma))$ convergent to 0 by the pointwise boundedness assumption. In particular, $(1/i)\mu_i(A_i) \to 0$ as desired.

It should be noted that a version of the theorem for countably additive set functions defined on σ-algebras with values in an LCTVS follows immediately from Theorem 4.59.

Corollary 4.60. *Let X be an LCTVS and let M be an family of countably additive set functions defined on Σ with values in X. If $\{\mu(E) : \mu \in M\}$ is bounded in X for every $E \in \Sigma$, then $\{\mu(E) : \mu \in M, E \in \Sigma\}$ is bounded.*

Proof: Let $x' \in X'$. Then the subset $\{x'\mu : \mu \in M\} \subset ca(\Sigma)$ is pointwise bounded on Σ so the set $\{x'\mu : \mu \in M\}$ is uniformly bounded on Σ by Theorem 4.59. That is, the set $\{x'\mu : \mu \in M, E \in \Sigma\}$ is $\sigma(X, X')$ bounded and, therefore, bounded in X.

A few remarks pertaining to the Nikodym Boundedness Theorem are in order. First, the local convex assumption in Corollary 4.60 is important. Turpin has given an example of a countably additive set function defined on a σ-algebra with values in a (non-locally convex) TVS which is unbounded ([Rol]). Theorem 4.59 actually holds for bounded, finitely additive set functions defined on σ-algebras; we will give a proof of this version of the Nikodym Boundedness Theorem in Chapter 7 based on the Hahn-Schur Theorem. As the following example shows, the conclusion of Theorem 4.59 is false for set functions defined on algebras.

Example 4.61. Let $\mathcal{A}$ be the algebra of finite/co-finite subsets of $\mathbb{N}$. Let δ_n be the Dirac measure concentrated at n; $\delta_n(E) = 1$ if $n \in E$ and $\delta_n(E) = 0$ if $n \notin E$. Define $\mu_n(E) = n(\delta_{n+1}(E) - \delta_n(E))$ if E is finite and $\mu_n(E) = -n(\delta_{n+1}(E) - \delta_n(E))$ if $E^c = \mathbb{N} \setminus E$ is finite. Then $\{\mu_n\}$ is pointwise bounded on $\mathcal{A}$ but not uniformly bounded on $\mathcal{A}$ $[\mu_n(\{n\}) = n]$.

Despite Example 4.61, there are algebras for which the conclusion of Theorem 4.59 holds. The treatise by Schachermeyer contains examples, references and other discussions concerning the Nikodym Boundedness Theorem ([Sm]); see also Diestel and Uhl ([DU]).

Finally, we present several versions and applications of the Orlicz-Pettis Theorem in an abstract setting. Let E, F be vector spaces such that there is a bilinear mapping from $\cdot : E \times F \to X$, $(x, y) \to x \cdot y, x \in E, y \in F$, where X is an LCTVS. Of course, an example of this situation is when E, F are two vector spaces in duality ; we give other examples in the applications which follow. Let $w(E, F)$ $[w(F, E)]$ be the weakest topology on E $[F]$ such that the linear maps $x \to x \cdot y$ $[y \to x \cdot y]$ from E into X $[F$ into $X]$ are continuous for all $y \in F$ $[x \in E]$. If E, F are 2 vector spaces in duality, then $w(E, F)$ $[w(F, E)]$ is just the weak topology $\sigma(E, F)$ $[\sigma(F, E)]$. A subset $K \subset F$ is said to be conditionally $w(F, E)$ sequentially compact if for every sequence $\{y_j\} \subset K$, there is a subsequence $\{y_{n_j}\}$ such that $\lim_j x \cdot y_{n_j}$ exists for every $x \in E$. In this setting we have the analogue of Corollary 4.10 for the topology $\gamma(X, X')$. Again, if E, F are 2 vector spaces in duality, this agrees with previous terminology.

We establish a version of Corollary 4.10 for the topology $\gamma(E, F)$ in this setting.

Theorem 4.62. *Let λ have signed-WGHP. If the series $\sum_j x_j$ is λ multiplier convergent in E with respect to $w(E, F)$, then for each $t \in \lambda$ and each conditionally $w(F, E)$ sequentially compact subset $K \subset F$, the series $\sum_{j=1}^{\infty} t_j x_j \cdot y$ converge uniformly for $y \in K$.*

Proof: If the conclusion fails to hold, there exists a neighborhood of 0, W, in X, $y_k \in K$ and an increasing sequence of intervals $\{I_k\}$ such that

$$(\#) \quad \sum_{l \in I_k} t_l x_l \cdot y_k \notin W$$

for every k. We may assume, by passing to a subsequence if necessary, that $\lim_k x \cdot y_k$ exists for every $x \in E$. Consider the matrix

$$M = [m_{ij}] = [\sum_{l \in I_j} t_l x_l \cdot y_i].$$

We claim that M is a signed $\mathcal{K}$-matrix (Appendix D). First, the columns of M converge. Next, given an increasing sequence of positive integers there is a subsequence $\{n_j\}$ and a sequence of signs $\{s_j\}$ such that $u = \sum_{j=1}^{\infty} s_j \chi_{I_{n_j}} t \in \lambda$. Then

$$\left\{ \sum_{j=1}^{\infty} s_j m_{in_j} \right\}_i = \left\{ \sum_{j=1}^{\infty} s_j \sum_{l \in I_{n_j}} t_l x_l \cdot y_i \right\}_i = \left\{ \sum_{l=1}^{\infty} u_l x_l \cdot y_i \right\}_i$$

converges. Hence, M is a signed $\mathcal{K}$-matrix so the diagonal of M converges to 0 by the signed version of the Antosik-Mikusinski Matrix Theorem (Appendix D.3). But, this contradicts $(\#)$.

We now derive an analogue of Corollary 4.10 for λ multiplier convergent series and the topology $\lambda(X, X')$.

Theorem 4.63. *Let λ have signed-WGHP. If the series $\sum_j x_j$ is λ multiplier convergent in E with respect to $w(E, F)$, then for each $w(F, E)$ compact (countably compact) subset $K \subset F$ and each $t \in \lambda$, the series $\sum_{j=1}^{\infty} t_j x_j \cdot y$ are convergent uniformly for $y \in K$.*

Proof: Let p be a continuous semi-norm on X. We need to show that the series $\sum_j t_j x_j \cdot y$ converge uniformly for $y \in K$ with respect to p. This will follow if we can show that this property holds in the quotient space X/p. Hence, we may assume that p is actually a norm. Define an equivalence relation $\sim$ on F by $y \sim z$ iff $x_j \cdot y = x_j \cdot z$ for all j. If

$E_0 = \{\sum_{j=1}^{\infty} s_j x_j : s \in \lambda,$ where $\sum_{j=1}^{\infty} s_j x_j$ is the $w(E, F)$ sum of the series$\}$, then $x \cdot y = x \cdot z$ for every $x \in E_0$ when $y \sim z$. Let y^- be the equivalence class of $y \in F$ and set $F^- = \{f^- : f \in F\}$. Define a metric d on F^- by

$$d(y^-, z^-) = \sum_{j=1}^{\infty} p(x_j \cdot (y - z))/2^j(1 + p(x_j \cdot (y - z)));$$

note that d is a metric since p is a norm. Define a bilinear mapping

$$\cdot : E_0 \times F^- \to (X, p)$$

by $x \cdot y^- = x \cdot y$ so we may consider the triple $E_0, F^-, (X, p)$ as above. The quotient map $F \to F^-$ is $w(F, E) - w(F^-, E_0)$ continuous and the inclusion $(F^-, w(F^-, E_0)) \subset (F^-, d)$ is continuous so K^- is compact (countably compact) with respect to $w(F^-, E_0)$ and d and, therefore, $w(F^-, E_0) = d$ on K^- and K^- is $w(F^-, E_0)$ sequentially compact. Since the series $\sum_j x_j$ is λ multiplier convergent with respect to $w(E, F)$, the series $\sum_j x_j$ is λ multiplier convergent with respect to $w(E_0, F^-)$ in the abstract triple $E_0, F^-, (X, p)$. Since K^- is sequentially compact in $w(F^-, E_0)$, by Theorem 4.62 the series $\sum_{j=1}^{\infty} t_j x_j \cdot y^- = \sum_{j=1}^{\infty} t_j x_j \cdot y$ converge uniformly for $y^- \in K^-$ with respect to p.

Theorems 4.62 and 4.63 have as immediate corollaries the results in Corollary 4.10. We now present another corollary related to Theorem 4.9.

Corollary 4.64. *Let λ have signed-WGHP. Then $(\lambda, \gamma(\lambda, \lambda^\beta))$ and $(\lambda, \lambda(\lambda, \lambda^\beta))$ are AK-spaces.*

Proof: The series $\sum_j e^j$ is λ multiplier convergent with respect to $\sigma(\lambda, \lambda^\beta)$ and, therefore, is λ multiplier convergent with respect to $\gamma(\lambda, \lambda^\beta)$ and $\lambda(\lambda, \lambda^\beta)$ by Theorems 4.62 and 4.63. The result is now immediate.

Note that the results above were derived in the other order previously. We now give several applications of Theorems 4.62 and 4.63.

Example 4.65. Let Σ be a σ-algebra of subsets of a set S and let $ca(\Sigma, X)$ be the space of all X valued countably additive set functions from Σ into X. If $E = \mathcal{S}(\Sigma)$ and $F = ca(\Sigma, X)$, then $f \cdot \mu = \int_S f d\mu$, $f \in E$, $\mu \in F$, defines a bilinear map from $E \times F$ into X (note that we are only integrating simple functions so no elaborate integration theory is involved). If $\{E_j\} \subset \Sigma$ is pairwise disjoint, then the series $\sum_j \chi_{E_j}$ is $w(\mathcal{S}(\Sigma), ca(\Sigma, X))$ subseries convergent. By Theorem 4.62 above, the series $\sum_{j=1}^{\infty} \mu(E_j)$ converge uniformly for μ belonging to any conditionally $w(ca(\Sigma, X), \mathcal{S}(\Sigma))$ sequentially compact subset of $ca(\Sigma, X)$. In particular, we have as a special case the Nikodym Convergence Theorem.

Theorem 4.66. *Let* $\{\mu_j\} \subset ca(\Sigma, X)$ *be such that* $\lim_j \mu_j(E) = \mu(E)$ *exists for every* $E \in \Sigma$. *Then* $\{\mu_j\}$ *is uniformly countably additive and* $\mu \in ca(\Sigma, X)$.

Proof: By the observation above, since $\{\mu_j\}$ is conditionally $w(ca(\Sigma, X), \mathcal{S}(\Sigma))$ sequentially compact, $\{\mu_j\}$ is uniformly countably additive. That $\mu \in ca(\Sigma, X)$ then follows.

From Theorem 4.63, we can also derive a result of Graves and Ruess ([GR]) Lemma 6).

Theorem 4.67. *If* $K \subset ca(\Sigma, X)$ *is* $w(ca(\Sigma, X), \mathcal{S}(\Sigma))$ *compact, then* K *is uniformly countably additive.*

Next, we derive a version of a theorem of Thomas ([Th]).

Example 4.68. Let S be a sequentially compact Hausdorff space. Let $E = SC(S, X)$ be the space of sequentially continuous functions from S into X and let $F = span\{\delta_t : t \in S\}$, where δ_t is the Dirac measure concentrated at t. Then $f \cdot t = f(t)$ defines a bilinear mapping from $E \times F$ into X. Note that S is conditionally $w(span\{\delta_t : t \in S\}, SC(S, X))$ sequentially compact since S is sequentially compact [here we are identifying t with δ_t]. Thus, from Theorem 4.62 above if λ has signed-WGHP and $\sum_j f_j$ is λ multiplier convergent in $SC(S, X)$ with respect the topology of pointwise convergence on S, then for each $t \in \lambda$ the series $\sum_j t_j f_j$ converges uniformly on S. Similarly, if S is compact, then S is $w(span\{\delta_t : t \in S\}, C(S, X))$ compact so from Theorem 4.63 if λ has signed-WGHP and the series $\sum_j f_j$ is λ multiplier convergent in $C(S, X)$ with respect to the topology of pointwise convergence on S, then for each $t \in \lambda$ the series $\sum_j t_j f_j$ converges uniformly on S. The subseries version of this result is due to Thomas ([Th]).

We can also use Theorems 4.62 and 4.63 above to derive a version of the Orlicz-Pettis Theorem for continuous linear operators.

Example 4.69. Let Z be an LCTVS. Set $E = L(Z, X)$ and $F = Z$ and define a bilinear mapping from $E \times F$ into X by $T \cdot x = Tx$. Then $w(E, F)$ is just the topology of pointwise convergence on Z or $L_s(Z, X)$. If $K \subset Z$ is sequentially compact (compact), then K is conditionally $w(Z, L(Z, X))$

sequentially compact $(w(Z, L(Z, X))$ compact) so if $\mathcal{K}$ $(\mathcal{C})$ denotes the set of all sequentially compact (compact) subsets of Z, from Theorem 4.62 (Theorem 4.63) above, we have

Theorem 4.70. *Let λ have signed-WGHP. If $\sum_j T_j$ is λ multiplier convergent in $L_s(Z, X)$, then $\sum_j T_j$ is λ multiplier convergent in $L_{\mathcal{K}}(Z, X)$ $(L_C(Z, X))$.*

We will obtain some similar results for operator valued series later in Chapter 6.

An operator $T \in L(Z, X)$ is completely continuous if T carries weakly convergent sequences into convergent sequences; denote all such operators by $CC(Z, X)$. Note that if T is completely continuous, then T carries weak Cauchy sequences into Cauchy sequences. Now consider the abstract triple $E = CC(Z, X)$, $F = Z$ and the bilinear map $\cdot : E \times F \to X$ defined by $\cdot : (T, z) \to T \cdot z = Tz$. If a subset $K \subset Z$ is conditionally weakly sequentially compact, then K is conditionally $w(CC(Z, X), Z)$ sequentially compact. If CW denotes the set of all conditionally weakly sequentially compact subsets of Z, then from Theorem 4.62 we have

Theorem 4.71. *Let λ have signed-WGHP. If the series $\sum_j T_j$ is λ multiplier convergent in $CC_s(Z, X)$, then $\sum_j T_j$ is λ multiplier convergent in $CC_{CW}(Z, X)$.*

An operator $T \in L(Z, X)$ is weakly compact if T carries bounded sets to relatively weakly compact sets; denote all such operators by $W(Z, X)$. The space Z has the Dunford-Pettis property if every weakly compact operator from Z into any locally convex space X carries weak Cauchy sequences into convergent sequences. Consider the abstract triple $E = W(Z, X), F = Z$ and the bilinear map $\cdot : E \times F \to X$ defined by $\cdot : (T, z) \to T \cdot z = Tz$. If $K \subset Z$ is conditionally weakly sequentially compact and Z has the Dunford-Pettis property, then K is conditionally $w(W(Z, X), Z)$ sequentially compact. If CW denotes the set of all conditionally weakly compact subsets of Z, then from Theorem 4.62 we have

Theorem 4.72. *Let λ have signed-WGHP and assume that Z has the Dunford-Pettis property. If the series $\sum_j T_j$ is λ multiplier convergent in $W_s(Z, X)$, then $\sum_j T_j$ is λ multiplier convergent in $W_{CW}(Z, X)$.*

A space Z is almost reflexive if every bounded sequence contains a weak Cauchy subsequence ([LW]). For example, Banach spaces with separable duals, quasi-reflexive Banach spaces and $c_0(S)$ are almost reflexive ([LW]). If Z is almost reflexive and has the Dunford-Pettis property, then every bounded set is conditionally $w(W(Z,X),Z)$ sequentially compact so from Theorem 4.62, we have

Theorem 4.73. *Let λ have signed-WGHP and assume that Z is almost reflexive with the Dunford-Pettis property. If the series $\sum_j T_j$ is λ multiplier convergent in $W_s(Z,X)$, then $\sum_j T_j$ is λ multiplier convergent in $W_b(Z,X)$.*

As another application of Theorem 4.63, we derive an Orlicz-Pettis result of Stiles for a locally convex TVS with a Schauder basis ([Sti]). Stiles' version of the Orlicz-Pettis Theorem is for subseries convergent series with values in an F-space with a Schauder basis and his proof uses the metric properties of the space. Other proofs of Stiles' result have been given in [Bs] and [Sw5]. We will establish a version of Stiles' result for multiplier convergent series which requires no metrizability assumptions. Later in Chapter 9 we will establish a version of the result for non-locally convex spaces using the Antosik Interchange Theorem.

Let X be an LCTVS with a Schauder basis $\{b_j\}$ and associated coordinate functionals $\{f_j\}$. That is, every $x \in X$ has a unique series representation $x = \sum_{j=1}^{\infty} t_j b_j$ and $f_j : X \to \mathbb{R}$ is defined by $\langle f_j, x \rangle = t_j$. We do not assume that the coordinate functionals are continuous although this is the case when X is an F-space ([Sw2] 10.1.13). Define $P_i : X \to X$ by $P_i x = \sum_{j=1}^{i} \langle f_j, x \rangle b_j$. Let $E = X$, $F = span\{P_i : i \in \mathbb{N}\}$ and let the bilinear mapping from $E \times F$ into X be the extension to F of the mapping $x \cdot P_i = P_i x$. Let $G = span\{f_i : i \in \mathbb{N}\}$.

Theorem 4.74. *Let λ have signed-WGHP. If $\sum_j x_j$ is λ multiplier convergent with respect to $\sigma(X,G)$, then $\sum_j x_j$ is λ multiplier convergent with respect to the original topology of X.*

Proof: Since a sequence in X is $\sigma(X,G)$ convergent iff the sequence is $w(E,F)$ convergent, the series $\sum_j x_j$ is λ multiplier convergent with respect to $w(E,F)$. Now $\{P_i : i \in \mathbb{N}\}$ is conditionally $w(F,E)$ sequentially compact since $P_i x \to x$ for every $x \in X$. By Theorem 4.62, for every $t \in \lambda$ the series $\sum_{j=1}^{\infty} t_j P_i x_j$ converge uniformly for $i \in \mathbb{N}$. Let U be a closed neighborhood of 0 in X. There exists N such that $\sum_{j=m}^{\infty} t_j P_i x_j = P_i(\sum_{j=m}^{\infty} t_j x_j) \in U$ for $m \geq N, i \in \mathbb{N}$. Let $i \to \infty$ gives $\sum_{j=m}^{\infty} t_j x_j \in U$ for $m \geq N$.

Note that we did not use the continuity of the coordinate functionals in the proof so the topology of X may not even be comparable to $\sigma(X,G)$.

We next consider a more general situation than that encountered in Stiles' result. Assume that there exists a sequence of linear operators P_i : $X \to X$ such that for each $x \in X$, $x = \sum_{i=1}^{\infty} P_i x$ [convergence in X]. When each P_i is continuous, $\{P_i\}$ is called a Schauder decomposition ([LT]). If X has a Schauder basis $\{b_i\}$ with coordinate functionals $\{f_i\}$, then $P_i x = \langle f_i, x \rangle b_i$ is an example of this situation. Let $E = X, F = span\{P_i : i \in \mathbb{N}\}$ and let the bilinear mapping from $E \times F$ into X be the extension of the map $x \cdot P_i = P_i x$.

Theorem 4.75. *Let λ have signed-WGHP and assume that each P_i is $w(E, F) - X$ continuous. If the series $\sum_j x_j$ is λ multiplier convergent with respect to $w(E, F)$, then the series $\sum_j x_j$ is λ multiplier convergent in X with respect to the original topology.*

Proof: Define $S_n : X \to X$ by $S_n = \sum_{i=1}^{n} P_i$. Then $\{S_n : n \in \mathbb{N}\}$ is conditionally $w(F, E)$ sequentially compact so by Theorem 4.62 for each $t \in \lambda$ the series $\sum_{j=1}^{\infty} t_j S_n x_j$ converge uniformly for $n \in \mathbb{N}$. Let U be a closed neighborhood of 0 in X. There exists N such that $\sum_{j=m}^{\infty} t_j S_n x_j = S_n(\sum_{j=m}^{\infty} t_j x_j) \in U$ for $m \geq N, n \in \mathbb{N}$. Letting $n \to \infty$ gives $\sum_{j=m}^{\infty} t_j x_j \in U$ for $m \geq N$.

We give an example where the theorem above is applicable.

Example 4.76. Let Y be an LCTVS and let X be a vector space of Y valued sequences containing the space of sequences which are eventually 0. Then X is an AK-space if the coordinate functionals $f_j : X \to Y$, $f_j(\{x_j\}) = x_j$ are continuous for every j and each $x = \{x_j\}$ has a representation $x = \sum_{j=1}^{\infty} e^j \otimes x_j$ [Appendix C; here $e^j \otimes x$ denotes the sequence with x in the jth coordinate and 0 in the other coordinates]. The space X has the property (I) if the injections $x \to e^j \otimes x$ are continuous from Y into X. If $P_j : X \to X$ is defined by $P_j(\{x_j\}) = e^j \otimes x_j$, then $\{P_j\}$ is a Schauder decomposition for X. If X has property (I), then the topology of coordinatewise convergence is equal to $w(E, F)$ so the result above applies and if λ has signed-WGHP, then any series which is λ multiplier convergent in the topology of coordinatewise convergence converges in the topology of X.

For examples where the result above applies let Y be a normed space. If $1 \leq p < \infty$, then $l^p(Y)$ and $c_0(Y)$ are AK-spaces satisfying the conditions in the example above.

We will consider non-locally convex versions of these results later in Chapter 9.

Finally, we show that the result in Theorem 2.26 can be obtained from Theorem 4.62.

Theorem 4.77. *Let λ have signed-WGHP. Assume that $\sum_j x_{ij}$ is λ multiplier convergent for every $i \in \mathbb{N}$ and that $\lim_i \sum_{j=1}^{\infty} t_j x_{ij}$ exists for every $t \in \lambda$ with $x_j = \lim_i x_{ij}$ for every j. Then for every $t \in \lambda$ the series $\sum_{j=1}^{\infty} t_j x_{ij}$ converge uniformly for $i \in \mathbb{N}$.*

Proof: For every $i \in \mathbb{N}$ define a linear map $f_i : \lambda \to X$ by $f_i(t) = \sum_{j=1}^{\infty} t_j x_{ij}$ and set $F = span\{f_i : i \in \mathbb{N}\}$. Consider the abstract triple $E = \lambda, F$ and X and let the bilinear mapping from $E \times F$ into X be the extension of the map $(t, f_i) \to t \cdot f_i = f_i(t)$. We first claim that the series $\sum_j e^j$ is λ multiplier convergent with respect to $w(E, F)$. For if $t \in \lambda$, $\sum_{j=1}^{\infty} t_j e^j \cdot f_i = \sum_{j=1}^{\infty} t_j f_i(e^j) = \sum_{j=1}^{\infty} t_j x_{ij}$ converges for every i. Now $\{f_i\}$ is conditionally $w(F, E)$ sequentially compact since $\{t \cdot f_i\} = \{\sum_{j=1}^{\infty} t_j x_{ij}\}$ converges for every $t \in \lambda$. Theorem 4.62 implies that the series $\sum_{j=1}^{\infty} t_j f_i(e^j) = \sum_{j=1}^{\infty} t_j x_{ij}$ converge uniformly for $i \in \mathbb{N}$.

Recall that Theorem 4.77 (Theorem 2.26) and the convergence result in Lemma 2.27 were used to derive Stuart's completeness result in Theorem 2.28.

We can also derive a version of Kalton's Theorem on subseries convergence in the space of compact operators. Let X and Y be normed spaces and let $K(X, Y)$ be the space of all compact operators from X into Y (an operator $T \in L(X, Y)$ is compact if T carries bounded sets into relatively compact sets). The space X has the DF property if every weak* subseries convergent series in X' is $\|\cdot\|$ subseries convergent ([DF]; Diestel and Faires have shown that for B-spaces this is equivalent to X' containing no subspace isomorphic to l^{∞}).

Theorem 4.78. *Let X and Y be normed spaces and let X have the DF property. If the series $\sum_j T_j$ is subseries convergent in the weak operator topology of $K(X, Y)$, then the series is subseries convergent in the norm topology of $K(X, Y)$.*

Proof: Each T_j has separable range so we may assume that Y is separable by replacing Y with $span \cup_{j=1}^{\infty} T_j X$. By Lemma A.6 of Appendix A or Lemma 4.4, it suffices to show that $\|T_j\| \to 0$ or, equivalently, $\|T_j'\| \to 0$. Pick $y_j' \in Y'$, $\|y_j'\| = 1$, such that $\|T_j'\| \leq \|T_j' y_j'\| + 1/j$. By the separability of Y there exists a subsequence $\{y_{n_j}'\}$ which is weak* convergent to some $y' \in Y'$; for convenience assume that the sequence $\{y_j'\}$ is weak* convergent to y'. Consider the abstract triple $E = \{T' : T \in$

$K(\dot{X}, Y)\}, F = Y'$ and $(X', \|\cdot\|)$ with the bilinear map $E \times F \to (X', \|\cdot\|)$ defined by $(T', y') \to T' \cdot y' = T'y'$. For each $z' \in Y'$, the series $\sum_j T'_j z'$ is weak* subseries convergent in X' and is, therefore, subseries convergent in $(X', \|\cdot\|)$ by the DF property. Hence, the series $\sum_j T'_j$ is $w(E, F)$ subseries convergent. The sequence $\{y'_j\}$ is $w(F, E)$ relatively sequentially compact since $\|\cdot\| - \lim T'y'_j = T'y'$ for every $T \in K(X, Y)^{\cdot}$ ([DS] VI.5.6). By Theorem 4.62 the series $\sum_{j=1}^{\infty} T'_j y'_i$ converge uniformly for $i \in \mathbb{N}$. In particular, $\|T'_j y'_j\| \to 0$ so $\|T'_j\| = \|T_j\| \to 0$ as desired.

Kalton's Theorem will also be considered in Chapter 6.

We next show that the conclusions in Theorems 4.62 and 4.63 can be strengthened if the multiplier space has signed-SGHP instead of signed-WGHP.

Theorem 4.79. *Let $\Lambda \subset \lambda$ have signed-SGHP. If $\sum_j x_j$ is λ multiplier convergent with respect to $w(E, F)$, then for each conditionally $w(F, E)$ sequentially compact ($w(F, E)$ compact, $w(F, E)$ countably compact) subset $K \subset F$ and each bounded subset $B \subset \Lambda$, the series $\sum_{j=1}^{\infty} t_j x_j \cdot y$ converge uniformly for $y \in K, t \in B$.*

Proof: If the conclusion fails to hold, there exist a neighborhood, W, in X , $y_k \in K, t^k \in B$ and an increasing sequence of intervals $\{I_k\}$ such that

$$(\#) \quad \sum_{l \in I_k} t^k_l x_l \cdot y_k \notin W$$

for every k. We may assume, by passing to a subsequence if necessary, that $\lim_k x \cdot y_k$ exists for every $x \in E$. Consider the matrix

$$M = [m_{ij}] = [\sum_{l \in I_j} t^j_l x_l \cdot y_i].$$

We claim that M is a signed $\mathcal{K}$ matrix as in Theorem 4.62 (Appendix D.3). First, the columns of M converge. Next given an increasing sequence of positive integers, there exist a sequence of signs $\{s_j\}$ and a subsequence $\{n_j\}$ such that $u = \sum_{k=1}^{\infty} s_k \chi_{I_{n_k}} t^{n_k} \in \Lambda$. Then

$$\{\sum_{j=1}^{\infty} s_j m_{in_j}\}_i = \{\sum_{j=1}^{\infty} s_j \sum_{l \in I_{n_j}} t^{n_j}_l x_l \cdot y_i\}_i = \{\sum_{l=1}^{\infty} u_l x_l \cdot y_i\}_i$$

converges. Hence, M is a signed $\mathcal{K}$ matrix so the diagonal of M converges to 0 by the signed version of the Antosik-Mikusinski Matrix Theorem (Appendix D.3). But, this contradicts $(\#)$.

The proof of the statements in parentheses follow as in the proof of Theorem 4.63.

If the multiplier space $\Lambda \subset \lambda$ has signed-SGHP, then the conclusion of Corollary 4.10 can be improved (see Theorem 2.16).

Corollary 4.80. *Let Λ have signed-SGHP and let E, F be in duality. If the series $\sum_j x_j$ is λ multiplier convergent with respect to $\sigma(E, F)$, then the series $\sum_{j=1}^{\infty} t_j x_j$ converge uniformly for t belonging to bounded subsets of Λ with respect to both $\lambda(E, F)$ and $\gamma(E, F)$.*

Corollary 4.80 covers the case of subseries convergent series ($\Lambda = \{\chi_\sigma : \sigma \subset \mathbb{N}\} \subset m_0 = \lambda$) and bounded multiplier convergent series (Λ the unit ball of l^∞).

Using Theorem 4.79 we can also obtain an improved conclusion in Theorem 4.66. In particular, if $\{E_j\}$ is a pairwise disjoint sequence from Σ, then the series $\sum_{j=1}^{\infty} \chi_E(j) \mu_i(E_j)$ converge uniformly for $i \in \mathbb{N}, E \in \Sigma$. That is, the series $\sum_{j=1}^{\infty} \mu_i(E_j)$ are uniformly unordered convergent for $i \in \mathbb{N}$.

Similarly, we can obtain an improvement to the statements in Example 4.68 if the multiplier space λ has signed-SGHP. If λ has signed-SGHP and the series $\sum_j f_j$ is λ multiplier convergent in $SC(S, X)$ $(C(S, X))$ with respect to the topology of pointwise convergence on S, then the series $\sum_{j=1}^{\infty} t_j f_j(s)$ converge uniformly for $s \in S$ and t belonging to bounded subsets of λ (Theorem 4.79).

We can also obtain a strengthened version of the result given in Theorem 4.77.

Theorem 4.81. *Let λ have signed-SGHP. Assume that $\sum_j x_{ij}$ is λ multiplier convergent for each $i \in \mathbb{N}$ and that $\lim_i \sum_{j=1}^{\infty} t_j x_{ij}$ exists for each $t \in \lambda$ with $x_j = \lim_i x_{ij}$ for every j. Then the series $\sum_{j=1}^{\infty} t_j x_{ij}$ converge uniformly for t belonging to bounded subsets of λ and $i \in \mathbb{N}$, and the series $\sum_j x_j$ is λ multiplier convergent.*

The proof of Theorem 4.77 carries forward using Theorem 4.79 in place of Theorem 4.62.

Chapter 5

Orlicz-Pettis Theorems for the Strong Topology

In this chapter we consider Orlicz-Pettis Theorems for the strong topology. As the following example shows, in general, an Orlicz-Pettis Theorem does not hold for the strong topology. Recall that if X, X' is a pair of vector spaces in duality, the strong topology $\beta(X, X')$ is the polar topology of uniform convergence on the family of $\sigma(X', X)$ bounded subsets of X' (Appendix A, Example A.2). As before, throughout this chapter λ will denote a scalar sequence space which contains c_{00}, the space of sequences which are eventually 0.

Example 5.1. The series $\sum_j e^j$ is subseries convergent in l^∞ with respect to the weak topology $\sigma(l^\infty, l^1)$ but is not subseries convergent in the strong topology $\beta(l^\infty, l^1) = \|\cdot\|_\infty$.

In order to obtain an Orlicz-Pettis Theorem for the strong topology, we will impose stronger conditions on the multiplier space λ. Before proceeding in this direction, we use the Nikodym Boundedness Theorem to show that although a weak subseries convergent series may fail to be subseries convergent in the strong topology, the partial sums of the series are strongly bounded.

Theorem 5.2. Let X be an LCTVS. If $\sum_j x_j$ is $\sigma(X, X')$ subseries convergent, then $P = \{\sum_{j \in \sigma} x_j : \sigma \subset \mathbb{N}\}$ is $\beta(X, X')$ bounded, where $\sum_{j \in \sigma} x_j$ is the $\sigma(X, X')$ sum of the series.

Proof: Let $\mathcal{P}$ be the power set of $\mathbb{N}$ and define $\mu : \mathcal{P} \to X$ by $\mu(\sigma) = \sum_{j \in \sigma} x_j$ [$\sigma(X, X')$ sum of the series]. Let $B \subset X'$ be $\sigma(X', X)$ bounded. The family $\mathcal{M} = \{x'\mu : x' \in B\}$ is a family of scalar valued, signed measures which is pointwise bounded on $\mathcal{P}$ since B is $\sigma(X', X)$ bounded. By the

Nikodym Boundedness Theorem [Theorem 4.59], $\mathcal{M}$ is uniformly bounded on $\mathcal{P}$. Since the range of μ is P, P is $\beta(X, X')$ bounded.

A similar result for barrelled AB-spaces λ is given in Corollary 2.6.

First, we recall an Orlicz-Pettis result relative to the strong topology which was established in Corollary 4.6.

Corollary 5.3. *Let* λ *have signed-WGHP. Let* X, X' *be in duality. If* $(X, \beta(X, X'))$ *is separable, then any series* $\sum_j x_j$ *which is* λ *multiplier convergent with respect to the weak topology* $\sigma(X, X')$ *is* λ *multiplier convergent with respect to the strong topology* $\beta(X, X')$.

Example 5.1, where $X = l^\infty$ and $X' = l^1$, shows that the separability condition in Corollary 5.3 is important.

We next establish an Orlicz-Pettis Theorem for the strong topology which requires strong topological assumptions on the multiplier space.

Theorem 5.4. *Assume that* λ *is a barrelled AK-space and* X *is an LCTVS. If* $\sum_j x_j$ *is* λ *multiplier convergent with respect to the weak topology* $\sigma(X, X')$, *then* $\sum_j x_j$ *is* λ *multiplier convergent with respect to the strong topology* $\beta(X, X')$.

Proof: By Proposition 2.5, $\lambda' = \lambda^\beta$ so the original topology of λ is $\beta(\lambda, \lambda^\beta)$ and λ is an AK-space with respect to $\beta(\lambda, \lambda^\beta)$. Since the strong topology is a Hellinger-Toeplitz topology, the result follows from Theorem 4.5.

Example 5.1 where $\lambda = m_0$ shows that the AK assumption in Theorem 5.4 is important.

Although the assumptions on the multiplier space in Theorem 5.4 are quite restrictive, the result covers a large number of multiplier spaces.

Example 5.5. If λ is a Banach [Frechet] AK-space, Theorem 5.4 applies. For example, $\lambda = c_0, l^p$ $(1 \le p < \infty), cs$ or bv_0 are Banach AK-spaces. Likewise, if $\lambda = s$ or if λ is a Köthe echelon space ([K1] 30.8), then λ is a Frechet AK-space. The spaces $(l^p, \|\cdot\|_1)$, $0 < p < 1$, are barrelled AK-spaces ([Be]) so Theorem 5.4 applies. Examples of barrelled subspaces of $(l^1, \|\cdot\|_1)$ are given in [RS] and more examples of barrelled AK-spaces are given in [BK].

Whereas the assumptions on the multiplier space in Theorem 5.4 are topological in nature, we now consider a gliding hump assumption which is purely algebraic.

We recall a gliding hump property which will be employed (Appendix B.36)

Definition 5.6. The space λ has the infinite gliding hump property (∞-GHP) if whenever $t \in \lambda$ and $\{I_j\}$ is an increasing sequence of intervals, there exist a subsequence $\{n_j\}$ and $a_{n_j} > 0, a_{n_j} \to \infty$ such that every subsequence of $\{n_j\}$ has a further subsequence $\{p_j\}$ such that the coordinate sum $\sum_{j=1}^{\infty} a_{p_j} \chi_{I_{p_j}} t \in \lambda$.

The term "infinite gliding hump" is used to suggest that the "humps", $\chi_{I_{p_j}} t$, are multiplied by a sequence of scalars which converges to ∞; there are other gliding hump properties where the humps are multiplied by elements of classical sequence spaces [see Appendix B for the μ-gliding hump property].

Examples of multiplier spaces with ∞-GHP are given in Appendix B. For example, $\lambda = l^p$, $0 < p < \infty$, or $\lambda = cs$ have ∞-GHP.

We now establish an Orlicz-Pettis Theorem for spaces with ∞-GHP.

Theorem 5.7. Let λ have ∞-GHP and let X be an LCTVS. If $\sum_j x_j$ is λ multiplier convergent with respect to the weak topology $\sigma(X, X')$, then $\sum_j x_j$ is λ multiplier convergent with respect to the strong topology $\beta(X, X')$.

Proof: If the conclusion fails to hold, there exist $t \in \lambda$, a $\sigma(X', X)$ bounded subset $B \subset X'$ and $\epsilon > 0$ such that for every k there exist $x'_k \in B, m_k > k$ such that $\left| \sum_{j=m_k}^{\infty} \langle x'_k, t_j x_j \rangle \right| > 2\epsilon$. For $k = 1$, let x'_1, m_1 satisfy the previous condition. There exists $n_1 > m_1$ such that $\left| \sum_{j=n_1+1}^{\infty} \langle x'_1, t_j x_j \rangle \right| < \epsilon$. Then

$$\left| \sum_{j=m_1}^{n_1} \langle x'_1, t_j x_j \rangle \right| = \left| \sum_{j=m_1}^{\infty} \langle x'_1, t_j x_j \rangle - \sum_{j=n_1+1}^{\infty} \langle x'_1, t_j x_j \rangle \right| > \epsilon.$$

Continuing this construction produces $\{x'_k\} \subset B$, increasing sequences $\{m_k\}, \{n_k\}$ with $m_k < n_k < m_{k+1}$ satisfying

$$(*) \qquad \left| \sum_{j=m_k}^{n_k} t_j \langle x'_k, x_j \rangle \right| > \epsilon.$$

Set $I_k = [m_k, n_k]$. Since λ has ∞-GHP, there exists $\{p_j\}$, $a_{p_j} > 0$, $a_{p_j} \to \infty$ such that every subsequence of $\{p_j\}$ has a further subsequence $\{q_j\}$ such that the coordinate sum $\sum_{j=1}^{\infty} a_{q_j} \chi_{I_{q_j}} t \in \lambda$. Define matrix

$$M = [m_{ij}] = [\sum_{l \in I_{p_j}} a_{p_j} t_l \langle x_i'/a_{p_i}, x_l \rangle].$$

We claim that M is a $\mathcal{K}$-matrix [Appendix D.2]. First, the columns of M converge to 0 since $\{x_i'\}$ is $\sigma(X', X)$ bounded and $a_{p_i} \to \infty$. Next, given a subsequence there exists a further subsequence $\{q_j\}$ such that $s = \sum_{j=1}^{\infty} a_{q_j} \chi_{I_{q_j}} t \in \lambda$. Then

$$\sum_{j=1}^{\infty} m_{iq_j} = \sum_{j=1}^{\infty} \sum_{l \in I_{q_j}} s_l \langle x_i'/a_{p_i}, x_l \rangle = \left\langle x_i'/a_{q_i}, \sum_{l=1}^{\infty} s_l x_l \right\rangle \to 0,$$

where $\sum_{l=1}^{\infty} s_l x_l$ is the $\sigma(X, X')$ sum of the series. Thus, M is a $\mathcal{K}$-matrix and by the Antosik-Mikusinski Matrix Theorem [Appendix D.2], the diagonal of M converges to 0. But, this contradicts (∗) and establishes the result.

The spaces $d = \{t : \sup_k |t_k|^{1/k} < \infty\}$ and $\delta = \{t : \lim |t_k|^{1/k} = 0\}$ furnish examples of spaces to which Theorem 5.7 applies but Theorem 5.4 does not [the natural metric on d does not give a vector topology ([KG] p.68)].

Diestel and Faires have established an interesting Orlicz-Pettis Theorem for the weak* topology on the dual of a Banach space. They have shown that if X is a Banach space such that X' contains no subspace isomorphic to l^{∞} and if $\sum_j x_j'$ is weak* subseries convergent, then $\sum_j x_j'$ is norm subseries convergent ([DU] I.4.7). We give statements of Theorems 5.4 and 5.7 for the weak* topology where the emphasis is on the multiplier space λ instead of topological assumptions on the dual space as in the Diestel-Faires Theorem.

Corollary 5.8. *Assume that λ either has ∞-GHP or is a barrelled AK-space. Let X be an LCTVS. If $\sum_j x_j'$ is λ multiplier convergent in the weak* topology $\sigma(X', X)$ of X', then $\sum_j x_j'$ is λ multiplier convergent with respect to the strong topology $\beta(X', X)$.*

Recall that if X is a Banach space, then the strong topology $\beta(X', X)$ is just the dual norm topology so Corollary 5.8 can be compared to the Diestel-Faires result in this case.

Finally, we observe that the proof of Theorem 5.7 shows directly that the strong topology $\beta(X, X')$ is weaker than the Dierolf topology $\mathcal{D}_\lambda(X, X')$ [Definition 4.38]. Theorem 5.7 will then follow from Theorem 4.41.

Theorem 5.9. *Let λ have ∞-GHP and let X be an LCTVS. Then the strong topology $\beta(X, X')$ is weaker than the Dierolf topology $\mathcal{D}_\lambda(X, X')$.*

Proof: Let B be $\sigma(X', X)$ bounded. If B does not belong to $\mathcal{D}_\lambda$, there exists $t \in \lambda$ and a λ multiplier convergent series $\sum_j x_j$ such that the series $\sum_{j=1}^\infty t_j \langle x', x_j \rangle$ do not converge uniformly for $x' \in B$ [Remark 4.40]. The proof of Theorem 5.7 then yields the result.

Theorem 5.9 and Theorem 4.41 then yield Theorem 5.7 as a corollary and furnishes an alternate proof of Theorem 5.7.

Chapter 6

Orlicz-Pettis Theorems for Linear Operators

In this chapter we consider multiplier convergent series of continuous linear operators and establish Orlicz-Pettis Theorems for such series. Throughout this chapter let X and $Y \neq \{0\}$ be LCTVS and $L(X, Y)$ the space of continuous linear operators from X into Y. We first describe the topologies on $L(X, Y)$ which will be considered.

Let $\mathcal{A}$ be a family of bounded subsets of X whose union is all of X and let $\mathcal{Y}$ be the family of all continuous semi-norms on Y. The pair $(\mathcal{A}, \mathcal{Y})$ generate a locally convex topology on $L(X, Y)$ defined by the family of semi-norms

$$(1) \quad p_{A,q}(T) = \sup\{q(Tx) : x \in A\}, q \in \mathcal{Y}, A \in \mathcal{A}.$$

We denote by $L_{\mathcal{A}}(X, Y)$ the locally convex topology on $L(X, Y)$ generated by the semi-norms in (1); this notation suppresses the dependence of the topology on the semi-norms in $\mathcal{Y}$. A net $\{T_\delta\}$ in $L(X, Y)$ converges to 0 in $L_{\mathcal{A}}(X, Y)$ iff for every $A \in \mathcal{A} \lim T_\delta x = 0$ uniformly for $x \in A$; for this reason the topology $L_{\mathcal{A}}(X, Y)$ is called the topology of uniform convergence on $\mathcal{A}$ (Appendix A).

If $\mathcal{A}$ is the family of all bounded subsets of X, the topology $L_{\mathcal{A}}(X, Y)$ is denoted by $L_b(X, Y)$. In the case when X and Y are normed spaces, the topology $L_b(X, Y)$ is called the uniform operator topology and is generated by the semi-norm

$$\|T\| = \sup\{\|Tx\| : \|x\| \leq 1\}.$$

If $\mathcal{A}$ is the family of all finite subsets of X, the topology $L_{\mathcal{A}}(X, Y)$ is denoted by $L_s(X, Y)$ and is just the topology of pointwise convergence on X. When Y has its original topology, the topology $L_s(X, Y)$ is called the strong operator topology of $L(X, Y)$. When Y has the weak topology

89

$\sigma(Y, Y')$, the topology $L_s(X, Y)$ is called the weak operator topology of $L(X, Y)$. Thus, a net $\{T_\delta\}$ converges to 0 in the strong operator topology (weak operator topology) iff $T_\delta x \to 0$ in Y for every $x \in X$ ($\langle y', T_\delta x \rangle \to 0$ for every $x \in X, y' \in Y'$). See Appendix A for more details.

Throughout this chapter let λ be a sequence space containing c_{00}, the space of all sequences which are eventually 0. We begin by establishing an Orlicz-Pettis Theorem for the weak and strong operator topologies. Recall that λ has the signed weak gliding hump property (signed-WGHP) if for every $t \in \lambda$ and every increasing sequence of intervals $\{I_j\}$ there exist a sequence of signs $\{s_j\}$ and a subsequence $\{n_j\}$ such that the coordinate sum $\sum_{j=1}^{\infty} s_j \chi_{I_{n_j}} t \in \lambda$; if the signs can all be chosen equal to 1, then λ has the weak gliding hump property (WGHP). Examples of spaces with these properties are given in Appendix B. For example, any monotone space has WGHP while bs has signed-WGHP but not WGHP.

Theorem 6.1. *Let λ have signed-WGHP. If the series $\sum_j T_j$ is λ multiplier convergent in $L(X, Y)$ with respect to the weak operator topology, then the series $\sum_j T_j$ is λ multiplier convergent in $L(X, Y)$ with respect to the strong operator topology.*

Proof: For every $x \in X$ the series $\sum_j T_j x$ is λ multiplier convergent in Y with respect to $\sigma(Y, Y')$. By the Orlicz-Pettis Theorem in Corollary 4.10, the series $\sum_j T_j x$ is λ multiplier convergent in Y with respect to the original topology of Y. That is, the series $\sum_j T_j$ is λ multiplier convergent in the strong operator topology.

Since subseries convergence is just m_0 multiplier convergence as a special case of Theorem 6.1, we have

Corollary 6.2. *If the series $\sum_j T_j$ is subseries convergent with respect to the weak operator topology of $L(X, Y)$, then the series is subseries convergent with respect to the strong operator topology.*

If the multiplier space λ does not satisfy the signed-WGHP, then the conclusion of Theorem 6.1 may fail.

Example 6.3. Let $\lambda = c_c$, the space of all sequences which are eventually constant (Appendix B). Then a series $\sum_j z_j$ is λ multiplier convergent in a TVS Z iff the series $\sum_j z_j$ converges in Z. Define continuous linear operators $T_j : \mathbb{R} \to c_0$ by $T_j s = s(e^{j+1} - e^j) = -s\left(e^j - e^{j+1}\right)$. Then for every $s \in \mathbb{R}$, the series $\sum_{j=1}^{\infty} T_j s$ converges in $\sigma(c_0, l^1)$ but does not converge in $(c_0, \|\cdot\|_\infty)$. That is, the series $\sum_j T_j$ is λ multiplier convergent in the weak operator topology of $L(\mathbb{R}, c_0)$ but is not λ multiplier convergent in the strong operator topology of $L(\mathbb{R}, c_0)$.

First, we have the analogue of Lemma A.6 of Appendix A for λ multiplier convergent series.

Lemma 6.4. *Let X be a vector space and σ and τ two vector topologies on X such that τ is linked to σ. Let λ have signed-WGHP. Suppose that every series $\sum_j x_j$ which is λ multiplier convergent with respect to σ is such that for every $t \in \lambda$ and increasing sequence of intervals $\{I_k\}$ there is a subsequence $\{n_k\}$ such that $\tau - \lim \sum_{j \in I_{n_k}} t_j x_j = 0$. Then every series $\sum_j x_j$ which is λ multiplier convergent with respect to σ is also λ multiplier convergent with respect to τ.*

Proof: By Lemma A.4, it suffices to show that if $t \in \lambda$, then the partial sums of the series $\sum_j t_j x_j$ are τ Cauchy. Suppose the partial sums of the series $\sum_j t_j x_j$ are not τ Cauchy. Then there exist a symmetric τ neighborhood of $0, U$, and an increasing sequence of intervals $\{I_k\}$ such that

$$(*) \quad \sum_{j \in I_k} t_j x_j \notin U.$$

Since λ has signed-WGHP, there exist a sequence of signs $\{s_k\}$ and a subsequence $\{n_k\}$ such that $u = \sum_{k=1}^\infty s_k \chi_{I_{n_k}} t$ [coordinate sum] belongs to λ. The series $\sum_{j=1}^\infty u_j x_j = \sum_{k=1}^\infty s_k \sum_{j \in I_{n_k}} t_j x_j$ is σ convergent so by hypothesis, $\{n_k\}$ has a subsequence $\{p_k\}$ such that $\tau - \lim \sum_{j \in I_{p_k}} t_j x_j = 0$. But, this contradicts $(*)$.

We now use Lemma 6.4 to establish a general Orlicz-Pettis Theorem for linear operators. Let

$$\mathcal{C} = \{C \subset X : \text{if } \{x_j\} \subset C, \text{ then } \lim T x_j \text{ exists for every } T \in L(X, Y)\}.$$

Theorem 6.5. *Let λ have signed-WGHP. If $\sum_j T_j$ is λ multiplier convergent in $L(X, Y)$ with respect to the weak operator topology, then $\sum_j T_j$ is λ multiplier convergent in $L_\mathcal{C}(X, Y)$.*

Proof: By Lemma 6.4 it suffices to show that if $t \in \lambda$ and $\{I_k\}$ is an increasing sequence of intervals and $\{x_i\} \in \mathcal{C}$, then

$$(\#) \quad \lim_k (\sum_{l \in I_k} t_l T_l)(x_i) = 0 \text{ uniformly for } i \in \mathbb{N}.$$

Define the matrix

$$M = [m_{ij}] = [\sum_{l \in I_j} (t_l T_l)(x_i)].$$

We claim that M is a signed $\mathcal{K}$-matrix (Appendix D.3). First, the columns of M converge since $\{x_i\} \in \mathcal{C}$. Next, if $\{p_j\}$ is an increasing sequence of integers, then there exist a subsequence $\{q_k\}$ of $\{p_k\}$ and a sequence of signs $\{s_k\}$ such that the coordinate sum of the series $u = \sum_{k=1}^{\infty} s_k \chi_{I_{q_k}} t$ belongs to λ. Then

$$\sum_{j=1}^{\infty} s_j m_{iq_j} = \sum_{j=1}^{\infty} s_j \sum_{l \in I_{q_j}} t_l T_l x_i = (\sum_{l=1}^{\infty} u_l T_l) x_i$$

converges as $i \to \infty$, where $\sum_{l=1}^{\infty} u_l T_l \in L(X,Y)$ is the weak operator sum of the series. Hence, M is a signed $\mathcal{K}$-matrix. By the signed version of the Antosik-Mikusinski Matrix Theorem the condition (#) is satisfied [Appendix D.3].

We use Theorem 6.5 to establish Orlicz-Pettis Theorems for two of the common topologies employed on $L(X,Y)$. If $\mathcal{A}$ is the family of all sequences in X which converge to 0, the topology $L_{\mathcal{A}}(X,Y)$ is denoted by $L_{\to 0}(X,Y)$[Appendix A]; this topology is studied in [GDS]. From Theorem 6.5, we have

Corollary 6.6. *Let λ have signed-WGHP. If $\sum_j T_j$ is λ multiplier convergent in $L(X,Y)$ with respect to the weak operator topology, then $\sum_j T_j$ is λ multiplier convergent in $L_{\to 0}(X,Y)$.*

Proof: If $x_j \to 0$ in X, then $\{x_j\} \in \mathcal{C}$.

We next consider the topology of uniform convergence on precompact subsets. If $\mathcal{A}$ is the family of all precompact subsets of X, the topology $L_{\mathcal{A}}(X,Y)$ is denoted by $L_{pc}(X,Y)$. In order to establish an Orlicz-Pettis Theorem for $L_{pc}(X,Y)$, we require the following representation theorem for precompact sets.

Proposition 6.7. *Let Z be a dense subspace of X and assume that X is metrizable. If K is a precompact subset of X, then there exists a null sequence $\{x_k\} \subset Z$ such that every $x \in K$ has a representation $x = \sum_{j=1}^{\infty} t_j x_j$ with $\sum_{j=1}^{\infty} |t_j| \leq 1$.*

Proof: Let $\|\cdot\|_1 \leq \|\cdot\|_2 \leq \dots$ be a sequence of semi-norms which generate the topology of X. Since Z is dense and K is precompact, there is a finite set $F_1 \subset Z$ such that for every $x \in K$ there exists $z^1(x) \in F_1$ such that

$$\left\|x - z^1(x)\right\|_1 \leq (1/2)(1/2^3).$$

Now $K - F_1$ is precompact so there is a finite set $F_2 \subset Z$ such that for every $x \in K$ there exists $z^2(x) \in F_2$ such that

$$\left\|x - z^1(x) - z^2(x)\right\|_2 \leq (1/3)(1/2^4).$$

Continuing this construction produces a sequence of finite subsets $F_1, F_2, \dots$ of Z such that every $x \in K$ there exist $z^i(x) \in F_i$ satisfying

$$(1) \quad \left\|x - z^1(x) - \dots - z^i(x)\right\|_i \leq (1/(i+1))(1/2^{i+2}).$$

Therefore,

$$(2) \quad \left\|z^i(x)\right\|_{i-1}$$
$$\leq \left\|x - z^1(x) - \dots - z^i(x)\right\|_{i-1} + \left\|x - z^1(x) - \dots - z^{i-1}(x)\right\|_{i-1}$$
$$\leq (1/i)(1/2^i)$$

Set $y^i(x) = 2^i z^i(x)$ for $x \in K, i \in \mathbb{N}$. Arrange the elements of $2F_1, 2^2 F_2, \dots$ in a sequence with the elements of F_1 first, those of F_2 second and so on. By (2) this sequence converges to 0 and by (1)

$$x = z^1(x) + z^2(x) + \dots = (1/2)y^1(x) + (1/2^2)y^2(x) + \dots$$

This gives the desired representation.

Theorem 6.8. *Let λ have signed-WGHP and let X be metrizable or the regular strict inductive limit of a sequence of metrizable LCTVS. If $\sum_j T_j$ is λ multiplier convergent in $L(X,Y)$ with respect to the weak operator topology, then $\sum_j T_j$ is λ multiplier convergent in $L_{pc}(X,Y)$.*

Proof: First suppose that X is metrizable and let $K \subset X$ be precompact. By Proposition 6.7 there exists $\{x_j\} \subset K, x_j \to 0$ such that every $x \in K$ has a representation $x = \sum_{j=1}^{\infty} t_j x_j$ with $\sum_{j=1}^{\infty} |t_j| \leq 1$. Let U be a closed, absolutely convex neighborhood of 0 in Y and let $s \in \lambda$. By Corollary 6.6 there exists N such that $\sum_{j=m}^{n} s_j T_j x_k \in U$ for $n > m \geq N$ and $k \in \mathbb{N}$. If $x \in K$ has the representation above, then for $n > m \geq N$ we have

$$\sum_{j=m}^{n} s_j T_j x = \sum_{j=m}^{n} s_j T_j \sum_{k=1}^{\infty} t_k x_k = \sum_{k=1}^{\infty} t_k \sum_{j=m}^{n} s_j T_j x_k \in U$$

. since U is closed and absolutely convex. By Theorem 6.1, $\sum_{j=m}^{\infty} s_j T_j x \in U$ for $m \geq N$ and $x \in K$. This establishes the result in the metrizable case.

If X is a regular strict inductive limit of metrizable spaces $\{X_k\}$ and $K \subset X$ is precompact, then $K \subset X_k$ for some k and K is precompact in X_k. Thus, the first part gives the result in this case.

A similar result was derived in Theorem 4.70.

We next consider Orlicz-Pettis Theorems for the topology $L_b(X, Y)$. As was the case for the strong topology $\beta(X, X')$ considered in Chapter 5 such results require strong hypotheses on the multiplier space λ. We present an example which shows that, in general, a series which is subseries convergent in the strong operator topology may not be subseries convergent in $L_b(X, Y)$. The example also suggests that to obtain Orlicz-Pettis theorems connecting the weak (strong) operator topology to the topology of $L_b(X, Y)$, one should consider the space $K(X, Y)$ of compact operators from X to Y.

Example 6.9. Let X be a Banach space with an unconditional Schauder basis $\{b_j\}$, i.e., every $x \in X$ has a unique expansion

$$x = \sum_{j=1}^{\infty} t_j b_j,$$

where the series is unconditionally (subseries) convergent. Let f_j be the jth coordinate functional associated with the basis $\{b_j\}$ defined by $\langle f_j, x \rangle = t_j$ in the expansion above. Each f_j is linear and continuous ([Sw2] 10.1.13). Let $P_j x = \langle f_j, x \rangle b_j$. If $T \in L(X, Y)$, where Y is a Banach space, then $Tx = \sum_{j=1}^{\infty} \langle f_j, x \rangle T b_j = \sum_{j=1}^{\infty} T P_j x$, where the series is norm convergent in Y. That is, the series $\sum_{j=1}^{\infty} T P_j$ is subseries convergent in the strong operator topology of $L(X, Y)$ to T. If $L(X, Y)$ has the property that any series which is subseries convergent in the strong operator topology (or weak operator topology) is subseries convergent in the norm topology of $L(X, Y)$, it follows that every $T \in L(X, Y)$ is compact, being the norm limit of a sequence of compact operators, $\{\sum_{j=1}^{n} T P_j\}$, with finite dimensional range. That is, if $L(X, Y)$ has this property, then $L(X, Y) = K(X, Y)$, where $K(X, Y)$ is the space of all compact operators from X into Y. In particular, if $X = Y$, then the identity operator is compact and X must be finite dimensional ([Sw2] 7.8).

Before considering Orlicz-Pettis theorems for $L_b(X, Y)$, we observe a necessary condition for such a result to hold for subseries convergent series of operators.

Definition 6.10. The space X has the DF property iff every series $\sum_j x_j'$ in X' which is subseries convergent in $\sigma(X', X)$ is also subseries convergent in $\beta(X', X)$.

For Banach spaces, Diestel and Faires have shown that X' has the DF property iff X' contains no subspace isomorphic to l^∞ ([DF]).

Theorem 6.11. *If every series $\sum_j T_j$ in $L(X, Y)$ which is subseries convergent in the weak operator topology is also subseries convergent in $L_b(X, Y)$, then X has the DF property.*

Proof: Let $\sum_j x_j'$ be subseries convergent in X' with respect to $\sigma(X', X)$. Pick $y \in Y, y \neq 0$. Define $T_j \in L(X, Y)$ by $T_j x = \langle x_j', x \rangle y$. Then $\sum_j T_j$ is subseries convergent in the strong operator topology so by hypothesis the series converges in $L_b(X, Y)$. Thus, the series $\sum_{j=1}^\infty \langle x_j', x \rangle$ converge uniformly for x belonging to bounded subsets of X. That is, the series $\sum_j x_j'$ converges in $(X', \beta(X', X))$.

For our first Orlicz-Pettis Theorem for $L_b(X, Y)$, we establish the analogue of Theorem 5.4 for operators.

Theorem 6.12. *Let λ be a barrelled AK space. If the series $\sum_j T_j$ is λ multiplier convergent in the weak operator topology of $L(X, Y)$, then $\sum_j T_j$ is λ multiplier convergent in $L_b(X, Y)$.*

Proof: If $x \in X$ and $y' \in Y'$, define a continuous linear functional $x \otimes y'$ on $L(X, Y)$ by $\langle x \otimes y', T \rangle = \langle y', Tx \rangle$. Let $X \otimes Y' = span\{x \otimes y' : x \in X, y' \in Y'\}$. Note that the weak operator topology on $L(X, Y)$ is just $\sigma(L(X, Y), X \otimes Y')$. From Theorem 5.4 it follows that the series $\sum_j T_j$ is λ multiplier convergent in the strong topology $\beta(L(X, Y), X \otimes Y')$. Thus, it suffices to show that the strong topology $\beta(L(X, Y), X \otimes Y')$ is stronger than $L_b(X, Y)$. Let $\{S_\delta\}$ be a net in $L(X, Y)$ which converges to 0 in $\beta(L(X, Y), X \otimes Y')$. Let $A \subset X$ be bounded, $B \subset Y'$ be equicontinuous and set $C = \{x \otimes y' : x \in A, y' \in B\}$. Since A is $\beta(X, X')$ bounded from the barrelledness assumption,

$$\sup\{|\langle x \otimes y', T \rangle| : x \in A, y' \in B\} < \infty$$

for every $T \in L(X, Y)$; that is, C is $\sigma(X \otimes Y', L(X, Y))$ bounded. Thus,

$$\sup\{|\langle x \otimes y', S_\delta \rangle| : x \in A, y' \in B\} \to 0$$

so $S_\delta \to 0$ in Y uniformly for $x \in A$ or $S_\delta \to 0$ in $L_b(X, Y)$. Thus, $\beta(L(X, Y), X \otimes Y')$ is stronger than $L_b(X, Y)$ as desired.

An extensive list of multiplier spaces satisfying the assumptions of Theorem 6.12 is given in Remark 4.7.

We next establish the analogue of Theorem 5.7 for linear operators.

Theorem 6.13. *Let λ have ∞-GHP. If the series $\sum_j T_j$ is λ multiplier convergent in the strong operator topology, then $\sum_j T_j$ is λ multiplier convergent in $L_b(X, Y)$.*

Proof: If the conclusion fails to hold, there exist $\epsilon > 0, A \subset X$ bounded,$t \in \lambda$, a continuous semi-norm p on Y and sequences $\{m_k\}, \{n_k\}$ with $m_1 < n_1 < m_2 < \dots$ and

$$\sup_{x \in A} p\left(\sum_{l=m_k}^{n_k} t_l T_l x \right) = p_A\left(\sum_{l=m_k}^{n_k} t_l T_l \right) > \epsilon.$$

For every k there exists $x_k \in A$ such that

$$(*) \quad p\left(\sum_{l=m_k}^{n_k} t_l T_l x_k \right) > \epsilon.$$

Set $I_k = [m_k, n_k]$. Since λ has ∞-GHP, there exist $\{p_k\}, a_{p_k} > 0, a_{p_k} \to \infty$ such that every subsequence of $\{p_k\}$ has a further subsequence $\{q_k\}$ such that $s = \sum_{k=1}^{\infty} a_{q_k} \chi_{I_{q_k}} t \in \lambda$. Define a matrix

$$M = [m_{ij}] = \left[\sum_{l \in I_j} (t_l a_{p_j}) T_l(x_i / a_{p_i}) \right].$$

We claim that M is a $\mathcal{K}$-matrix [Appendix D.2]. First, the columns of M converge to 0 since $x_i / a_{p_i} \to 0$ and each T_l is continuous. Next, given a subsequence there is a further subsequence $\{q_k\}$ such that $s = \sum_{k=1}^{\infty} a_{q_k} \chi_{I_{q_k}} t \in \lambda$. The series $\sum_{l=1}^{\infty} s_l T_l$ converges in the strong operator topology to an operator $T \in L(X, Y)$. Hence,

$$\sum_{j=1}^{\infty} m_{iq_j} = \sum_{j=1}^{\infty} \sum_{l \in I_{q_j}} s_l T_l(x_i / a_{p_i}) = T(x_i / a_{p_i}) \to 0.$$

Hence, M is a $\mathcal{K}$-matrix and by the Antosik-Mikusinski Matrix Theorem [Appendix D.2], the diagonal of M converges to 0. But, this contradicts $(*)$.

Remark 6.14. If λ also has signed-WGHP in Theorem 6.13, we may replace the assumption that the series $\sum_j T_j$ is λ multiplier convergent in the strong operator topology with the assumption that the series is λ multiplier convergent in the weak operator topology (Theorem 6.1).

Example 6.9 suggests that if one wishes to establish Orlicz-Pettis theorems for λ multiplier convergent series with respect to $L_b(X, Y)$ without imposing strong conditions on the multiplier space λ, one should consider the space $K(X, Y)$ of compact operators. The major result in this area is a result of Kalton. Kalton has shown that if X has the DF property (or, if X' contains no subspace isomorphic to l^∞) and if $\sum_j T_j$ is a series of compact operators from a Banach space X into a Banach space Y which is subseries convergent in the weak operator topology of $K(X, Y)$, then the series $\sum_j T_j$ is subseries convergent in the uniform operator topology of $K(X, Y)$ ([Ka]).

We first establish a result of Wu and Lu which characterizes the Orlicz-Pettis property for the space of compact operators ([WL]). Their result contains Kalton's result as a special case. Let $K_b(X, Y)$ be the topology on $K(X, Y)$ induced by $L_b(X, Y)$.

Theorem 6.15. *Let λ have signed-WGHP. The following are equivalent:*

(i) *Every series $\sum_j T_j$ which is λ multiplier convergent in the weak operator topology of $K(X, Y)$ is λ multiplier convergent in $K_b(X, Y)$.*

(ii) *Every continuous linear operator $S : X \to (\lambda^\beta, \sigma(\lambda^\beta, \lambda))$ is sequentially compact (an operator is sequentially compact if it carries bounded sets into relatively sequentially compact sets).*

Proof: Suppose (ii) holds. Let $\sum_j T_j$ be λ multiplier convergent in the weak operator topology of $K(X, Y)$. By Theorem 6.1 the series is λ multiplier convergent in the strong operator topology of $K(X, Y)$. Suppose there exists $t \in \lambda$ such that the series $\sum_j t_j T_j$ is not convergent in $K_b(X, Y)$. Then there exist $T \in K(X, Y)$ and a bounded set $A \subset X$ such that $\sum_{j=1}^\infty t_j T_j x = Tx$ for every $x \in X$ but the series do not converge uniformly for $x \in A$. Thus, there exist a continuous semi-norm p on Y, increasing sequences $\{m_k\}$ and $\{n_k\}$ with $m_k < n_k < m_{k+1}, x_k \in A$ and $\epsilon > 0$ such that

$$p\left(\sum_{l=m_k}^{n_k} t_l T_l x_k\right) > \epsilon$$

for all k. By the Hahn-Banach Theorem there is a sequence $\{y_k'\} \subset Y'$ such that

$$(*) \quad \left\langle y_k', \sum_{l=m_k}^{n_k} t_l T_l x_k \right\rangle > \epsilon$$

and
$$\sup\{|\langle y_k', y\rangle| : p(y) \le 1\} \le 1.$$

Let Y_0 be the closure in Y of $span\{T_i x_j : i, j \in \mathbb{N}\}$. Then (Y_0, p) is a separable semi-norm space. By the Banach-Alaoglu Theorem for separable semi-norm spaces, $\{y_k'\}$ has a subsequence $\{y_{n_k}'\}$ and $y' \in Y'$ such that
$$\lim \langle y_{n_k}', y\rangle = \langle y', y\rangle$$
for every $y \in Y_0$ and
$$\sup\{|\langle y', y\rangle| : p(y) \le 1\} \le 1.$$

For notational convenience, assume that $n_k = k$.

Define a semi-norm q on X' by $q(x') = \sup\{|\langle x', x_k\rangle| : k \in \mathbb{N}\}$. We claim that if $U \in K(X, Y)$ satisfies $U x_k \in Y_0$, then
$$(**) \qquad \lim q(U' y_k' - U' y') = 0.$$

If $(**)$ fails to hold, there exist $\delta > 0$, a subsequence $\{y_{n_k}'\}$ and a subsequence $\{x_{n_k}\}$ such that
$$(***) \qquad |\langle U' y_{n_k}' - U' y', x_{n_k}\rangle| > \delta.$$

Since U is compact, $\{U x_{n_k}\}$ is a relatively compact subset of Y and, therefore, a relatively compact subset of (Y_0, p). Without loss of generality, we may assume that there exists $y \in Y_0$ such that $p(U x_{n_k} - y) \to 0$. Then
$$|\langle y_{n_k}' - y', U x_{n_k}\rangle| \le |\langle y_{n_k}' - y', U x_{n_k} - y\rangle| + |\langle y_{n_k}' - y', y\rangle|$$
$$\le \sup\{|\langle y_{n_k}' - y', z\rangle| : p(z) \le 1\} p(U x_{n_k} - y)$$
$$+ |\langle y_{n_k}' - y', y\rangle|$$
$$\le 2p(U x_{n_k} - y) + |\langle y_{n_k}' - y', y\rangle| \to 0.$$

This contradicts $(***)$ and establishes the claim.

If $s \in \lambda$, $x \in X$ and $z' \in Y'$, the series $\sum_{j=1}^{\infty} s_j \langle z', T_j x\rangle$ converges so we define a linear operator $S(= S_{z'}) : X \to (\lambda^\beta, \sigma(\lambda^\beta, \lambda))$ by $Sx = \{\langle z', T_j x\rangle\}$. Since S is obviously continuous, S is sequentially compact by condition (ii). Thus, SA is sequentially compact with respect to $\sigma(\lambda^\beta, \lambda)$. By Corollary 2.29, if $s \in \lambda$, then the series $\sum_{j=1}^{\infty} s_j \langle z', T_j x\rangle$ converge uniformly for $x \in A$ or, equivalently, the series $\sum_{j=1}^{\infty} s_j \langle T_j' z', x\rangle$ converge uniformly for $x \in A$.

Now consider the matrix
$$M = [m_{ij}] = \left[\sum_{l=m_j}^{n_j} t_l T_l' y_i' \right].$$

We show that M is a signed $\mathcal{K}$-matrix with values in the semi-norm space (X', q) [Appendix D.3]. First, the columns of M converge by condition $(**)$. Next, given any subsequence $\{p_j\}$ there is a further subsequence $\{q_j\}$ and a sequence of signs $\{\epsilon_j\}$ such that $s = \sum_{l=1}^{\infty} \epsilon_l \sum_{j=m_{q_l}}^{n_{q_l}} t_j \in \lambda$. There exist $U \in K(X,Y)$ such that $\sum_{j=1}^{\infty} s_j T_j = \sum_{l=1}^{\infty} \epsilon_l \sum_{j=m_{q_l}}^{n_{q_l}} t_j T_j$ converges to U in the strong operator topology. By the paragraph above $\sum_{l=1}^{\infty} \epsilon_l \sum_{j=m_{q_l}}^{n_{q_l}} t_j T_j' y_k'$ converges to $U' y_k'$ uniformly for $x \in A$. In particular,

$$q\left(\sum_{l=1}^{\infty} \epsilon_l \sum_{j=m_{q_l}}^{n_{q_l}} t_j T_j' y_k' - U' y_k' \right) \to 0.$$

Thus, M is a signed $\mathcal{K}$-matrix (with respect to (X', q). By the signed version of the Antosik-Mikusinski Matrix Theorem (Appendix D.3), the diagonal of M converges to 0 in (X', q). This contradicts $(*)$ and establishes that (ii) implies (i).

Suppose that (i) holds. Let $S : X \to (\lambda^{\beta}, \sigma(\lambda^{\beta}, \lambda))$ be linear and continuous. So $Sx = \{Sx \cdot e^j\}$. Let $y \in Y, y \neq 0$. Define $T_j \in K(X,Y)$ by $T_j x = (Sx \cdot e^j) y$. Let $t \in \lambda$. Define $T \ (= T_t) \in K(X,Y)$ by $Tx = (Sx \cdot t) y$. Then $\sum_{j=1}^{\infty} t_j T_j x = Tx$ for every $x \in X$, i.e., the series $\sum_j t_j T_j$ converges to T in the strong operator topology of $K(X,Y)$. By (i) the series

$$\sum_{j=1}^{\infty} t_j T_j x = \sum_{j=1}^{\infty} t_j (Sx \cdot e^j) y = Tx = (Sx \cdot t) y$$

converge uniformly for x belonging to bounded subsets of X or the series $\sum_{j=1}^{\infty} t_j (Sx \cdot e^j) = Sx \cdot t$ converge uniformly for x belonging to bounded subsets of X. Now to show S is sequentially compact, let $\{x_k\}$ be a bounded sequence in X. Then $\{Sx_k\}$ is coordinatewise bounded in λ^{β} since S is bounded. By the diagonal method ([DeS] 26.10), there is a subsequence $\{n_k\}$ such that $\lim_k Sx_{n_k} \cdot e^j$ exists for every j and since the series $\sum_{j=1}^{\infty} t_j (Sx_{n_k} \cdot e^j)$ converge uniformly for $k \in \mathbb{N}$, $\lim_k t \cdot Sx_{n_k}$ exists. Thus, $\{Sx_{n_k}\}$ is $\sigma(\lambda^{\beta}, \lambda)$ Cauchy. By Corollary 2.28 there exists $u \in \lambda^{\beta}$ such that $Sx_{n_k} \to u$ in $\sigma(\lambda^{\beta}, \lambda)$. Therefore, $\{Sx_{n_k}\}$ is relatively sequentially compact in $(\lambda^{\beta}, \sigma(\lambda^{\beta}, \lambda))$.

Remark 6.16. Wu Junde has shown that subsets of λ^{β} are $\sigma(\lambda^{\beta}, \lambda)$ sequentially compact iff they are $\sigma(\lambda^{\beta}, \lambda)$ compact so condition (ii) of Theorem 6.15 can be replaced with the hypothesis that the operator S is compact ([Wu]).

For the case of subseries convergent series, that is, when $\lambda = m_0$, we have

Theorem 6.17. *Let X be a barrelled LCTVS. The following are equivalent:*

(i) *Every series $\sum_j T_j$ which is subseries convergent in the weak operator topology of $K(X,Y)$ is subseries convergent in $K_b(X,Y)$.*

(ii) *Every continuous linear operator $S : X \to (l^1, \sigma(l^1, m_0))$ is compact.*

(iii) *Every continuous linear operator $S : X \to (l^1, \|\cdot\|_1)$ is compact.*

(iv) *$(X', \beta(X', X))$ contains no subspace isomorphic to c_0.*

(v) *X has the DF property.*

Proof: Since subsets of l^1 are $\sigma(l^1, m_0)$ $[\|\cdot\|_1]$ sequentially compact iff they are compact (Proposition 4.18), (i), (ii) and (iii) are equivalent by Theorem 6.15. Since X is barrelled (iii), (iv) and (v) are equivalent by Theorem 3.20.

Remark 6.18. If X and Y are Banach spaces, the equivalence of (i) and (v) is Kalton's result except that Kalton uses the hypothesis that X' contains no subspace isomorphic to l^∞ which is equivalent to the DF property by the Diestel/Faires result ([DF]). For Banach spaces the equivalence of (i) and (iv) was established by Bu and Wu ([BW]).

We used the abstract set-up preceding Theorem 4.62 to establish a version of Kalton's theorem for normed spaces (Theorem 4.78) which we restate.

Theorem 6.19. *Let X and Y be normed spaces and let X have the DF property. If the series $\sum_j T_j$ is subseries convergent in the weak operator topology of $K(X,Y)$, then the series is subseries convergent in the norm topology of $K(X,Y)$.*

Chapter 7

The Hahn-Schur Theorem

In this chapter we establish vector versions of the classical Hahn-Schur Theorem for multiplier convergent series. For later reference, we first give a statement of one version of the Schur Theorem for absolutely convergent scalar valued series.

Theorem 7.1. *(Schur) For each $i \in \mathbb{N}$, let $\sum_j t_{ij}$ be an absolutely convergent series of scalars. Assume*

(h) $\lim_i \sum_{j=1}^{\infty} s_j t_{ij}$ *exists for every $\{s_j\} \in l^\infty$ and let $\lim_i t_{ij} = t_j$ for every $j \in \mathbb{N}$.*

Then

(i) $\sum_j t_j$ *is absolutely convergent and $\lim_i \sum_{j=1}^{\infty} s_j t_{ij} = \sum_{j=1}^{\infty} s_j t_j$ for every $\{s_j\} \in l^\infty$,*
(ii) $\lim_i \sum_{j=1}^{\infty} |t_{ij} - t_j| = 0$,
(iii) *the series $\sum_{j=1}^{\infty} |t_{ij}|$ converge uniformly for $i \in \mathbb{N}$.*

The statement in Theorem 7.1 is often referred to as the Schur Lemma ([Sr]). In particular, Theorem 7.1 implies that a sequence in l^1 which is weak ($\sigma(l^1, l^\infty)$) Cauchy is norm ($\|\cdot\|_1$) convergent.

The Hahn version of Theorem 7.1 relaxes the hypothesis (h) and retains conclusions (ii) and (iii) with a slight restatement of condition (i). Again for later reference, we give a statement of Hahn's theorem ([Ha]).

Theorem 7.2. *(Hahn) For each $i \in \mathbb{N}$, let $\sum_j t_{ij}$ be an absolutely convergent series of scalars. Assume*

(h') $\lim_i \sum_{j \in \sigma} t_{ij}$ *exists for every* $\sigma \subset \mathbb{N}$ *and let* $\lim_i t_{ij} = t_j$ *for every*
$j \in \mathbb{N}$.

Then

(i) $\sum_j t_j$ *is absolutely convergent and* $\lim_i \sum_{j \in \sigma} t_{ij} = \sum_{j \in \sigma} t_j$ *for every*
$\sigma \subset \mathbb{N}$,
(ii) *and (iii).*

In particular, the Hahn version in Theorem 7.2 implies that a sequence in l^1 which is $\sigma(l^1, m_0)$ Cauchy is $\|\cdot\|_1$ convergent ([Ha]). Hahn's theorem can also be used to establish an important result from summability. Namely,

(S) A matrix $T = [t_{ij}]$ maps the sequence space m_0 into the space c of convergent sequences iff the matrix T maps l^∞ into c.

([Sw2] 9.5.3; see Theorem 7.29 for a vector version of statement (S)).

Both the Schur and Hahn theorems have numerous applications to various topics in analysis; in particular, Schur's theorem was used in the proofs of the versions of the Orlicz-Pettis Theorem given by both Orlicz and Pettis ([Or], [Pe]).

The conclusions in (ii) and (iii) (as well as those in the hypothesis and condition (i)) involve absolutely convergent series and, therefore, do not represent interesting suggestions for vector valued generalizations of either Theorem 7.1 or Theorem 7.2. However, conditions (ii) and (iii) can be restated in forms which do not involve absolute convergence and which do suggest possible vector valued generalizations. We first give a restatement of conditions (ii) and (iii) for hypothesis (h).

Proposition 7.3. *Let* $\sum_j t_{ij}$ *be absolutely convergent for every* $i \in \mathbb{N}$.
Condition (ii) is equivalent to:

(ii)' $\lim_i \sum_{j=1}^\infty s_j t_{ij} = \sum_{j=1}^\infty s_j t_j$ *uniformly for* $\|\{s_j\}\|_\infty \le 1$.

Condition (iii) is equivalent to:

(iii)' *the series* $\sum_{j=1}^\infty s_j t_{ij}$ *converge uniformly for* $i \in \mathbb{N}$ *and* $\|\{s_j\}\|_\infty \le 1$.

Proof: Assume (ii). If $\{s_j\} \in l^\infty$, then

$$\left| \sum_{j=1}^\infty s_j(t_{ij} - t_j) \right| \le \|\{s_j\}\|_\infty \sum_{j=1}^\infty |t_{ij} - t_j|$$

so (ii)' follows immediately.

Assume (ii)'. Fix $i \in \mathbb{N}$. Define $s_j = sign(t_{ij} - t_j)$. Then $\sum_{j=1}^{\infty} s_j(t_{ij} - t_j) = \sum_{j=1}^{\infty} |t_{ij} - t_j|$. Since $\|\{s_j\}\|_{\infty} \leq 1$, (ii) follows.

The equivalence of (iii) and (iii)' are established similarly.

We next give a restatement of conditions (ii) and (iii) for hypothesis (h)'. For this we use Lemma 3.37.

Proposition 7.4. *Let $\sum_j t_{ij}$ be absolutely convergent for every $i \in \mathbb{N}$.*

Condition (ii) is equivalent to:

Proposition 7.5.

(ii)" $\lim_i \sum_{j \in \sigma} t_{ij} = \sum_{j \in \sigma} t_j$ uniformly for $\sigma \subset \mathbb{N}$.

Condition (iii) is equivalent to:

(iii)" for every $\epsilon > 0$ there exist N such that $\sigma \subset \mathbb{N}, \min \sigma \geq N$ implies $\left| \sum_{j \in \sigma} t_{ij} \right| < \epsilon$ for all $i \in \mathbb{N}$.

Proof: Clearly (ii) implies (ii)" and (iii) implies (iii)".

Assume (ii)" holds. Let $\epsilon > 0$. There exists N such that $i \geq N$ implies $\left| \sum_{j \in \sigma} (t_{ij} - t_j) \right| < \epsilon/2$ for $\sigma \subset \mathbb{N}$. By Lemma 3.37, $\sum_{j=1}^{\infty} |t_{ij} - t_j| \leq \epsilon$ for $i \geq N$. Thus, (ii) holds.

Similarly, (iii)" implies (iii).

The hypothesis (h) [(h)'] and conclusions (ii)' and (iii)' [(ii)" and (iii)"] suggest generalizations of the Schur and Hahn theorems for multiplier convergent series with values in a TVS. Hypothesis (h) and conditions (ii)' and (iii)' [(h)' and conditions (ii)" and (iii)"] use bounded multiplier convergent series [subseries convergent series] or Λ multiplier convergent series where $\Lambda = l^{\infty}$ [where $\Lambda = \{\chi_{\sigma} : \sigma \subset \mathbb{N}\}$]. Thus, it would be natural to seek versions of Theorems 7.1 and 7.2 for Λ multiplier convergent series. We pursue these versions in this chapter. Theorems 7.1 and 7.2 will follow directly from our general results for multiplier convergent series.

Let λ be a scalar sequence space which contains c_{00}, the space of sequences which are eventually 0 and let $\Lambda \subset \lambda$. Let X be a TVS. The analogue of hypotheses (h) and (h)' for Λ multiplier convergent series would then be:

- (H) Let $\sum_j x_{ij}$ be Λ multiplier convergent in X for every $i \in \mathbb{N}$. Assume that $\lim_i \sum_{j=1}^{\infty} t_j x_{ij}$ exists for every $t \in \Lambda$ and assume that $\lim_i x_{ij} = x_j$ exists for every $j \in \mathbb{N}$.

The analogues of conditions (ii)' [(ii)''], and (iii)' [(iii)''] for multiplier convergent series require topological assumptions on the multiplier space Λ. Assume that λ is a K-space. The analogues of conditions (i), (ii)' [(ii)''], and (iii)' [(iii)''] for multiplier convergent series would then be:

- (C1) the series $\sum_j x_j$ is Λ multiplier convergent and $\lim_i \sum_{j=1}^{\infty} t_j x_{ij} = \sum_{j=1}^{\infty} t_j x_j$ for every $t \in \Lambda$.
- (C2) $\lim_i \sum_{j=1}^{\infty} t_j x_{ij} = \sum_{j=1}^{\infty} t_j x_j$ uniformly for t belonging to bounded subsets of Λ.
- (C3) the series $\sum_{j=1}^{\infty} t_j x_{ij}$ converge uniformly for t belonging to bounded subsets of Λ.

We first consider conclusion (C1). Under the hypothesis (H) conclusion (C1) follows from Lemma 2.27 and Theorem 2.26 preceding Stuart's weak completeness result. Recall that Λ has the signed weak gliding hump property (signed-WGHP) if whenever $t \in \Lambda$ and $\{I_j\}$ is an increasing sequence of intervals, there exist a sequence of signs $\{s_j\}$ and a subsequence $\{n_j\}$ such that the coordinate sum of the series $\sum_{j=1}^{\infty} s_j \chi_{I_{n_j}} t$ belongs to Λ [Appendix B.6].

Theorem 7.6. *Assume that Λ has signed-WGHP. Then condition*

(H) Let $\sum_j x_{ij}$ be Λ multiplier convergent in X for every $i \in \mathbb{N}$. Assume that $\lim_i \sum_{j=1}^{\infty} t_j x_{ij}$ exists for every $t \in \Lambda$ and assume that $\lim_i x_{ij} = x_j$ exists for every $j \in \mathbb{N}$

implies the conclusion

(C1) the series $\sum_j x_j$ is Λ multiplier convergent and $\lim_i \sum_{j=1}^{\infty} t_j x_{ij} = \sum_{j=1}^{\infty} t_j x_j$ for every $t \in \Lambda$.

The result in Theorem 7.5 may fail if λ does not have signed-WGHP.

Example 7.7. Let $\lambda = c$. Define x_{ij} by $x_{ij} = 1$ if $i = j$ and $x_{ij} = 0$ otherwise. If $t \in c$, then $\lim_i \sum_{j=1}^{\infty} t_j x_{ij} = \lim_i t_i$ and $x_j = \lim_i x_{ij} = 0$. But, $\lim_i \sum_{j=1}^{\infty} t_j x_{ij} = \lim_i t_i \neq \sum_{j=1}^{\infty} t_j x_j = 0$ if $\lim_i t_i \neq 0$.

This example shows that if $\{e^j\} \subset l^1$, then $\{e^j\}$ is $\sigma(l^1, c)$ Cauchy but does not have a $\sigma(l^1, c)$ limit. That is, $\sigma(l^1, c)$ is not sequentially complete.

We next consider conclusions (C2) and (C3). This will require stronger assumptions on the multiplier space λ. Let λ be a K-space and let $\Lambda \subset \lambda$. Recall that Λ has the signed strong gliding hump property (signed-SGHP) if whenever $\{t^k\} \subset \Lambda$ is bounded and $\{I_k\}$ is an increasing sequence of intervals, there exist a sequence of signs $\{s_k\}$ and a subsequence $\{n_k\}$ such that the coordinate sum of the series $\sum_{k=1}^{\infty} s_k \chi_{I_{n_k}} t^{n_k}$ belongs to Λ. If all of the signs can be chosen equal to 1, then Λ is said to have the strong gliding hump property (SGHP)(Appendix B.17). For example, the space l^{∞} has SGHP while the subset $\Lambda = M_0 = \{\chi_{\sigma} : \sigma \subset \mathbb{N}\} \subset m_0$ has SGHP but the space m_0 does not have SGHP; the space bs of bounded series has signed-SGHP but not SGHP [see Appendix B for these and additional examples].

We first establish a lemma which is a special case of condition (C2). The proof of the lemma uses a property of TVS which we now establish.

Lemma 7.8. *Let X be a TVS. If $\lim x_j = 0$ in X, then $\lim t x_j = 0$ uniformly for $|t| \leq 1$.*

Proof: Let U be a balanced neighborhood of 0 in X. There exists N such that $k \geq N$ implies that $x_k \in U$. Therefore, if $k \geq N$ and $|t| \leq 1$, $t x_k \in t U = U$.

Remark 7.9. There is another proof of Lemma 7 in Yosida ([Y] I.2.2) which uses Egoroff's Theorem and properties of Lebesgue measure and one in [Sw1] 8.2.4 which uses the Banach-Steinhaus Theorem.

Lemma 7.10. *Assume that $\Lambda \subset \lambda$ and Λ has signed-SGHP and $B \subset \Lambda$ is bounded. If $\sum_j x_{ij}$ is Λ multiplier convergent for every $i \in \mathbb{N}$, $\lim_i \sum_{j=1}^{\infty} t_j x_{ij} = 0$ for every $\{t_j\} \in \Lambda$ and $\lim_i x_{ij} = 0$ for every j, then $\lim_i \sum_{j=1}^{\infty} t_j x_{ij} = 0$ uniformly for $t \in B$.*

Proof: It suffices to show that $\lim_i \sum_{j=1}^{\infty} t_j^i x_{ij} = 0$ for any sequence $\{t^i\} \subset B$. Let U be a neighborhood of 0 in X and pick a symmetric neighborhood of $0, V$, in X such that $V + V + V \subset U$. Set $n_1 = 1$ and pick N_1 such that $\sum_{j=N_1}^{\infty} t_j^{n_1} x_{n_1 j} \in V$. Since $\lim_i x_{ij} = 0$ for every j and $\{t_j^i : i \in \mathbb{N}\}$ is bounded from the K-space assumption, $\lim_i t_j^i x_{ij} = 0$ for every j by Lemma 7.7. Therefore, there exists $n_2 > n_1$ such that $\sum_{j=1}^{N_1 - 1} t_j^i x_{ij} \in V$ for every $i \geq n_2$. Pick $N_2 > N_1$ such that $\sum_{j=N_2}^{\infty} t_j^{n_2} x_{n_2 j} \in V$. Continuing this construction produces increasing sequences $\{n_k\}, \{N_k\}$ such that

$$\sum_{l=N_j}^{\infty} t_l^{n_j} x_{n_j l} \in V \ and \ \sum_{l=1}^{N_j-1} t^i x_{il} \in V \ for \ i \geq n_j.$$

Set $I_j = \{l : N_{j-1} \leq l < N_j\}$. Define a matrix

$$M = [m_{ij}] = [\sum_{l \in I_j} t_l^{n_j} x_{n_i l}].$$

We show that M is a signed $\mathcal{K}$-*matrix* (Appendix D.3). First, the columns of M converge to 0 since $\lim_i x_{il} = 0$ for every l. Given an increasing sequence $\{p_j\}$, there is a sequence of signs $\{s_j\}$ and a subsequence $\{q_j\}$ of $\{p_j\}$ such that $t = \sum_{j=1}^{\infty} s_j \chi_{I_{q_j}} t_j^{q_j} \in \Lambda$ (coordinate sum). Then

$$\sum_{j=1}^{\infty} s_j m_{iq_j} = \sum_{j=1}^{\infty} s_j \sum_{l \in I_{q_j}} t_l^{q_j} x_{n_i l} = \sum_{j=1}^{\infty} t_j x_{n_i j}$$

and $\sum_{j=1}^{\infty} t_j x_{n_i j} \to 0$ by hypothesis. Hence, M is a signed $\mathcal{K}$-*matrix* and by the signed version of the Antosik-Mikusinski Matrix Theorem the diagonal of M converges to 0 (Appendix D.3). Thus, there exists N such that $m_{ii} \in V$ for $i \geq N$. If $i \geq N$, then

$$\sum_{l=1}^{\infty} t_l^{n_i} x_{n_i l} = \sum_{l=1}^{N_{i-1}-1} t_l^{n_i} x_{n_i l} + \sum_{l \in I_i} t_l^{n_i} x_{n_i l} + \sum_{l=N_i}^{\infty} t_l^{n_i} x_{n_i l} \in V + V + V \subset U$$

so $\lim_i \sum_{l=1}^{\infty} t_l^{n_i} x_{n_i l} = 0$. Since the same argument can be applied to any subsequence, it follows that $\lim_i \sum_{j=1}^{\infty} t_j^i x_{ij} = 0$.

We can now establish the result with conclusions (C2) and (C3) under hypothesis (H).

Theorem 7.11. *Assume that $\Lambda \subset \lambda$ and Λ has signed-SGHP. If*

(H) *Let $\sum_j x_{ij}$ be Λ multiplier convergent in X for every $i \in \mathbb{N}$. Assume that $\lim_i \sum_{j=1}^{\infty} t_j x_{ij}$ exists for every $t \in \Lambda$ and assume that $\lim_i x_{ij} = x_j$ exists for every $j \in \mathbb{N}$,*

then the following conclusions hold:

(C1) *the series $\sum_j x_j$ is Λ multiplier convergent and $\lim_i \sum_{j=1}^{\infty} t_j x_{ij} = \sum_{j=1}^{\infty} t_j x_j$ for every $t \in \Lambda$,*

(C2) $\lim_i \sum_{j=1}^{\infty} t_j x_{ij} = \sum_{j=1}^{\infty} t_j x_j$ *uniformly for t belonging to bounded subsets of Λ and*

(C3) *the series $\sum_{j=1}^{\infty} t_j x_{ij}$ converge uniformly for t belonging to bounded subsets of Λ*

Proof: Since Λ has signed-WGHP, conclusion (C1) holds by Theorem 7.5.

Since $\lim_i \sum_{j=1}^{\infty} t_j(x_{ij} - x_j) = 0$ by conclusion (C1), Lemma 7.9 now applies and gives conclusion (C2) immediately.

Suppose that (C3) fails to hold. Then there exist a closed, symmetric neighborhood of $0, U$, in X and a bounded set $B \subset \Lambda$ such that for every i there exist $k_i > i$, $n_i > i$, $t^i \in B$ with $\sum_{j=n_i}^{\infty} t_j^i x_{k_i j} \notin U$. There exists $m_i > n_i$ such that $\sum_{j=n_i}^{m_i} t_j^i x_{k_i j} \notin U$. Set $I = [n_i, m_i]$ so $\sum_{j \in I_i} t_j^i x_{k_i j} \notin U$.

By the condition above for $i_1 = 1$, there exist k_1, a finite interval I_1 with $\min I_1 > i_1, t^1 \in B$ with $\sum_{j \in I_1} t_j^1 x_{k_1 j} \notin U$. By Theorem 2.35, there exists j_1 such that $\sum_{k=j}^{\infty} t_k x_{ik} \in U$ for every $t \in B, 1 \le i \le k_1, j \ge j_1$. Set $i_2 = \max\{I_1 + 1, j_1\}$. Again by the condition above there exist $k_2 > i_2$, a finite interval I_2 with $\min I_2 > i_2, t^2 \in B$ such that $\sum_{k \in I_2} t_k^2 x_{k_2 k} \notin U$. Note that $k_2 > k_1$ by the definition of i_2. Continuing this construction produces an increasing sequence $\{k_i\}$, an increasing sequence of intervals $\{I_i\}$ and $\{t^i\} \subset B$ such that

$$(*) \qquad \sum_{k \in I_i} t_k^i x_{k_i k} \notin U.$$

Define a matrix

$$M = [m_{ij}] = [\sum_{k \in I_j} t_k^j x_{k_i k}].$$

We claim that M is a signed $\mathcal{K}$-matrix (Appendix D.3). First, the columns of M converge by hypothesis. Next, given any increasing sequence $\{p_j\}$, there exist a sequence of signs $\{s_j\}$ and a subsequence $\{q_j\}$ of $\{p_j\}$ such that the coordinate sum $t = \sum_{j=1}^{\infty} s_j \chi_{I_{q_j}} t_j^{q_j} \in \Lambda$. Then the sequence

$$\sum_{j=1}^{\infty} s_j m_{iq_j} = \sum_{j=1}^{\infty} s_j \sum_{l \in I_{q_j}} t_l^{q_j} x_{k_i l} = \sum_{j=1}^{\infty} t_j x_{k_i j}$$

converges by hypothesis. Hence, M is a signed $\mathcal{K}$-matrix so by the signed version of the Antosik-Mikusinski Matrix Theorem (Appendix D.3), the diagonal of M converges to 0. But, this contradicts $(*)$.

If the multiplier space Λ does not have signed-SGHP, then the hypothesis (H) may not imply (C2) and (C3) even if Λ has WGHP or 0-GHP.

Example 7.12. Let $1 \leq p < \infty$ and $\lambda = \Lambda = l^p$. Define $x_{ij} = e^j$ if $1 \leq j \leq i$ and $x_{ij} = 0$ if $j > i$. Then $\sum_j x_{ij}$ is λ multiplier convergent for every i, $\lim_i x_{ij} = e^j = x_j$ for every j and

$$\lim_i \sum_{j=1}^{\infty} t_j x_{ij} = \lim_i \sum_{j=1}^{i} t_j e^j = \sum_{j=1}^{\infty} t_j e^j = \sum_{j=1}^{\infty} t_j x_j$$

in l^p for every $t \in l^p$ so (H) holds and (C1) holds. However, both (C2) and (C3) fail to hold. [Take $t^k = e^k$ so $\{t^k\}$ is bounded in l^p but $\sum_{j=1}^{\infty} t_j^k x_{ij} = \sum_{j=1}^{i} t_j^k e^j = e^k$ if $i \geq k$ so (C2) and (C3) fail.]

We next consider the hypothesis (H) as a conclusion.

Proposition 7.13. *Let* $\sum_j x_{ij}$ *be* Λ *multiplier convergent for every* $i \in \mathbb{N}$ *and assume that* $\lim_i x_{ij} = x_j$ *exists for every* $j \in \mathbb{N}$.

(1) If for every $t \in \Lambda$ *the series* $\sum_{j=1}^{\infty} t_j x_{ij}$ *converge uniformly for* $i \in \mathbb{N}$, *then for every* $t \in \Lambda$ *the sequence* $\{\sum_{j=1}^{\infty} t_j x_{ij}\}_i$ *is Cauchy.*

(2) If the series $\sum_{j=1}^{\infty} t_j x_{ij}$ *converge uniformly for* $i \in \mathbb{N}$ *and* t *belonging to bounded subsets of* Λ, *then the sequences* $\{\sum_{j=1}^{\infty} t_j x_{ij}\}_i$ *satisfy a Cauchy condition uniformly for* t *belonging to bounded subsets of* Λ.

Proof: (1): Let $t \in \Lambda$. Let U be a neighborhood of 0 in X. Pick a symmetric neighborhood of 0, V, such that $V + V + V \subset U$. There exists N such that $n \geq N$ implies $\sum_{j=n}^{\infty} t_j x_{ij} \in V$ for every $i \in \mathbb{N}$. There exists $n > N$ such that $i, k \geq n$ implies $\sum_{j=1}^{N} t_j(x_{ij} - x_{kj}) \in V$. If $i, k \geq n$, then

$$(*) \quad \sum_{j=1}^{\infty} t_j x_{kj} - \sum_{j=1}^{\infty} t_j x_{ij} =$$

$$\sum_{j=1}^{N} t_j(x_{kj} - x_{ij}) + \sum_{j=N+1}^{\infty} t_j x_{kj} - \sum_{j=N+1}^{\infty} t_j x_{ij} \in V + V + V \subset U.$$

(2): Let $B \subset \Lambda$ be bounded. Let U be a neighborhood of 0 in X. Pick a symmetric neighborhood of 0, V, such that $V + V + V \subset U$. There exists N such that $n \geq N$ implies $\sum_{j=n}^{\infty} t_j x_{ij} \in V$ for every $i \in \mathbb{N}, t \in B$. By the K-space assumption $\{t_j : t \in B\}$ is bounded for every j. By Lemma 7.7 there exists $n > N$ such that $\sum_{j=1}^{N} t_j(x_{ij} - x_{kj}) \in V$ for every $t \in B$. If $i, k \geq n$, then $(*)$ holds.

Proposition 7.14. *Assume that λ is an AB-space (Appendix B.3). Let $\sum_j x_{ij}$ be Λ multiplier convergent for every $i \in \mathbb{N}$ and assume that $\lim_i x_{ij} = x_j$ exists for every $j \in \mathbb{N}$. If $\lim_i \sum_{j=1}^{\infty} t_j x_{ij}$ exists for every $t \in \Lambda$ and the series $\sum_{j=1}^{\infty} t_j x_{ij}$ converge uniformly for $i \in \mathbb{N}$ and t belonging to bounded subsets of Λ, then $\sum_j x_j$ is Λ multiplier convergent and $\lim_i \sum_{j=1}^{\infty} t_j x_{ij} = \sum_{j=1}^{\infty} t_j x_j$ uniformly for t belonging to bounded subsets of Λ.*

Proof: Let $t \in \Lambda$. Put $z = \lim_i \sum_{j=1}^{\infty} t_j x_{ij}$. Let U be a neighborhood of 0 in X. Pick a closed, symmetric neighborhood of 0, V, such that $V + V + V \subset U$. There exists k such that $\sum_{j=1}^{\infty} t_j x_{kj} - z \in V$ and by Proposition 7.12(2) since V is closed and $\{P_m t : m \in \mathbb{N}\}$ is bounded by the AB-assumption (P_m is the section map $P_m t = \sum_{j=1}^{m} t_j e^j$), $\sum_{j=1}^{m} t_j(x_j - x_{kj}) \in V$ for every m. There exists M such that $m \geq M$ implies $\sum_{j=m+1}^{\infty} t_j x_{kj} \in V$. Then if $m \geq M$, we have

$$\sum_{j=1}^{m} t_j x_j - z = \sum_{j=1}^{\infty} t_j x_{kj} - z + \sum_{j=1}^{m} t_j(x_j - x_{kj}) - \sum_{j=m+1}^{\infty} t_j x_{kj} \in V + V + V \subset U.$$

Hence, $z = \lim_m \sum_{j=1}^{m} t_j x_j = \sum_{j=1}^{\infty} t_j x_j$.

Let $B \subset \Lambda$ be bounded. Since $\sum_{j=1}^{\infty} t_j x_{ij}$ converges uniformly for $i \in \mathbb{N}, t \in B$, the series $\sum_{j=1}^{\infty} t_j x_{ij}$ satisfy a Cauchy condition uniformly for $i \in \mathbb{N}, t \in B$. Therefore, there exists N such that $n > m \geq N$ implies that $\sum_{j=m}^{n} t_j x_{ij} \in V$ for $i \in \mathbb{N}, t \in B$. Hence, $\sum_{j=m}^{\infty} t_j x_j \in V$ for $m \geq N, t \in B$. Since $\{t_j : t \in B\}$ is bounded for every j, by Lemma 7.7 there exists M such that $\sum_{j=1}^{N} t_j(x_{ij} - x_j) \in V$ for $i \geq M, t \in B$. If $i \geq M$, then

$$\sum_{j=1}^{\infty} t_j(x_{ij} - x_j) = \sum_{j=1}^{N} t_j(x_{ij} - x_j) + \sum_{j=N+1}^{\infty} t_j x_{ij} - \sum_{j=N+1}^{\infty} t_j x_j \in V + V + V \subset U$$

for every $t \in B$.

Corollary 7.15. *Let Λ have signed-WGHP and let X be sequentially complete. Let $\sum_j x_{ij}$ be Λ multiplier convergent for every $i \in \mathbb{N}$ and assume that $\lim_i x_{ij} = x_j$ exists for every $j \in \mathbb{N}$. The following are equivalent:*

(i) $\lim_i \sum_{j=1}^{\infty} t_j x_{ij}$ *exists for every* $t \in \Lambda$,
(ii) $\sum_j x_j$ *is Λ multiplier convergent and* $\lim_i \sum_{j=1}^{\infty} t_j x_{ij} = \sum_{j=1}^{\infty} t_j x_j$,
(iii) for every $t \in \Lambda$ *the series* $\sum_{j=1}^{\infty} t_j x_{ij}$ *converge uniformly for* $i \in \mathbb{N}$.

Proof: That (ii) implies (i) is clear; (iii) implies (i) by Proposition 7.12(1); (i) implies (ii) by Lemma 2.27; (ii) implies (iii) by Theorem 2.26.

Corollary 7.16. *Let Λ have signed-SGHP and let X be sequentially complete. Let $\sum_j x_{ij}$ be Λ multiplier convergent for every $i \in \mathbb{N}$ and assume that $\lim_i x_{ij} = x_j$ exists for every $j \in \mathbb{N}$. The following are equivalent:*

(I) $\lim_i \sum_{j=1}^{\infty} t_j x_{ij}$ exists for every $t \in \Lambda$,

(II) $\sum_j x_j$ is Λ multiplier convergent and $\lim_i \sum_{j=1}^{\infty} t_j x_{ij} = \sum_{j=1}^{\infty} t_j x_j$ uniformly for t belonging to bounded subsets of Λ,

(III) the series $\sum_{j=1}^{\infty} t_j x_{ij}$ converge uniformly for $i \in \mathbb{N}, t$ belonging to bounded subsets of Λ.

(IV) for every $t \in \Lambda$ the series $\sum_{j=1}^{\infty} t_j x_{ij}$ converge uniformly for $i \in \mathbb{N}$.

Proof: Clearly (II) implies (I); (I) implies (II) and (III) by the Hahn-Schur Theorem 7.10; that (III) implies (IV) is clear; (IV) implies (I) by Proposition 7.12(2).

We also obtain a boundedness result.

Proposition 7.17. *Let $\sum_j x_{ij}$ be Λ multiplier convergent for every $i \in \mathbb{N}$ and assume that $\lim_i x_{ij} = x_j$ exists for every $j \in \mathbb{N}$. If $B \subset \Lambda$ is bounded and the series $\sum_{j=1}^{\infty} t_j x_{ij}$ converge uniformly for $i \in \mathbb{N}, t \in B$, then*

$$S = \left\{ \sum_{j=1}^{\infty} t_j x_{ij} : i \in \mathbb{N}, t \in B \right\}$$

is bounded.

Proof: Let U be a balanced neighborhood of 0 in X and pick a balanced neighborhood of 0, V, such that $V + V \subset U$. There exists N such that $\sum_{j=N+1}^{\infty} t_j x_{ij} \in V$ for $i \in \mathbb{N}, t \in B$. Since $\{x_{ij} : i \in \mathbb{N}\}$ and $\{t_j : t \in B\}$ are bounded for every j, there exists $t > 1$ such that $\{\sum_{j=1}^{N} t_j x_{ij} : i \in \mathbb{N}, t \in B\} \subset tV$. Hence, $S \subset V + tV \subset tU$ and S is bounded.

Since both $M_0 = \{\chi_\sigma : \sigma \subset \mathbb{N}\} \subset m_0 = \bar{span} M_0$ and l^∞ have SGHP, the previous results hold for both M_0 multiplier (=subseries) convergent series and l^∞ multiplier (=bounded multiplier) convergent series. We record these special cases for the previous results. We begin with the subseries case. From Theorem 7.10, we have

Theorem 7.18. *Let $\sum_j x_{ij}$ be subseries convergent for every $i \in \mathbb{N}$. Assume that $\lim_i \sum_{j \in \sigma} x_{ij}$ exists for every $\sigma \subset \mathbb{N}$ and $\lim_i x_{ij} = x_j$ for every j. Then*

(1) $\sum_j x_j$ is subseries convergent,
(2) $\lim_i \sum_{j \in \sigma} x_{ij} = \sum_{j \in \sigma} x_j$ uniformly for $\sigma \subset \mathbb{N}$ and
(3) the series $\sum_{j \in \sigma} x_{ij}$ converge uniformly for $i \in \mathbb{N}, \sigma \subset \mathbb{N}$.

Note that the scalar case of Theorem 7.17 gives the scalar version of the Hahn Theorem stated in Theorem 7.2.

From Proposition 7.12, we have

Proposition 7.19. *Let $\sum_j x_{ij}$ be subseries convergent for every $i \in \mathbb{N}$ and $\lim_i x_{ij} = x_j$ for every j.*

(1) If for every $\sigma \subset \mathbb{N}$ the series $\sum_{j \in \sigma} x_{ij}$ converge uniformly for $i \in \mathbb{N}$, then for every $\sigma \subset \mathbb{N}$ the sequence $\{\sum_{j \in \sigma} x_{ij}\}_i$ is Cauchy.
(2) If the series $\sum_{j \in \sigma} x_{ij}$ converge uniformly for $i \in \mathbb{N}, \sigma \subset \mathbb{N}$, then the sequences $\{\sum_{j \in \sigma} x_{ij}\}_i$ satisfy a Cauchy condition uniformly for $\sigma \subset \mathbb{N}$.

From Corollary 7.15, we have

Corollary 7.20. *Let X be sequentially complete. Let $\sum_j x_{ij}$ be subseries convergent for every $i \in \mathbb{N}$ and $\lim_i x_{ij} = x_j$ for every j. The following are equivalent:*

(1) $\lim_i \sum_{j \in \sigma} x_{ij}$ exists for every $\sigma \subset \mathbb{N}$,
(2) $\sum_j x_j$ is subseries convergent and $\lim_i \sum_{j \in \sigma} x_{ij} = \sum_{j \in \sigma} x_j$ uniformly for $\sigma \subset \mathbb{N}$,
(3) the series $\sum_{j \in \sigma} x_{ij}$ converge uniformly for $i \in \mathbb{N}, \sigma \subset \mathbb{N}$,
(4) for every $\sigma \subset \mathbb{N}$ the series $\sum_{j \in \sigma} x_{ij}$ converge uniformly for $i \in \mathbb{N}$.

From Proposition 7.16, we have

Proposition 7.21. *Let $\sum_j x_{ij}$ be subseries convergent for every $i \in \mathbb{N}$ and $\lim_i x_{ij} = x_j$ for every j. If the series $\sum_{j \in \sigma} x_{ij}$ converge uniformly for $i \in \mathbb{N}, \sigma \subset \mathbb{N}$, then*

$$S = \left\{ \sum_{j \in \sigma} x_{ij} : i \in \mathbb{N}, \sigma \subset \mathbb{N} \right\}$$

is bounded.

We now show that Theorem 7.17 can be improved for sequentially complete spaces by replacing the family of all subsets of $\mathbb{N}$ by a smaller family of subsets. Recall that a family of subsets $\mathcal{F}$ of $\mathbb{N}$ is an FQσ family if $\mathcal{F}$ contains the finite subsets and whenever $\{I_k\}$ is a pairwise disjoint sequence of finite subsets there is a subsequence $\{I_{n_k}\}$ such that $\cup_{k=1}^{\infty} I_{n_k} \in \mathcal{F}$ (see Appendix B for examples). If $\Lambda = \{\chi_\sigma : \sigma \in \mathcal{F}\} \subset m_0$ where $\mathcal{F}$ is an FQσ family, then Λ has SGHP (Appendix B) so Theorem 7.10 applies. We first establish a lemma.

Lemma 7.22. *Let X be sequentially complete and $\mathcal{F}$ be an FQσ family with $\Lambda = \{\chi_\sigma : \sigma \in \mathcal{F}\} \subset m_0$. If $\sum_j x_j$ is Λ multiplier convergent in X, then $\sum_j x_j$ is subseries convergent.*

Proof: Let $\{n_j\}$ be a subsequence and $\sigma = \{n_j : j \in \mathbb{N}\}$. If $\sum_{j=1}^{\infty} x_{n_j}$ does not converge, there exist a neighborhood of 0, U, in X and an increasing sequence of intervals $\{I_k\}$ such that $\sum_{j \in I_k} x_{n_j} = \sum_{j \in I_k \cap \sigma} x_j \notin U$. By the FQ$\sigma$ property there is a subsequence $\{m_k\}$ such that $I = \cup_{k=1}^{\infty} I_{m_k} \cap \sigma \in \mathcal{F}$. But, then $\sum_{j \in I} x_j$ does not converge since the series fails the Cauchy condition.

We now give the improvement of Theorem 7.17 for sequentially complete spaces.

Theorem 7.23. *Let X be sequentially complete and $\mathcal{F}$ be an FQσ family with $\Lambda = \{\chi_\sigma : \sigma \in \mathcal{F}\} \subset m_0$. Assume that $\sum_j x_{ij}$ is Λ multiplier convergent for every i and that $\lim_i \sum_{j \in \sigma} x_{ij}$ exists for every $\sigma \in \mathcal{F}$ with $x_j = \lim_i x_{ij}$ for every j. Then $\lim_i \sum_{j \in \sigma} x_{ij}$ exists for every $\sigma \subset \mathbb{N}$ so conclusions (1),(2) and (3) of Theorem 7.17 hold.*

Proof: We first claim that for each $\sigma \subset \mathbb{N}$, the series $\sum_{j \in \sigma} x_j$ converges. We show that $\sum_{j=1}^{\infty} x_j$ converges; the same argument can be applied to any subseries $\sum_{j \in \sigma} x_j$. Let U be a neighborhood of 0 in X and pick a closed, symmetric neighborhood of 0, V, such that $V+V \subset U$. By (C2) of Theorem 7.10 there exists n such that $\sum_{j \in \sigma}(x_{ij} - x_j) \in V$ for $i \geq n, \sigma \in \mathcal{F}$. By Lemma 7.21 there exists m such that $\sum_{j=k}^{l} x_{nj} \in V$ for $l \geq k \geq m$. If $l \geq k \geq m$, then

$$\sum_{j=1}^{l} x_j - \sum_{j=1}^{k} x_j = \sum_{j=k+1}^{l} (x_j - x_{nj}) + \sum_{j=k+1}^{l} x_{nj} \in V + V \subset U$$

so $\sum_{j=1}^{\infty} x_j$ converges by the sequential completeness of X.

Since $\sum_{j \in \sigma}(x_{ij} - x_j) \in V$ for $i \geq n$ and finite σ, Lemma 7.21 gives that $\sum_{j \in \sigma}(x_{ij} - x_j) \in V$ for $i \geq n$ and any $\sigma \subset \mathbb{N}$. Therefore, the hypothesis of Theorem 7.17 holds and (1), (2) and (3) of Theorem 7.17 follow.

We next give a scalar corollary of Theorem 7.22 due to Samaratanga and Sember ([SaSe]) which will be used later.

Corollary 7.24. *Let $\mathcal{F}$ be an $FQ\sigma$ family with $\Lambda = \{\chi_\sigma : \sigma \in \mathcal{F}\} \subset m_0$ and set $\lambda = \mathrm{span}\Lambda$. Assume that $t^i \in l^1$ and $\lim_i t^i \cdot s$ exists for each $s \in \lambda$ with $t_j = \lim_i t^i_j$. Then $t = \{t_j\} \in l^1$ and $\left\| t^i - t \right\|_1 \to 0$. In particular, if $t^i \to 0$ in $\sigma(l^1, \lambda)$, then $\left\| t^i \right\|_1 \to 0$.*

We next show that we can relax the hypothesis in Theorem 7.22 and retain part of the conclusion of the theorem. Recall that a family $\mathcal{F}$ of subsets of $\mathbb{N}$ is an $IQ\sigma$ family if $\mathcal{F}$ contains the finite subsets of $\mathbb{N}$ and whenever $\{I_k\}$ is an increasing sequence of intervals there is a subsequence $\{I_{n_k}\}$ such that $\cup_{k=1}^\infty I_{n_k} \in \mathcal{F}$ (see Appendix B for examples).

If $\mathcal{F}$ is an $IQ\sigma$ family and $\Lambda = \{\chi_\sigma : \sigma \in \mathcal{F}\} \subset m_0$, then Λ has SGHP so Theorem 7.10 applies and gives

Theorem 7.25. *Let $\mathcal{F}$ be an $IQ\sigma$ family which contains $\mathbb{N}$, let $\Lambda = \{\chi_\sigma : \sigma \in \mathcal{F}\} \subset m_0$ and assume that $\sum_j x_{ij}$ is Λ multiplier convergent for every i. If $\lim_i \sum_{j \in \sigma} x_{ij}$ exists for every $\sigma \in \mathcal{F}$ and $x_j = \lim_i x_{ij}$, then $\sum_j x_j$ is Λ multiplier convergent and $\lim_i \sum_{j=1}^\infty x_{ij} = \sum_{j=1}^\infty x_j$. [Note that $\sum_{j=1}^\infty x_{ij}$ converges since $\mathbb{N} \in \mathcal{F}$.]*

We use this result later in Chapter 9 when we consider iterated series.

For the bounded multiplier case we have the following result as a special case of Theorem 7.10.

Theorem 7.26. *Let $\sum_j x_{ij}$ be bounded multiplier convergent for every $i \in \mathbb{N}$. Assume that $\lim_i \sum_{j=1}^\infty t_j x_{ij}$ exists for every $t \in l^\infty$ and $\lim_i x_{ij} = x_j$ for every j. Then*

(1) $\sum_j x_j$ is bounded multiplier convergent,
(2) $\lim_i \sum_{j=1}^\infty t_j x_{ij} = \sum_{j=1}^\infty t_j x_j$ uniformly for $\left\| \{t_j\} \right\|_\infty \leq 1$ and
(3) the series $\sum_{j=1}^\infty t_j x_{ij}$ converge uniformly for $i \in \mathbb{N}, \left\| \{t_j\} \right\|_\infty \leq 1$.

Note that the scalar version of Theorem 7.25 gives the scalar version of the Schur Theorem stated in Theorem 7.1.

From Proposition 7.12, we have

Proposition 7.27. *Let $\sum_j x_{ij}$ be bounded multiplier convergent for every $i \in \mathbb{N}$ and $\lim_i x_{ij} = x_j$ for every j.*

(1) If for every $t \in l^\infty$, the series $\sum_{j=1}^\infty t_j x_{ij}$ converge uniformly for $i \in \mathbb{N}$, then for every $t \in l^\infty$ the sequence $\{\sum_{j=1}^\infty t_j x_{ij}\}_i$ is Cauchy.

(2) If the series $\sum_{j=1}^\infty t_j x_{ij}$ converge uniformly for $i \in \mathbb{N}, \|\{t_j\}\|_\infty \le 1$, then the sequences $\{\sum_{j=1}^\infty t_j x_{ij}\}_i$ satisfy a Cauchy condition uniformly for $\|\{t_j\}\|_\infty \le 1$.

From Corollary 7.15, we have

Corollary 7.28. *Let X be sequentially complete. Let $\sum_j x_{ij}$ be bounded multiplier convergent for every $i \in \mathbb{N}$ and $\lim_i x_{ij} = x_j$ for every j. The following are equivalent:*

(1) $\lim_i \sum_{j=1}^\infty t_j x_{ij}$ exists for every $t \in l^\infty$,

(2) $\sum_j x_j$ is bounded multiplier convergent and $\lim_i \sum_{j=1}^\infty t_j x_{ij} = \sum_{j=1}^\infty t_j x_j$ uniformly for $\|\{t_j\}\|_\infty \le 1$,

(3) the series $\sum_{j=1}^\infty t_j x_{ij}$ converge uniformly for $i \in \mathbb{N}, \|\{t_j\}\|_\infty \le 1$,

(4) for every $t \in l^\infty$ the series $\sum_{j=1}^\infty t_j x_{ij}$ converge uniformly for $i \in \mathbb{N}$.

From Proposition 7.16, we have

Proposition 7.29. *Let $\sum_j x_{ij}$ be bounded multiplier convergent for every $i \in \mathbb{N}$ and $\lim_i x_{ij} = x_j$ for every j. If the series $\sum_{j=1}^\infty t_j x_{ij}$ converge uniformly for $i \in \mathbb{N}, \|\{t_j\}\|_\infty \le 1$, then*

$$S = \left\{ \sum_{j=1}^\infty t_j x_{ij} : i \in \mathbb{N}, \|\{t_j\}\|_\infty \le 1 \right\}$$

is bounded.

In a sequentially complete LCTVS we can obtain a stronger conclusion in Theorem 7.17 for the subseries convergent version of the Hahn-Schur Theorem. Recall that the inequality of McArthur/Rutherford (Lemma 2.53) implies that a series in a sequentially complete LCTVS is subseries convergent iff the series is bounded multiplier convergent (Theorem 2.54). Also, from the inequality, we obtain

Theorem 7.30. *Let X be a sequentially complete LCTVS. Assume that $\sum_j x_{ij}$ is subseries convergent for every j and $\lim_i x_{ij} = x_j$ exists for every j. If $\lim_i \sum_{j \in \sigma} x_{ij}$ exists for every $\sigma \subset \mathbb{N}$, then*

(1) $\sum_j x_j$ *is bounded multiplier convergent,*

(2) $\lim_i \sum_{j=1}^{\infty} t_j x_{ij} = \sum_{j=1}^{\infty} t_j x_j$ *uniformly for* $\|\{t_j\}\|_{\infty} \leq 1$ *and*

(3) the series $\sum_{j=1}^{\infty} t_j x_{ij}$ *converge uniformly for* $i \in \mathbb{N}, \|\{t_j\}\|_{\infty} \leq 1$.

Proof: The first statement in (1) follows from Theorems 7.17(1) and 2.54.

For (2) let p be a continuous semi-norm on X. By Theorem 7.17(2), for $\epsilon > 0$ there exists n such that $p(\sum_{j \in \sigma}(x_{ij} - x_j)) < \epsilon$ for $i \geq N, \sigma \subset \mathbb{N}$. If $\{t_j\} \in l^{\infty}$ and $i \geq n$, then from Lemma 2.53 for $\sigma \subset \mathbb{N}$ finite

$$p(\sum_{j \in \sigma} t_j(x_{ij} - x_j)) \leq 2 \|\{t_j\}\|_{\infty} \sup_{\sigma' \subset \sigma} p(\sum_{j \in \sigma'}(x_{ij} - x_j)) \leq \|\{t_j\}\|_{\infty} \epsilon.$$

Therefore, if $i \geq n$, then

$$p(\sum_{j=1}^{\infty} t_j(x_{ij} - x_j)) \leq \|\{t_j\}\|_{\infty} \epsilon$$

and (2) follows.

(3) follows from the McArthur/Rutherford inequality in a similar fashion.

Theorem 7.29 gives a generalization of the summability result stated in (S) following Theorem 7.2. Namely, we have:

(S)' If X is sequentially complete, the vector valued matrix $[x_{ij}]$ maps m_0 into $c(X)$, the space of X valued convergent sequences, iff $[x_{ij}]$ maps l^{∞} into $c(X)$.

Corollaries 7.19 and 7.27 give necessary and sufficient conditions for a vector valued matrix $[x_{ij}]$ to map m_0 or l^{∞} into $c(X)$ analogous to the scalar case (see [Sw2] 9.5.3).

We next consider a generalization of the compactness result in Theorem 2.45. For this we require a preliminary lemma.

Lemma 7.31. *Let S be a compact Hausdorff space and $g_i : S \to X$ continuous functions for $i = 0, 1, 2, \ldots$. Suppose that $\lim g_i(t) = g_0(t)$ uniformly for $t \in S$. Then $R = \cup_{i=0}^{\infty} \mathcal{R}g_i$ is compact, where $\mathcal{R}g_i$ is the range of g_i.*

Proof: Let $\mathcal{G}$ be an open cover of R. For each $x \in R$ there exists $U_x \in \mathcal{G}$ such that $x \in U_x$. Then $-x + U_x$ is an open neighborhood of 0 so there is an open neighborhood of 0, V_x, such that $V_x + V_x \subset -x + U_x$. Then $\mathcal{G}' = \{x + V_x : x \in R\}$ is an open cover of R.

Since $\mathcal{R}g_0$ is compact, there exist finite $x_1 + V_{x_1}, ..., x_k + V_{x_k}$ covering $\mathcal{R}g_0$. Put $V = \cap_{j=1}^k V_{x_j}$ so V is an open neighborhood of 0. There exists n such that $g_i(t) - g_0(t) \in V$ for $i \geq n, t \in S$. For $t \in S$ there exists j such that $g_0(t) \in x_j + V_{x_j}$ so $g_i(t) \in g_0(t) + V \subset x_j + V_{x_j} + V_{x_j} \subset U_{x_j}$ for $i \geq n$. Hence, $U_{x_1}, ..., U_{x_k}$ covers $\cup_{i=n}^\infty \mathcal{R}g_i$.

Since $\mathcal{R}g_i, i = 0, ...n - 1$, are compact, a finite subcover of $\mathcal{G}$ covers the union of these sets, and, hence, $\mathcal{G}$ has a finite subcover covering R.

From Theorem 7.10, Corollary 2.43 and Lemma 7.30, we have

Theorem 7.32. *Assume that $\Lambda \subset \lambda$ is bounded and has signed-SGHP and is compact with respect to p, the topology of pointwise convergence on Λ. If*

(H) *Let $\sum_j x_{ij}$ be Λ multiplier convergent in X for every $i \in \mathbb{N}$. Assume that $\lim_i \sum_{j=1}^\infty t_j x_{ij}$ exists for every $t \in \Lambda$ and assume that $\lim_i x_{ij} = x_j$ exists for every $j \in \mathbb{N}$,*
    then $B = \{\sum_{j=1}^\infty t_j x_{ij} : i \in \mathbb{N}, t \in \Lambda\} \cup \{\sum_{j=1}^\infty t_j x_j : t \in \Lambda\}$ is compact.

Proof: Let S_i (S_0) be the summing operator with respect to $\sum_j x_{ij}$ ($\sum_j x_j$). If (C2) holds, then S_i and S_0 are continuous with respect to p and the topology of X (Corollary 2.43). If (C3) holds, then $S_i \to S_0$ uniformly on Λ so it follows from Lemma 7.30 that B is compact.

In particular, if $\Lambda = \{t \in l^\infty : \|\{t_j\}\|_\infty \leq 1\} \subset \lambda = l^\infty$ or if $\Lambda = \{\chi_\sigma : \sigma \subset \mathbb{N}\} \subset \lambda = m_0$, Theorem 7.31 applies.

We next consider another property of the multiplier space which implies vector versions of the Hahn-Schur Theorem.

Definition 7.33. Let λ be a K-space. The multiplier space λ has the Hahn-Schur property if $s^i \in \lambda^\beta$ and $s^i \cdot t \to 0$ for every $t \in \lambda$ implies that $\lim s^i \cdot t = 0$ uniformly for t belonging to bounded subsets of λ.

From Corollary 7.23 we have the following example of a multiplier space with the Hahn-Schur property.

Example 7.34. Let $\mathcal{F}$ be an FQσ family with $\Lambda = \{\chi_\sigma : \sigma \in \mathcal{F}\} \subset m_0$ and set $\lambda = span\Lambda$. Then λ with the sup-norm has the Hahn-Schur property.

For multiplier spaces with the Hahn-Schur property we have a vector Hahn-Schur Theorem.

Theorem 7.35. *Assume that λ has the Hahn-Schur property. If $\{x^k\} \subset \lambda^{\beta X}$ is such that $\lim x^k \cdot t = 0$ for every $t \in \lambda$, then $\lim x^k \cdot t = 0$ uniformly for t belonging to bounded subsets of λ.*

Proof: Let $B \subset \lambda$ be bounded. Suppose that the conclusion of the theorem fails for B. Then there exist $\delta > 0$, a continuous semi-norm p on X, $t^k \in B$ and an increasing sequence $\{n_k\}$ such that

$$p(x^{n_k} \cdot t^k) > \delta.$$

By the Hahn-Banach Theorem, for every k there exist $x'_k \in X'$ such that $\sup\{|\langle x'_k, x\rangle| : p(x) \le 1\} \le 1$ and

$$(*) \qquad \left\langle x'_k, \sum_{j=1}^{\infty} t^k_j x^{n_k}_j \right\rangle > \delta.$$

Since $p(\sum_{j=1}^{\infty} t_j x^{n_k}_j) \to 0$ for every $t \in \lambda$ and $\sup\{|\langle x'_k, x\rangle| : p(x) \le 1\} \le 1$,

$$\lim_k \sum_{j=1}^{\infty} t_j \langle x'_k, x^{n_k}_j \rangle = 0$$

uniformly for t belonging to bounded subsets of λ by the Hahn-Schur property. This contradicts $(*)$.

Corollary 7.36. *Assume that λ has the Hahn-Schur property, $(\lambda^{\beta X}, \omega(\lambda^{\beta X}, \lambda))$ is sequentially complete and $\{x^k\} \subset \lambda^{\beta X}$. If $\lim x^k \cdot t$ exists for every $t \in \lambda$ and $x_j = \lim_k x^k_j$, then $x \in \lambda^{\beta X}$ and $\lim x^k \cdot t = x \cdot t$ uniformly for t belonging to bounded subsets of λ.*

Proof: By the sequential completeness assumption, there exists $y \in \lambda^{\beta X}$ such that $x^k \to y$ in $\omega(\lambda^{\beta X}, \lambda)$. Since $x^k_j \to y_j$ for every j, $y = \{y_j\} = \{x_j\} = x$. If $t \in \lambda$, then $x^k \cdot t \to x \cdot t$ so the result follows from Theorem 7.34.

We now consider some Hahn-Schur type results for normed spaces due to the Spanish school in Cadiz ([AP1])).

Let X be a Banach space. The space X is a Grothendieck space if every weak* convergent sequence in X' is weakly convergent. For example, the space l^∞ is a Grothendieck space by Phillips' Lemma (Lemma 7.52 or [Sw2] 15.16).

Definition 7.37. Let M be a subspace of X' such that $X \subset M \subset X''$. Then X is an M Grothendieck space if every $\sigma(X', X)$ convergent sequence in X' is $\sigma(X', M)$ convergent.

Thus, X is a Grothendieck space iff X is an X'' Grothendieck space. There are closed subspaces λ of l^∞ with $\lambda \neq l^\infty$ such that λ is a Grothendieck space not containing a subspace isomorphic to l^∞. For example, if $\mathcal{H}$ is the algebra of Hayden (Appendix B.21), and if H is the Stone space of $\mathcal{H}$, then $C(H)$ can be isometrically identified with a closed subspace of l^∞ such that $c_0 \subset C(H)$ and which is a Grothendieck space not containing a copy of l^∞ (see [AP1]).

When X is a Banach space and $c_0 \subset \lambda \subset l^\infty$, we define a norm on $\lambda^{\beta X}$ by

$$\|x\|' = \sup\{\|t \cdot x\| : t \in \lambda, \|t\|_\infty \leq 1\}.$$

Note that $\|\cdot\|'$ is finite by Proposition 3.8 since for any $x = \{x_i\} \in \lambda^{\beta X}$, the series $\sum_j x_j$ is c_0 multiplier convergent. Note that we also have

$$\|x\|' = \sup\{\|t \cdot x\| : t \in c_0, \|t\|_\infty \leq 1\}$$

$$= \sup\{\|t \cdot x\| : t \in c_{00}, \|t\|_\infty \leq 1\} = \sup\{\sum_{j=1}^\infty |\langle x', x_j \rangle| : \|x'\| \leq 1\}.$$

Theorem 7.38. *Let λ be a subspace of l^∞ containing c_0 which is an l^∞ Grothendieck space. Let $\sum_j x_{ij}$ be λ multiplier convergent for every i. If $\lim_i \sum_{j=1}^\infty t_j x_{ij}$ exists for every $t \in \lambda$ and $x_j = \lim_i x_{ij}$ for every j, then $\sum_j x_j$ is λ multiplier convergent and $\lim_i \sum_{j=1}^\infty t_j x_{ij} = \sum_{j=1}^\infty t_j x_j$ for every $t \in \lambda$.*

Proof: Put $x^i = \{x_{ij}\}_j \in \lambda^{\beta X}$ and $x = \{x_j\}$. We claim that $\{x^i\}$ is $\|\cdot\|'$ Cauchy in $\lambda^{\beta X}$. If not, there exist a subsequence $\{n_k\}$ and $\delta > 0$ such that $\|x^{n_{k+1}} - x^{n_k}\|' > \delta$. Put $z^k = x^{n_{k+1}} - x^{n_k}$ so $\|z^k\|' > \delta$ and $z^k \cdot t \to 0$ for every $t \in \lambda$. For each k pick $x_k' \in X'$ such that $\|x_k'\| \leq 1$ and

$$(*) \quad \sum_{j=1}^\infty |\langle x_k', z_j^k \rangle| > \delta.$$

The series $\sum_j z_j^k$ is λ multiplier convergent so let S_k be the summing operator with respect to the series $\sum_j z_j^k$, $S_k t = \sum_{j=1}^\infty t_j z_j^k$ for $t \in \lambda$ (Theorem 2.2). Then $x_k' S_k \in \lambda'$ and $x_k' S_k(t) = \langle x_k', \sum_{j=1}^\infty t_j z_j^k \rangle \to 0$ for $t \in \lambda$. That is, $\{x_k' S_k\}$ is $\sigma(\lambda', \lambda)$ convergent to 0. Since λ is an l^∞ Grothendieck space, $\{x_k' S_k\}$ is $\sigma(\lambda', l^\infty)$ convergent to 0. Thus, $x_k' S_k(t) = \langle x_k', \sum_{j=1}^\infty t_j z_j^k \rangle \to 0$ for $t \in l^\infty$. By the classical Hahn-Schur Theorem 7.1, $\{\langle x_k', z_j^k \rangle\}_j \to 0$ in $\|\cdot\|_1$ or $\sum_{j=1}^\infty |\langle x_k', z_j^k \rangle| \to 0$. This contradicts $(*)$.

We claim that $\sum_j x_j$ is λ multiplier convergent and $x^i \cdot t \rightarrow x \cdot t$ for every $t \in \lambda$. It will then follow that $\|x^i - x\|' \rightarrow 0$ and the result follows. Let $\epsilon > 0$ and $t \in \lambda$. There exists N such that $k, l \geq N$ implies $\left\| \sum_{j=m}^{n} t_j(x_j^k - x_j^l) \right\| < \epsilon$ for all $n > m$ by the part above. Let $k \rightarrow \infty$ to obtain

$$(**) \qquad \left\| \sum_{j=m}^{n} t_j(x_j - x_j^l) \right\| \leq \epsilon \text{ for all } n > m, l \geq N.$$

This shows that the series $\sum_j t_j(x_j - x_j^N)$ is Cauchy so $\sum_j t_j x_j$ is Cauchy and, therefore, convergent since X is complete. Condition $(**)$ also shows that $x^l \cdot t \rightarrow x \cdot t$ as desired.

Example 7.11 shows that even when the multiplier space λ has the 0-GHP and the signed-WGHP, the summing operators in Theorem 7.37 may not converge uniformly on bounded subsets of λ. However, we show that in this case we do have uniform convergence on null sequences.

Theorem 7.39. *Assume that λ has 0-GHP and signed-WGHP. Let $\sum_j x_{ij}$ be λ multiplier convergent for every i. If $\lim_i \sum_{j=1}^{\infty} t_j x_{ij}$ exists for every $t \in \lambda$ and $\lim_i x_{ij} = x_j$ for every j, then*

(C1) the series $\sum_j x_j$ is λ multiplier convergent and $\lim_i \sum_{j=1}^{\infty} t_j x_{ij} = \sum_{j=1}^{\infty} t_j x_j$ and
(C4) if $t^k \rightarrow 0$ in λ, then $\lim_i \sum_{j=1}^{\infty} t_j^k x_{ij} = \sum_{j=1}^{\infty} t_j^k x_j$ uniformly for $k \in \mathbb{N}$.

Proof: (C1) follows from Theorem 7.5.

For (C4) let U be a neighborhood of 0 in X and pick a neighborhood of 0, V, such that $V + V \subset U$. By Theorem 2.39 there exists n such that

$$\sum_{j=n+1}^{\infty} t_j^k(x_{ij} - x_j) \in V$$

for $i, k \in \mathbb{N}$. Since $\lim_i(x_{ij} - x_j) = 0$ for every j and $\{t_j^k : k \in \mathbb{N}\}$ is bounded for every j, then for every j $\lim_i t_j^k(x_{ij} - x_j) = 0$ uniformly for $k \in \mathbb{N}$ by Lemma 7.7. Therefore, there exists m such that $i \geq m$ implies $\sum_{j=1}^{n} t_j^k(x_{ij} - x_j) \in V$ for every $k \in \mathbb{N}$. If $i \geq m$, then

$$\sum_{j=1}^{\infty} t_j^k(x_{ij} - x_j) = \sum_{j=1}^{n} t_j^k(x_{ij} - x_j) + \sum_{j=n+1}^{\infty} t_j^k(x_{ij} - x_j) \in V + V \subset U$$

and (C4) follows.

As noted earlier Example 7.11 shows that conclusion (C4) cannot be improved to uniform convergence on bounded subsets of λ (conclusion (C2)). The following example shows that the signed-WGHP in Theorem 7.38 cannot be dropped even in the presence of 0-GHP.

Example 7.40. Let $\lambda = c$ and $X = \mathbb{R}$. Then $\sum_j \delta_{ij}$ is λ multiplier convergent for every i and $\lim_i \delta_{ij} = 0$ for every j. If $t \in c$, then $\lim_i \sum_{j=1}^{\infty} t_j \delta_{ij} = \lim_i t_i$ exists. However, if $t^k = e^k$, then $\sum_{j=1}^{\infty} t_j^k \delta_{ij} = 1$ so (C4) fails. Note that $\lambda = c$ has 0-GHP but not signed-WGHP.

We now establish a Hahn-Schur Theorem in the spirit of Li's Lemma 3.29. These theorems are useful in treating operator valued series with vector valued multipliers. Let Ω be a non-empty set and G be an Abelian topological group. Let $f_{ij} : \Omega \to G$ for $i, j \in \mathbb{N}$ and assume that Ω has a distinguished element w_0 such that $f_{ij}(w_0) = 0$ for every i, j.

Theorem 7.41. *Assume that the series $\sum_{j=1}^{\infty} f_{ij}(w_j)$ converges for every i and every sequence $\{w_j\} \subset \Omega$ and that $\lim_i \sum_{j=1}^{\infty} f_{ij}(w_j)$ exists for every sequence $\{w_j\} \subset \Omega$. Then*

(1) $\lim_i f_{ij}(w) = f_j(w)$ exists for every $w \in \Omega, j \in \mathbb{N}$,
(2) the series $\sum_{j=1}^{\infty} f_{ij}(w_j)$ converge uniformly for $i \in \mathbb{N}$ and all sequences $\{w_j\} \subset \Omega$,
(3) $\lim_i \sum_{j=1}^{\infty} f_{ij}(w_j) = \sum_{j=1}^{\infty} f_j(w_j)$ for every sequence $\{w_j\} \subset \Omega$.

Proof: Let $w \in \Omega$ and $j \in \mathbb{N}$. Define a sequence in Ω by $w_j = w$ and $w_i = w_0$ for $i \neq j$. Then $f_{ij}(w_j) = 0$ if $i \neq j$ so $\lim_i \sum_{j=1}^{\infty} f_{ij}(w_j) = \lim_i f_{ij}(w)$ exists by hypothesis and (1) holds.

We first show that for each $\{w_j\}$ the series $\sum_{j=1}^{\infty} f_{ij}(w_j)$ converge uniformly for $i \in \mathbb{N}$. If this fails to hold, there exists a neighborhood of 0, U, in G such that

$(*)$ for every k there exist $p > k$ and q such that $\displaystyle\sum_{j=p}^{\infty} f_{qj}(w_j) \notin U$.

Hence, there exist $n_1 > 1, i_1$ such that

$$\sum_{j=n_1}^{\infty} f_{i_1 j}(w_j) \notin U.$$

Pick a neighborhood of 0, V, such that $V + V \subset U$. There exists $m_1 > n_1$ such that

$$\sum_{j=m_1+1}^{\infty} f_{i_1 j}(w_j) \in V.$$

Hence,

$$\sum_{j=n_1}^{m_1} f_{i_1 j}(w_j) \notin V.$$

Since $\sum_{j=1}^{\infty} f_{ij}(w_j)$ converge for $i = 1, ..., i_1$ by $(*)$ there exist $n_2 > m_1, i_2 > i_1$ such that

$$\sum_{j=n_2}^{\infty} f_{i_2 j}(w_j) \notin U$$

and as above there exists $m_2 > n_2$ such that

$$\sum_{j=n_2}^{m_2} f_{i_2 j}(w_j) \notin V.$$

Continuing this construction produces increasing sequences $\{i_p\}, \{m_p\}$ and $\{n_p\}$ with $n_{p+1} > m_p > n_p$ such that

$$(**) \quad \sum_{j=n_p}^{m_p} f_{i_p j}(w_j) \notin V.$$

Now consider the matrix

$$M = [m_{pq}] = \left[\sum_{j=n_q}^{m_q} f_{i_p j}(w_j) \right].$$

We claim that M is a $\mathcal{K}$-matrix (Appendix D.2). The columns of M converge by (1). If $\{k_q\}$ is an increasing sequence, set $v_j = w_j$ if $n_{k_q} \leq j \leq m_{k_q}$ and $v_j = w_0$ otherwise. Then

$$\lim_p \sum_{q=1}^{\infty} m_{pk_q} = \lim_p \sum_{j=1}^{\infty} f_{i_p j}(v_j)$$

exists by hypothesis. Hence, M is a $\mathcal{K}$-matrix so by the Antosik-Mikusinski Matrix Theorem (Appendix D.2) the diagonal of M converges to 0. But, this contradicts $(**)$.

If (2) fails to hold, then as above there exist increasing sequences $\{i_k\}, \{m_k\}$ and $\{n_k\}$ with $n_k < m_k < n_{k+1}$, a matrix $\{w_{ij}\} \subset \Omega$ and a neighborhood, V, with

$$(***) \quad \sum_{j=n_k}^{m_k} f_{i_k j}(w_{kj}) \notin V.$$

Now define a sequence $\{w_j\} \subset \Omega$ by $w_j = w_{kj}$ if $n_k \leq j \leq m_k$ and $w_j = w_0$ otherwise. But, then the series $\sum_{j=1}^{\infty} f_{ij}(w_j)$ do not satisfy the Cauchy condition uniformly for $i \in \mathbb{N}$ by $(***)$ and, therefore, violates the condition established above.

For (3), let U be a neighborhood of 0 and $\{w_j\} \subset \Omega$. Pick a neighborhood of $0, V$, such that $V + V + V \subset U$. Put $g = \lim_i \sum_{j=1}^{\infty} f_{ij}(w_j)$. We show that the series $\sum_{j=1}^{\infty} f_j(w_j)$ converges to g. By (2) there exists n such that $\sum_{j=m}^{\infty} f_{ij}(w_j) \in V$ for $m \geq n$ and $i \in \mathbb{N}$. Suppose $m > n$. Then by (1) there exists i such that $\sum_{j=1}^{n}(f_{ij}(w_j) - f_j(w_j)) \in V$ and $g - \sum_{j=1}^{\infty} f_{ij}(w_j) \in V$. So

$$g - \sum_{j=1}^{n} f_j(w_j) = g - \sum_{j=1}^{\infty} f_{ij}(w_j) + \sum_{j=n+1}^{\infty} f_{ij}(w_j) + \sum_{j=1}^{n}(f_{ij}(w_j) - f_j(w_j))$$
$$\in V + V + V \subset U.$$

Concerning the converse of Theorem 7.40, we have

Theorem 7.42. *Assume that the series $\sum_{j=1}^{\infty} f_{ij}(w_j)$ converges for every i and every sequence $\{w_j\} \subset \Omega$ and that $\lim_i f_{ij}(w) = f_j(w)$ exists for every j and $w \in \Omega$. If for every $\{w_j\} \subset \Omega$ the series $\sum_{j=1}^{\infty} f_{ij}(w_j)$ converge uniformly for $i \in \mathbb{N}$, then $\{\sum_{j=1}^{\infty} f_{ij}(w_j)\}_i$ is Cauchy. If G is sequentially complete, then the stronger conclusion (2) of Theorem 7.40 holds.*

Proof: Let $\{w_j\} \subset \Omega$ and let U be a neighborhood of 0. Pick a symmetric neighborhood of 0, V, such that $V + V + V \subset U$. By hypothesis there exists n such that $\sum_{j=n}^{\infty} f_{ij}(w_j) \in V$ for all i. Since $\lim_i f_{ij}(w) = f_j(w)$ exists for every j and $w \in \Omega$ there exists m such that $\sum_{j=1}^{n-1}(f_{ij}(w_j) - f_{kj}(w_j)) \in V$ for all $i, k \geq m$. Then for all $i, k \geq m$,

$$\sum_{j=1}^{\infty} f_{ij}(w_j) - \sum_{j=1}^{\infty} f_{kj}(w_j)$$

$$= \sum_{j=1}^{n-1}(f_{ij}(w_j) - f_{kj}(w_j)) + \sum_{j=n}^{\infty} f_{ij}(w_j) - \sum_{j=n}^{\infty} f_{kj}(w_j)$$

$$\in V + V + V \subset U.$$

The last statement follows from Theorem 7.40.

Under stronger assumptions we establish a stronger convergence conclusion than condition (3) in Theorem 7.40.

Theorem 7.43. *Assume that the series $\sum_{j=1}^{\infty} f_{ij}(w_j)$ converges for every i and every sequence $\{w_j\} \subset \Omega$. If for each $j \in \mathbb{N}$ $\lim_i f_{ij}(w) = f_j(w)$ converges uniformly for $w \in \Omega$ and if the series $\sum_{j=1}^{\infty} f_{ij}(w_j)$ converge uniformly for all sequences $\{w_j\} \subset \Omega$ and $i \in \mathbb{N}$, then the sequences $\{\sum_{j=1}^{\infty} f_{ij}(w_j)\}_i$ satisfy a Cauchy condition uniformly for all sequences $\{w_j\} \subset \Omega$. If G is sequentially complete, then $\lim_i \sum_{j=1}^{\infty} f_{ij}(w_j) = \sum_{j=1}^{\infty} f_j(w_j)$ uniformly for all sequences $\{w_j\} \subset \Omega$.*

Proof: Let U be a closed neighborhood of 0 in G and pick a symmetric neighborhood of $0, V$, such that $V + V + V \subset U$.

There exists n such that $\sum_{j=n}^{\infty} f_{ij}(w_j) \in V$ for all $\{w_j\} \subset \Omega$ and $i \in \mathbb{N}$. There exists m such that $\sum_{j=1}^{n-1} (f_{ij}(w) - f_{kj}(w)) \in V$ for all $i, k \geq m$ and $w \in \Omega$ by the uniform convergence assumption. Hence, if $i, k \geq m$ and $\{w_j\} \subset \Omega$, we have

$$(*) \quad \sum_{j=1}^{\infty} f_{ij}(w_j) - \sum_{j=1}^{\infty} f_{kj}(w_j) =$$

$$\sum_{j=1}^{n-1} (f_{ij}(w_j) - f_{kj}(w_j)) + \sum_{j=n}^{\infty} f_{ij}(w_j) - \sum_{j=n}^{\infty} f_{kj}(w_j) \in V + V + V \subset U$$

so the first part of the statement is established.

If G is sequentially complete, then $\lim_i \sum_{j=1}^{\infty} f_{ij}(w_j)$ exists by (*). The last statement then follows from (3) of Theorem 7.40 and (*) above.

We can also obtain a version of Lemma 2.42.

Proposition 7.44. *Let Ω be a topological space with $g_j : \Omega \to G$ continuous and assume that the series $\sum_{j=1}^{\infty} g_j(w_j)$ converges for every $\{w_j\} \subset \Omega$. If $F : \Omega^{\mathbb{N}} \to G$ is defined by $F(\{w_j\}) = \sum_{j=1}^{\infty} g_j(w_j)$, then F is continuous with respect to the product topology.*

Proof: Let $w^k = \{w_j^k\}$ be a net in $\Omega^{\mathbb{N}}$ which converges to $w = \{w_j\}$ in the product topology. Let U be a neighborhood of 0 in G and pick a symmetric neighborhood, V, such that $V + V + V \subset U$. By Lemma 3.29

there exists n such that $\sum_{j=n}^{\infty} g_j(v_j) \in V$ for all $\{v_j\} \subset \Omega$. There exists k_0 such that $k \geq k_0$ implies $\sum_{j=1}^{n-1}(g_j(w_j^k) - g_j(w_j)) \in V$. If $k \geq k_0$, then

$$F(w^k) - F(w)$$

$$= \sum_{j=1}^{n-1}(g_j(w_j^k) - g_j(w_j)) + \sum_{j=n}^{\infty} g_j(w_j^k) - \sum_{j=n}^{\infty} g_j(w_j) \in V + V + V \subset U.$$

Thus, F is continuous.

From Proposition 7.43, we have

Corollary 7.45. *Let Ω be a compact topological space with $g_j : \Omega \to G$ continuous and assume that the series $\sum_{j=1}^{\infty} g_j(w_j)$ converges for every $\{w_j\} \subset \Omega$. Then*

$$S = \left\{ \sum_{j=1}^{\infty} g_j(w_j) : \{w_j\} \subset \Omega \right\}$$

is compact.

From Theorem 7.42, Lemma 7.30 and Proposition 7.43, we also obtain

Corollary 7.46. *Let Ω be a compact topological space. Assume that each f_{ij} is continuous, the series $\sum_{j=1}^{\infty} f_{ij}(w_j)$ converge uniformly for $\{w_j\} \subset \Omega$ and $i \in \mathbb{N}$ and for each $j \in \mathbb{N}$ $\lim_i f_{ij}(w) = f_j(w)$ converges uniformly for $w \in \Omega$, then*

$$S = \left\{ \sum_{j=1}^{\infty} f_{ij}(w_j) : \{w_j\} \subset \Omega, i \in \mathbb{N} \right\}$$

is compact.

Proof: As in Proposition 7.43 define $F_i : \Omega^{\mathbb{N}} \to G$ ($F_0 : \Omega^{\mathbb{N}} \to G$) by $F_i(\{w_j\}) = \sum_{j=1}^{\infty} f_{ij}(w_j)$ ($F_0(\{w_j\}) = \sum_{j=1}^{\infty} f_j(w_j)$). By Proposition 7.43 each F_i is continuous and by Theorem 7.42, $F_i \to F_0$ uniformly on $\Omega^{\mathbb{N}}$. The result follows from Lemma 7.30.

The results above cover the cases of subseries convergent series and bounded multiplier convergent series given in Theorems 7.17 and 7.25. In the subseries case, we take $\Omega = \{0, 1\}$ and in the bounded multiplier convergent case, we take $\Omega = [0, 1]$. If $\sum_j x_{ij}$ are the series in these statements, we define $f_{ij}(t) = tx_{ij}$ and take for the distinguished element $w_0 = 0$. That $\lim_i f_{ij}(w) = f_j(w)$ converges uniformly for $w \in \Omega$ follows from Lemma 7.7.

Throughout the remainder of this chapter we give applications of the Hahn-Schur results to various topics in functional analysis and measure theory.

The original proofs of the Orlicz-Pettis Theorem for normed spaces given by both Orlicz and Pettis used the version of the Schur Theorem stated in Theorem 7.1. We indicate how this version of the Orlicz-Pettis Theorem can easily be obtained from Hahn's Theorem 7.2.

Theorem 7.47. *Let X be a normed space. If the series $\sum_j x_j$ is subseries convergent in the weak topology of X, then the series is subseries convergent in the norm topology.*

Proof: By replacing X by the span of $\{x_j : j \in \mathbb{N}\}$, we may assume that X is separable. For every j pick $x_j' \in X'$ such that $\|x_j'\| = 1$ and $\langle x_j', x_j \rangle = \|x_j\|$. Since X is separable, $\{x_j'\}$ has a subsequence $\{x_{n_j}'\}$ which is weak* convergent to some $x' \in X'$. We have $\lim_i \sum_{j \in \sigma} \langle x_{n_i}', x_{n_j} \rangle = \langle x', \sum_{j \in \sigma} x_{n_j} \rangle$ for every $\sigma \subset \mathbb{N}$, where $\sum_{j \in \sigma} x_{n_j}$ is the weak sum of the series. By Theorem 7.2, the series $\sum_{j=1}^{\infty} \langle x_{n_i}', x_{n_j} \rangle$ converge uniformly for $i \in \mathbb{N}$. In particular, $\langle x_{n_j}', x_{n_j} \rangle = \|x_{n_j}\| \to 0$. Since the same argument can be applied to any subsequence of $\{x_j\}$, it follows that $\|x_j\| \to 0$ so $\sum_j x_j$ is norm subseries convergent by Lemma 4.4.

We next indicate several applications of the Hahn-Schur results to topics in vector valued measure theory. Let Σ be a σ-algebra of subsets of a set S. A set function $\mu : \Sigma \to X$ is countably additive if $\sum_{j=1}^{\infty} \mu(A_j) = \mu(\cup_{j=1}^{\infty} A_j)$ for every pairwise disjoint sequence $\{A_j\} \subset \Sigma$. A family of countably additive set functions $\{\mu_a : a \in I\}$ is uniformly countably additive if for every pairwise disjoint sequence $\{A_j\} \subset \Sigma$, the series $\sum_{j=1}^{\infty} \mu_a(A_j)$ converge uniformly for $a \in I$. We have the following result due to Nikodym.

Theorem 7.48. *(Nikodym Convergence Theorem) Let $\mu_j : \Sigma \to X$ be countably additive for every $j \in \mathbb{N}$. If $\lim \mu_j(A) = \mu(A)$ exists for every $A \in \Sigma$, then*

(1) μ is countably additive and
(2) $\{\mu_j\}$ is uniformly countably additive.

Proof: Let $\{A_j\} \subset \Sigma$ be pairwise disjoint. For any $\sigma \subset \mathbb{N}$, we have

$$\sum_{j \in \sigma} \mu_i(A_j) = \mu_i(\cup_{j \in \sigma} A_j) \to \mu(\cup_{j \in \sigma} A_j).$$

By Theorem 7.17 it follows that the series $\sum_{j=1}^{\infty} \mu_i(A_j)$ converge uniformly for $i \in \mathbb{N}$ so $\{\mu_j\}$ is uniformly countably additive. Also, from Theorem 7.17,

$$\sum_{j=1}^{\infty} \mu(A_j) = \lim_i \sum_{j=1}^{\infty} \mu_i(A_j) = \lim_i \mu_i(\cup_{j=1}^{\infty} A_j) = \mu(\cup_{j=1}^{\infty} A_j)$$

so μ is countably additive.

A theorem closely related to the Nikodym Convergence Theorem is the Vitali-Hahn-Saks Theorem which we now derive. Let $\nu : \Sigma \to [0, \infty]$ be a measure. If $\mu : \Sigma \to X$ is countably additive, then μ is ν continuous if $\lim_{\nu(A) \to 0} \mu(A) = 0$. If $\{\mu_j\}$ is a sequence of countably additive set functions, then $\{\mu_j\}$ is uniformly ν continuous if $\lim_{\nu(A) \to 0} \mu_j(A) = 0$ uniformly for $j \in \mathbb{N}$. We have the following result which connects uniform countable additivity and uniform ν continuity.

Theorem 7.49. *Let $\{\mu_j\}$ be countably additive, $\mu_j : \Sigma \to X$, such that each μ_j is ν continuous. If $\{\mu_j\}$ is uniformly countably additive, then $\{\mu_j\}$ is uniformly ν continuous.*

Proof: If the conclusion fails to hold, there exists a neighborhood, U, of 0 such that for every $\delta > 0$ there exist $k \in \mathbb{N}, E \in \Sigma$ such that $\mu_k(E) \notin U$ and $\nu(E) < \delta$. In particular, there exists $E_1 \in \Sigma, n_1$ such that $\mu_{n_1}(E_1) \notin U$ and $\nu(E_1) < 1$. Pick a neighborhood of 0, V, such that $V + V \subset U$. There exists $\delta_1 > 0$ such that $\mu_{n_1}(E) \in V$ when $\nu(E) < \delta_1$. There exist $E_2 \in \Sigma, n_2 > n_1$ such that $\mu_{n_2}(E_2) \notin U$ and $\nu(E_2) < \delta_1/2$. Continuing this construction produces sequences $\{E_k\} \subset \Sigma, \delta_{k+1} < \delta_k/2$, $\{n_k\}$ such that $\mu_{n_k}(E_k) \notin U, \nu(E_{k+1}) < \delta_k/2$ and $\mu_{n_k}(E) \in V$ when $\nu(E) < \delta_k$. Note that

$$\nu(\cup_{j=k+1}^{\infty} E_j) \leq \sum_{j=k+1}^{\infty} \nu(E_j) < \delta_k/2 + \delta_{k+1}/2 + ... < \delta_k/2 + \delta_k/2^2 + ... = \delta_k$$

so that

$$\mu_{n_k}(E_k \cap \cup_{j=k+1}^{\infty} E_j) \in V.$$

Now set $A_k = E_k \setminus \cup_{j=k+1}^{\infty} E_j$. The $\{A_k\}$ are pairwise disjoint and

$$\mu_{n_k}(A_k) = \mu_{n_k}(E_k) - \mu_{n_k}(E_k \cap_{j=k+1}^{\infty} E_j) \notin V.$$

However, by the uniform countable additivity of $\{\mu_j\}$ we have $\lim_k \mu_j(A_k) = 0$ uniformly for $j \in \mathbb{N}$. This gives the desired contradiction.

From the Nikodym Convergence Theorem, we can now obtain the Vitali-Hahn-Saks Theorem.

Theorem 7.50. *(Vitali-Hahn-Saks) Let $\{\mu_j\}$ be countably additive, $\mu_j :$ $\Sigma \to X$, such that each μ_j is ν continuous. If $\lim \mu_j(A) = \mu(A)$ exists for every $A \in \Sigma$, then*

(1) $\{\mu_j\}$ is uniformly ν continuous and
(2) μ is countably additive and ν continuous.

Proof: The result is an immediate consequence of Theorems 7.47 and 7.48.

As noted earlier there is a notion between the concepts of finite additivity and countable additivity called strong boundedness. If $\mu : \Sigma \to X$ is finitely additive, then μ is strongly bounded (strongly additive, exhaustive) if $\mu(A_j) \to 0$ whenever $\{A_j\}$ is a pairwise disjoint sequence from Σ (3.36). A family $\{\mu_a : a \in A\}$ of finitely additive set functions is uniformly strongly bounded (strongly additive, exhaustive) if whenever $\{A_j\}$ is a pairwise disjoint sequence from Σ, $\lim_j \mu_a(A_j) = 0$ uniformly for $a \in A$. We have the analogue of Theorem 3.42 for uniformly strong bounded set functions.

Theorem 7.51. *For $a \in A$ let $\mu_a : \Sigma \to X$ be finitely additive. The following are equivalent:*

(i) $\{\mu_a : a \in A\}$ is uniformly strongly bounded,
(ii) for any pairwise disjoint sequence $\{A_j\} \subset \Sigma$ the series $\sum_{j=1}^{\infty} \mu_a(A_j)$ satisfy a Cauchy condition uniformly for $a \in A$.

Proof: Clearly (ii) implies (i).

If (ii) fails to hold, there exist a neighborhood of $0, U$, an increasing sequence of intervals $\{I_k\}$ and a sequence $a_k \in A$ such that $\sum_{j \in I_k} \mu_{a_k}(A_j) \notin U$. Set $B_k = \cup_{j \in I_k} A_j$. Then $\{B_k\}$ is pairwise disjoint and $\mu_{a_k}(B_j) \notin U$ so (i) fails.

We now establish a version of the Nikodym Convergence Theorem for strongly bounded set functions.

Theorem 7.52. *Let $\mu_i : \Sigma \to X$ be strongly bounded for every $i \in \mathbb{N}$. If $\lim \mu_i(A) = \mu(A)$ exists for every $A \in \Sigma$, then*

(1) μ is strongly bounded and
(2) $\{\mu_i\}$ is uniformly strongly bounded.

Proof: Suppose that (2) fails. Then there exist a pairwise disjoint sequence $\{A_j\}$ and a neighborhood of 0, U, such that for every k there exist $j_k > k, i_k$ such that $\mu_{i_k}(A_{j_k}) \notin U$. For $k = 1$ there exist $j_1 > 1, i_1$ such that $\mu_{i_1}(A_{j_1}) \notin U$. There exist $J_1 > j_1$ such that $\mu_i(A_j) \in U$ for $1 \leq i \leq i_1$ and $j \geq J_1$. For $k = J_1$ there exist $j_2 > J_1$ and i_2 such that $\mu_{i_2}(A_{j_2}) \notin U$. Note that $i_2 > i_1$. Continuing this construction produces increasing sequences $\{j_k\}, \{i_k\}$ such that

$$(*) \quad \mu_{i_k}(A_{j_k}) \notin U.$$

By Drewnowski's Lemma (Appendix E.2) there is a subsequence $\{n_k\}$ such that each μ_{i_k} is countably additive on the σ-algebra Σ_0 generated by the $\{A_{j_{n_k}}\}$. By the Nikodym Convergence Theorem 7.47, $\{\mu_{i_{n_k}}\}$ is uniformly countably additive on Σ_0. In particular, $\lim \mu_{i_{n_k}}(A_{j_{n_k}}) = 0$. This contradicts $(*)$.

(1) follows from (2) since if $\{A_j\} \subset \Sigma$ is pairwise disjoint,

$$\lim_j \mu(A_j) = \lim_j \lim_i \mu_i(A_j) = \lim_i \lim_j \mu_i(A_j) = 0$$

by the uniform convergence of $\lim_j \mu_i(A_j) = 0$.

We next establish a vector version of a lemma due to Phillips which he used to show that there is no continuous projection from l^∞ onto c_0. We first state the scalar version of Phillips' Lemma. Let ba be the space of all finitely additive, bounded real valued set functions defined on $2^{\mathbb{N}}$ equipped with the variation norm, $\|\nu\| = var(\nu)(\mathbb{N})$. ba equipped with this norm is the dual space of $(l^\infty, \|\cdot\|_\infty)$ ([DS] IV.5.1, [SW3] 6.3). If $j \in \mathbb{N}$, we write $\nu(\{j\}) = \nu(j)$ for $\nu \in ba$.

Lemma 7.53. *(Phillips) Let $\nu_k \in ba$ for every $k \in \mathbb{N}$ and suppose that $\lim \nu_k(E) = 0$ for every $E \subset \mathbb{N}$. Then $\lim_k \sum_{j=1}^\infty |\nu_k(j)| = 0$.*

Phillips' Lemma has the following duality interpretation. Let J be the canonical imbedding of c_0 into its bidual l^∞. Then the transpose operator $J' : (l^\infty)' = ba \to (c_0)' = l^1$ is given by $J'\nu = \{\nu(j)\}$. Phillips' Lemma asserts that if $\{\nu_i\}$ converges to 0 in the weak topology $\sigma(ba, m_0)$, then $\{J'\nu_i\}$ converges to 0 in $\|\cdot\|_1$. In particular, if $\{\nu_i\}$ converges to 0 in the weak* topology $\sigma(ba, l^\infty)$, then $\|J'\nu_i\|_1 \to 0$. This also shows that l^∞ is a Grothendieck space. We next show how Phillips' Lemma can be used to show that there is no continuous projection of l^∞ onto c_0.

Theorem 7.54. *There is no continuous projection of l^∞ onto c_0.*

Proof: If P were such a projection, then for $y \in l^\infty$, $Py \in c_0$ so $e^k \cdot Py = P'e^k \cdot y \to 0$. Hence, $P'e^k \to 0$ weak* in ba. By the observation above,

$$\left\| J'P'e^k \right\|_1 = \sup\{ \left| J'P'e^k \cdot x \right| : x \in c_0, \|x\|_\infty \leq 1 \}$$
$$= \sup\{ \left| e^k \cdot PJx \right| : x \in c_0, \|x\|_\infty \leq 1 \}$$
$$= \sup\{ \left| e^k \cdot x \right| : x \in c_0, \|x\|_\infty \leq 1 \} = \left\| e^k \right\|_1 = 1 \to 0$$

an obvious contradiction.

We now establish a vector version of Phillips' Lemma which yields Lemma 7.52 as a special case.

Theorem 7.55. *Let X be sequentially complete and let $\mu_i : \Sigma \to X$ be strongly bounded for every $i \in \mathbb{N}$. If $\lim \mu_i(E) = 0$ for every $E \in \Sigma$, then for every pairwise disjoint sequence $\{E_j\}$ from Σ, $\lim_i \sum_{j \in \sigma} \mu_i(E_j) = 0$ uniformly for $\sigma \subset \mathbb{N}$.*

Proof: By Theorem 7.17 it suffices to show that $\lim_i \sum_{j \in \sigma} \mu_i(E_j) = 0$ for every $\sigma \subset \mathbb{N}$. If this fails to hold, we may assume, by passing to a subsequence if necessary, that there exists a closed neighborhood of 0, U, such that $\sum_{j=1}^\infty \mu_i(E_j) \notin U$ for every i. Pick a symmetric neighborhood of 0, V, such that $V + V \subset U$. There exists n_1 such that $\sum_{j=1}^{n_1} \mu_1(E_j) \notin U$. There exists m_1 such that $\sum_{j=1}^{n_1} \mu_i(E_j) \in V$ for $i \geq m_1$. There exists $n_2 > n_1$ such that $\sum_{j=1}^{n_2} \mu_{m_1}(E_j) \notin U$. Hence,

$$\sum_{j=n_1+1}^{n_2} \mu_{m_1}(E_j) = \sum_{j=1}^{n_2} \mu_{m_1}(E_j) - \sum_{j=1}^{n_1} \mu_{m_1}(E_j) \notin V.$$

Continuing this construction produces increasing sequences $\{m_i\}, \{n_i\}$ such that $\sum_{j=n_i+1}^{n_{i+1}} \mu_{m_i}(E_j) \notin V$. Set $F_i = \cup_{j=n_i+1}^{n_{i+1}} E_j$ so

$$(*) \quad \mu_{m_i}(F_i) \notin V$$

and $\{F_i\}$ is pairwise disjoint.

Consider the matrix

$$M = [m_{ij}] = [\mu_{m_i}(F_j)].$$

We claim that M is a $\mathcal{K}$-matrix (Appendix D.2). First, the columns of M converge to 0 by hypothesis. If $\{r_j\}$ is an increasing sequence, by Drewnowski's Lemma (Appendix E.2), there is a subsequence $\{s_j\}$ of $\{r_j\}$ such that each μ_i is countably additive on the σ-algebra generated by $\{F_{s_j}\}$. Thus,

$$\sum_{j=1}^\infty \mu_{m_i}(F_{s_j}) = \mu_{m_i}(\cup_{j=1}^\infty F_{s_j}) \to 0.$$

Hence, M is a $\mathcal{K}$-matrix so by the Antosik-Mikusinski Matrix Theorem (Appendix D.2) the diagonal of M converges to 0. But, this contradicts $(*)$.

We indicate how Theorem 7.54 yields Phillips' Lemma 7.52 and, therefore, can be viewed as a vector version of Phillips' Lemma. Let $\epsilon > 0$ and let the notation be as in Lemma 7.52. By Theorem 7.54 there exists N such that $\left|\sum_{j \in \sigma} \nu_i(j)\right| < \epsilon$ for every $\sigma \subset \mathbb{N}, i \geq N$. By Lemma 3.37, $\sum_{j=1}^{\infty} |\nu_i(j)| \leq 2\epsilon$ for $i \geq N$.

We next give a generalization of Theorem 7.54 which in turn yields a generalization of Phillips' Lemma 7.52.

Theorem 7.56. *Let X be sequentially complete and let $\mu_i : \Sigma \to X$ be strongly bounded for every $i \in \mathbb{N}$. If $\lim \mu_i(E) = \mu(E)$ exists for every $E \in \Sigma$, then for every pairwise disjoint sequence $\{E_j\} \subset \Sigma$, $\lim \sum_{j \in \sigma} \mu_i(E_j) = \sum_{j \in \sigma} \mu(E_j)$ uniformly for $\sigma \subset \mathbb{N}$. [In particular, μ is strongly bounded.]*

Proof: By Theorem 7.17 it suffices to show that $\lim_i \sum_{j \in \sigma} \mu_i(E_j)$ exists for every $\sigma \subset \mathbb{N}$. Since this is trivial for finite σ, assume that σ is infinite with $\sigma = \{m_1 < m_2 < ...\}$. We claim that $\{\sum_{j=1}^{\infty} \mu_i(E_{m_j})\}$ is a Cauchy sequence in X. For this, assume that $\{p_i\}$ and $\{q_i\}$ are increasing sequences with $p_i < q_i < p_{i+1}$. Then $\lim(\mu_{p_i}(E) - \mu_{q_i}(E)) = 0$ for every $E \in \Sigma$ so by Theorem 7.54, $\lim_i \sum_{j=1}^{\infty} (\mu_{p_i}(E_{m_j}) - \mu_{q_i}(E_{m_j})) = 0$ so $\{\sum_{j=1}^{\infty} \mu_i(E_{m_j})\}$ is a Cauchy sequence and the result follows.

The scalar case of Theorem 7.55 gives an improvement to Phillips' Lemma 7.52. In particular, as in the proof of Phillips' Lemma from Theorem 7.54 indicated above, we have

Corollary 7.57. *Let $\nu_k \in ba$ for every $k \in \mathbb{N}$ and suppose that $\lim \nu_k(E) = \nu(E)$ exists for every $E \subset \mathbb{N}$. Then $\nu \in ba$ and $\sum_{j=1}^{\infty} |\nu_k(j) - \nu(j)| \to 0$.*

We show that Theorem 7.55 can be used to derive a version of the Nikodym Boundedness Theorem for strongly bounded set functions.

Theorem 7.58. *(Nikodym Boundedness Theorem) Let $\nu_i : \Sigma \to \mathbb{R}$ be bounded and finitely additive. If $\{\nu_i(E) : i \in \mathbb{N}\}$ is bounded for every $E \in \Sigma$, then $\{\nu_i(E) : i \in \mathbb{N}, E \in \Sigma\}$ is bounded.*

Proof: Let $\{E_j\} \subset \Sigma$ be pairwise disjoint. By Lemma 4.58 it suffices to show that $\{\nu_i(E_i) : i \in \mathbb{N}\}$ is bounded.

Let $t_i \to 0$ in $\mathbb{R}$. Then $\lim_i t_i \nu_i(E) = 0$ for every $E \in \Sigma$. Since each $\nu \in ba$ is strongly bounded (Proposition 3.38) Theorem 7.55 implies that $\{t_i \nu_i\}$ is uniformly strongly bounded so $\lim_i t_i \nu_i(E_i) = 0$. Therefore, $\{\nu_i(E_i) : i \in \mathbb{N}\}$ is bounded.

From Theorem 7.57 and the Uniform Boundedness Principle, we can immediately obtain a version of the Nikodym Boundedness Theorem for LCTVS (see the proof of Corollary 4.60).

Theorem 7.59. *(Nikodym Boundedness Theorem) Let X be an LCTVS. Let $\nu_i : \Sigma \to X$ be strongly bounded. If $\{\nu_i(E) : i \in \mathbb{N}\}$ is bounded for every $E \in \Sigma$, then $\{\nu_i(E) : i \in \mathbb{N}, E \in \Sigma\}$ is bounded.*

Recall the local convex assumption in Theorem 7.58 cannot be dropped; see Remark 4.61.

We can use the version of the Nikodym Boundedness Theorem in Theorem 7.57 to show that $(m_0, \|\cdot\|_\infty)$ is barrelled. More generally, let $\mathcal{S}(\Sigma)$ be the space of all real valued Σ simple functions equipped with the sup-norm. Then the dual of $\mathcal{S}(\Sigma)$ is the space $ba(\Sigma)$ of all bounded, finitely additive, real valued set functions ν defined on Σ with the variation norm, $\|\nu\| = var(\nu)(S)$; the pairing between $f \in \mathcal{S}(\Sigma)'$ and $\nu \in ba(\Sigma)$ is given by integration, $\langle f, g \rangle = \int_S g d\nu, g \in \mathcal{S}(\Sigma)$ [no elaborate integration theory is used since we are only integrating simple functions] (see [DS], [Sw3] 6.3). A norm equivalent to the variation norm is given by $\|\nu\|' = \sup\{|\nu(E)| : E \in \Sigma\}$ ([DS], [Sw3] 2.2.1.7).

From the Nikodym Boundedness Theorem, we have

Theorem 7.60. *$(\mathcal{S}(\Sigma), \|\cdot\|_\infty)$ is barrelled. That is, if $\{\nu_j\} \subset ba(\Sigma)$ is $\sigma(ba(\Sigma), \mathcal{S}(\Sigma))$ bounded, then $\{\nu_j\}$ is norm bounded. In particular, $(m_0, \|\cdot\|_\infty)$ is barrelled.*

Proof: Since $\{\nu_j\}$ is pointwise bounded on $\mathcal{S}(\Sigma)$, $\{\nu_j(E)\}$ is bounded for each $E \in \Sigma$. Theorem 7.57 then implies that $\{\|\nu_j\|' : j \in \mathbb{N}\}$ is bounded so $\{\|\nu_j\| : j \in \mathbb{N}\}$ is bounded.

Chapter 8

Spaces of Multiplier Convergent Series and Multipliers

In this chapter we consider topological properties of the space of λ multiplier convergent series. Throughout this chapter let λ be a sequence space containing c_{00} and let X be an LCTVS. Recall that $\lambda^{\beta X}$ is the space of all X valued λ convergent series. If $x = \{x_j\} \in \lambda^{\beta X}$ and $t = \{t_j\} \in \lambda$, we write $x \cdot t = \sum_{j=1}^{\infty} t_j x_j$. We define a locally convex topology on $\lambda^{\beta X}$ induced by X and λ when λ is a K-space.

Assume that λ is a K-space and let

$$\mathcal{B} = \{B \subset \lambda : B \text{ is bounded and } \{x \cdot t : t \in B\} \text{ is bounded in } X \ \forall \ t \in \lambda\}.$$

Let $\mathcal{X}$ be the family of all continuous semi-norms on X. For $B \in \mathcal{B}$ and $p \in \mathcal{X}$, define a semi-norm on $\lambda^{\beta X}$ by

$$p_B(x) = \sup\{p(x \cdot t) : t \in B\}.$$

Let $\tau_{\mathcal{B}}$ be the locally convex topology on $\lambda^{\beta X}$ generated by the semi-norms p_B for $B \in \mathcal{B}$ and $p \in \mathcal{X}$.

Remark 8.1. The family $\mathcal{B}$ is equal to the family of all bounded subsets of λ if either λ has 0-GHP (Corollary 2.12) or if $\lambda^\beta \subset \lambda'$ (Corollary 2.4). If the maps $t \to x \cdot t$ from λ into X are continuous for all $x \in \lambda^{\beta X}$, then $\lambda^{\beta X} \subset L(\lambda, X)$ and the topology $\tau_{\mathcal{B}}$ is just the relative topology from $L_b(\lambda, X)$.

We consider the sequential completeness of $\tau_{\mathcal{B}}$. For this let P_i be the sectional operator on λ defined by $P_i t = \sum_{j=1}^{i} t_j e^j$. Recall the following property from Appendix B.4.

Definition 8.2. The space λ has the sections uniformly bounded property (SUB) if $\{P_i t : i \in \mathbb{N}, t \in B\}$ is bounded for every bounded subset B of λ.

Theorem 8.3. *Assume that X is sequentially complete, λ has SUB and for each $y \in \lambda^{\beta X}$ the operator $t \to y \cdot t$ from λ into X is bounded. Then τ_B is sequentially complete.*

Proof: Let $\{x^k\}$ be τ_B Cauchy, $B \in \mathcal{B}$, $p \in \mathcal{X}$ and $\epsilon > 0$.

First, put $t = e^j$. Then $p_{\{t\}}(x^k - x^l) = p(x_j^k - x_j^l)$ so $\{x_j^k\}_k$ is Cauchy in X for each j. Let $x_j = \lim_k x_j^k$. Set $x = \{x_j\}$.

Set $B' = B \cup \{P_i B : i \in \mathbb{N}\}$ so B' is bounded by the SUB assumption. Also, $B' \in \mathcal{B}$ by the boundedness assumption of the maps $t \to y \cdot t$ for $y \in \lambda^{\beta X}$. There exists N such that $k, l \geq N$ implies $p_{B'}(x^k - x^l) < \epsilon$. Thus, if $k, l \geq N$ and $n > m$, then $p(\sum_{i=m}^{n} t_i(x_i^k - x_i^l)) < \epsilon$ for $t \in B$. Hence, $p(\sum_{i=m}^{n} t_i(x_i^k - x_i)) \leq \epsilon$ for $t \in B, k \geq N, m > n$. This implies that $x \in \lambda^{\beta X}$ and $p_B(x^k - x) \leq \epsilon$ for $k \geq N$.

Remark 8.4. The boundedness assumption on the maps $t \to y \cdot t$ for $y \in \lambda^{\beta X}$ means that $\mathcal{B}$ is equal to the family of all bounded subsets of λ. This condition is satisfied, for example, if λ has 0-GHP (Corollary 2.12) or if $\lambda^\beta \subset \lambda'$ (Corollary 2.4).

The completeness assumption in Theorem 8.3 is necessary.

Proposition 8.5. *If $\lambda^{\beta X}$ is τ_B sequentially complete, then X is sequentially complete.*

Proof: Let $\{x_k\}$ be Cauchy in X. Put $x_j^k = x_k$ if $j = 1$ and $x_j^k = 0$ otherwise and set $x^k = \{x_j^k\}_j$. Then $\{x^k\}$ is τ_B Cauchy. Therefore, there exists $y = \{y_j\} \in \lambda^{\beta X}$ such that $x^k \to y$ in τ_B. In particular, $x_1^k = x_k \to y_1$ in X.

From the uniform convergence result in Theorem 2.16, we have

Theorem 8.6. *Let λ have signed-SGHP. Then for every $x \in \lambda^{\beta X}$,*

$$\tau_B - \lim \sum_{j=1}^{n} x_j e^j = x,$$

i.e., $\lambda^{\beta X}$ is a vector valued AK-space (Appendix C).

Recall that $\omega(\lambda^{\beta X}, \lambda)$ is the weakest topology on $\lambda^{\beta X}$ such that the maps $x \to x \cdot t$ from $\lambda^{\beta X}$ into X are continuous for all $t \in \lambda$. From Corollary 2.28, we have

Theorem 8.7. *If λ has signed-WGHP and X is sequentially complete, then $\omega(\lambda^{\beta X}, \lambda)$ is sequentially complete.*

We now give conditions which guarantee sequential convergence in $\omega(\lambda^{\beta X}, \lambda)$.

Proposition 8.8. *Let $\{x^k\} \subset \lambda^{\beta X}$.*

(1) If $x^k \to 0$ in $\omega(\lambda^{\beta X}, \lambda)$, then $\lim_k x_j^k = 0$ for every j.
(2) If $\lim_k x_j^k = 0$ for every j and if for every $t \in \lambda$ the series $\sum_j t_j x_j^k$ converge uniformly for $k \in \mathbb{N}$, then $x^k \to 0$ in $\omega(\lambda^{\beta X}, \lambda)$.
(3) If λ has signed-WGHP, the converse of (2) holds.

Proof: (1) follows since $\lim_k x^k \cdot e^j = \lim_k x_j^k = 0$.
For (2), let $t \in \lambda$ and consider

$$(*) \quad x^k \cdot t = \sum_{j=1}^n t_j x_j^k + \sum_{j=n+1}^\infty t_j x_j^k.$$

Let U be a neighborhood of 0 in X and pick a neighborhood of $0, V$, such that $V + V \subset U$. By hypothesis, there exists n such that the last term in $(*)$ belongs to V for every k. Since $\lim_k x_j^k = 0$ for every j, for large k the first term on the right hand side of $(*)$ belongs to V. Therefore, for large k, $x^k \cdot t \in U$.
The statement in (3) follows from (1) and Theorem 2.26.

One of the scalar versions of the Hahn-Schur theorem asserts that if the sequence $\{t^i\}$ in l^1 is $\sigma(l^1, l^\infty)$ convergent, then the sequence $\{t^i\}$ is $\|\cdot\|_1$ convergent. The vector version of the Hahn-Schur Theorem given in Theorem 7.10 can be given a similar interpretation.

Theorem 8.9. *Let λ have signed-SGHP and let X be sequentially complete. If $\{x^k\}$ is $\omega(\lambda^{\beta X}, \lambda)$ Cauchy, then there exists $x \in \lambda^{\beta X}$ such that $\lim x^k \cdot t = x \cdot t$ uniformly for t belonging to bounded subsets of λ.*

The scalar version of Theorem 8.9 gives a generalization of the classical scalar version of the Hahn-Schur Theorem for l^1 described above.

Corollary 8.10. *Let λ have signed-SGHP. If $\{t^k\}$ is $\sigma(\lambda^\beta, \lambda)$ Cauchy, then there exists $t \in \lambda^\beta$ such that $\beta(\lambda^\beta, \lambda) - \lim t^k = t$.*

For $\lambda = l^\infty$, this is the Hahn-Schur result for l^1 described above.
We next establish a Banach-Steinhaus equicontinuity type result for $\lambda^{\beta X}$. This will lead to another sequential completeness result for $\omega(\lambda^{\beta X}, \lambda)$.

Theorem 8.11. *Let λ have 0-GHP and $\{x^k\} \subset \lambda^{\beta X}$. If $\lim x^i \cdot t = Lt$ exists for every $t \in \lambda$, then $\{x^i\}$ is sequentially equicontinuous, i.e., if $t^j \to 0$ in λ, then $\lim_j x^i \cdot t^j = 0$ uniformly for $i \in \mathbb{N}$.*

Proof: If the conclusion fails to hold, we may assume that there exist $\epsilon > 0$, $p \in \mathcal{X}$ and $t^j \to 0$ in λ such that $p(x^j \cdot t^j) > \epsilon$ for all j. Set $m_1 = 1$ and pick n_1 such that

$$p(\sum_{k=1}^{n_1} t_k^{m_1} x_k^{m_1}) > \epsilon.$$

Note that $\lim_i x_k^i = \lim_i x^i \cdot e^k$ exists for every k so $\{x_k^i\}_i$ is bounded for each k. Thus, since $\lim_i t_k^i = 0$ for every k, $\lim_i t_k^i x_k^i = 0$ (Lemma 7.7) so there exists $m_2 > m_1$ such that

$$p(\sum_{k=1}^{n_1} t_k^{m_2} x_k^{m_2}) < \epsilon/2.$$

There exists $n_2 > n_1$ such that

$$p(\sum_{k=1}^{n_2} t_k^{m_2} x_k^{m_2}) > \epsilon.$$

Hence,

$$p(\sum_{k=n_1+1}^{n_2} t_k^{m_2} x_k^{m_2}) > \epsilon/2.$$

Continuing this construction produces increasing sequences $\{m_k\}, \{n_k\}$ such that

$$(*) \quad p(\sum_{k \in I_j} t_k^{m_j} x_k^{m_j}) > \epsilon/2$$

for all j, where $I_j = [n_{j-1} + 1, n_j]$.

Define a matrix

$$M = [m_{ij}] = [\sum_{k \in I_j} t_k^{m_j} x_k^{m_i}].$$

We claim that M is a $\mathcal{K}$-matrix (Appendix D.2). First, the columns of M converge by the observation above. Next, if $\{p_j\}$ is an increasing sequence, by 0-GHP, there is a further subsequence, still denoted by $\{p_j\}$, such that $s = \sum_{j=1}^{\infty} \chi_{I_{p_j}} t^{m_{p_j}} \in \lambda$. Then

$$\sum_{j=1}^{\infty} m_{ip_j} = \sum_{j=1}^{\infty} \sum_{k \in I_{p_j}} t_k^{m_{p_j}} x_k^{m_i} = x^{m_i} \cdot s$$

so $\lim_i x^{m_i} \cdot s = Ls$ exists. Thus, M is a $\mathcal{K}$-matrix and by the Antosik-Mikusinski Matrix Theorem (Appendix D.2) the diagonal of M converges to 0. But, this contradicts $(*)$.

We can now use Theorem 8.11 to obtain another sequential completeness result for $\omega(\lambda^{\beta X}, \lambda)$.

Corollary 8.12. *Let λ be an AK-space with 0-GHP and X be sequentially complete. If $\{x^i\}$ is $\omega(\lambda^{\beta X}, \lambda)$ Cauchy, then there exists $x \in \lambda^{\beta X}$ such that $\lim x^i \cdot t = x \cdot t$ for every $t \in \lambda$, i.e., $\omega(\lambda^{\beta X}, \lambda)$ is sequentially complete.*

Proof: If $Lt = \lim x^i \cdot t$ for $t \in \lambda$, then by Theorem 8.11, $L : \lambda \to \lambda$ is linear and sequentially continuous. Set $x_k = \lim_i x^i \cdot e^k$ and $x = \{x_k\}$. If $t \in \lambda$, by the AK-property, $t = \sum_{j=1}^{\infty} t_j e^j$ so $Lt = \sum_{j=1}^{\infty} t_j Le^j = \sum_{j=1}^{\infty} t_j x^j = x \cdot t$.

We next establish a uniform boundedness result for $\lambda^{\beta X}$.

Theorem 8.13. *Let λ have 0-GHP. If $\Gamma \subset \lambda^{\beta X}$ is pointwise bounded on λ, then Γ is uniformly bounded on bounded subsets of λ.*

Proof: Suppose the conclusion fails to hold. Then there exist $p \in \mathcal{X}$, $\epsilon > 0$, $\{x^k\} \subset \Gamma$, a bounded sequence $\{t^k\} \subset \lambda$ and $s_k \to 0, s_k > 0$, such that

$$p(s_k x^k \cdot t^k) > \epsilon$$

for all k. Put $k_1 = 1$ and pick m_1 such that

$$p(s_{k_1} \sum_{j=1}^{m_1} t_j^{k_1} x_j^{k_1}) > \epsilon.$$

Since λ is a K-space, $\{t_j^k : k \in \mathbb{N}\}$ is bounded for each j and $\{x_j^k : k \in \mathbb{N}\}$ is bounded for each j by hypothesis. Therefore, $\lim_k s_k t_j^k x_j^k = 0$ for every j (Lemma 7.7) so there exists $k_2 > k_1$ such that

$$p(s_{k_2} \sum_{j=1}^{m_1} t_j^{k_2} x_j^{k_2}) < \epsilon/2.$$

Pick $m_2 > m_1$ such that

$$p(s_{k_2} \sum_{j=1}^{m_2} t_j^{k_2} x_j^{k_2}) > \epsilon.$$

Set $I_2 = [m_1 + 1, m_2]$ and note

$$p(s_{k_2} \sum_{j \in I_2} t_j^{k_2} x_j^{k_2}) > \epsilon/2.$$

Continuing this construction produces an increasing sequence $\{k_p\}$ and an increasing sequence of intervals $\{I_p\}$ such that

$$(*) \quad p(s_{k_p} \sum_{j \in I_p} t_j^{k_p} x_j^{k_p}) > \epsilon/2$$

for all p.

Define a matrix

$$M = [m_{pq}] = [\sqrt{s_{k_p}} x^{k_p} \cdot \sqrt{s_{k_q}} \chi_{I_q} t^{k_q}].$$

We claim that M is a $\mathcal{K}$-matrix (Appendix D.2). First, the columns of M converge to 0 since $\{x^k\}$ is pointwise bounded on λ. Next, since $\sqrt{s_k} t^k \to 0$, by 0-GHP if $\{r_q\}$ is any subsequence, there is a further subsequence, still denoted by $\{r_q\}$, such that $t = \sum_{q=1}^{\infty} \sqrt{s_{k_{r_q}}} \chi_{I_{r_q}} t^{k_{r_q}} \in \lambda$. Hence,

$$\sum_{q=1}^{\infty} m_{pr_q} = \sqrt{s_{k_p}} x^{k_p} \cdot t \to 0.$$

Hence, M is a $\mathcal{K}$-matrix and by the Antosik-Mikusinski Matrix Theorem (Appendix D.2) the diagonal of M converges to 0. But, this contradicts $(*)$.

Recall that a pair of vector spaces X, X' in duality is called a Banach Mackey pair if $\sigma(X, X')$ bounded subsets are $\beta(X, X')$ bounded. An LCTVS X is a Banach-Mackey space if X, X' form a Banach-Mackey pair ([Wi] 10.4.3). The scalar version of Theorem 8.13 has the following corollary.

Corollary 8.14. *Let λ have 0-GHP.*

(i) λ, λ^β is a Banach-Mackey pair.
(ii) If $\lambda' \subset \lambda^\beta$, then λ, λ' is a Banach-Mackey pair.
(iii) If $\lambda' = \lambda^\beta$ and λ is quasi-barrelled, then λ is barrelled.

Conditions which guarantee that the hypotheses in (i) and (ii) are satisfied are given in Proposition 2.5.

We consider another uniform boundedness result which requires another type of gliding hump property (Appendix B.30 and B.31). Let μ be a sequence space containing c_{00}.

Definition 8.15. The K-space λ has the strong μ gliding hump property (strong μ-GHP) if whenever $\{I_k\}$ is an increasing sequence of intervals and $\{t^k\}$ is a bounded sequence in λ, then for every $s \in \mu$ the coordinate sum of the series $\sum_{j=1}^{\infty} s_j \chi_{I_j} t^j \in \lambda$.

Definition 8.16. The K-space λ has the weak μ gliding hump property (weak μ-GHP) if whenever $\{I_k\}$ is an increasing sequence of intervals and $\{t^k\}$ is a bounded sequence in λ, there is a subsequence $\{n_k\}$ such that for every $s \in \mu$ the coordinate sum $\sum_{j=1}^{\infty} s_j \chi_{I_{n_j}} t^{n_j} \in \lambda$.

The elements $s \in \mu$ are called multipliers since their coordinates multiply the blocks $\{\chi_{I_j} t^j\}$ determined by the $\{I_k\}$ and the $\{t^k\}$. The signed-WGHP and signed -SGHP are somewhat similar in that the "humps" are multiplied by ± 1 in these cases.

Examples of spaces with the strong μ-GHP and weak μ-GHP are given in Appendix B. For example, any locally complete LCTVS has strong l^1-GHP, l^{∞} and c_0 have strong c_0-GHP and $(l^2, \sigma(l^2, l^2))$ has strong l^1-GHP but not 0-GHP.

We next establish a basic lemma. If $A \subset \lambda$ and $B \subset \lambda^{\beta}$, we write

$$|B \cdot A| = \sup\{|s \cdot t| : s \in B, t \in A\}.$$

Lemma 8.17. *Suppose $A \subset \lambda$ is coordinate bounded and $B \subset \lambda^{\beta}$ is coordinate bounded. If $|B \cdot A| = \infty$, then there exist an increasing sequence of intervals $\{I_k\}$, $\{t^k\} \subset A$ and $\{s^k\} \subset B$ such that*

$$\left| s^k \cdot \chi_{I_k} t^k \right| > k^2$$

for all k.

Proof: There exist $s^k \in B, t^k \in A$ such that $\left| s^k \cdot t^k \right| > k^2 + k$. Set $k_1 = 1$ and pick n_1 such that $\left| \sum_{j=1}^{n_1} s_j^{k_1} t_j^{k_1} \right| > k_1^2 + 1$. By hypothesis for every j, $\{s_j^k : k \in \mathbb{N}\}$ and $\{t_j^k : k \in \mathbb{N}\}$ are bounded so there exists $k_2 > k_1$ such that $\frac{1}{k_2} \sum_{j=1}^{n_1} \left| s_j^{k_2} t_j^{k_2} \right| < 1$. Hence,

$$\left| \sum_{j=n_1+1}^{\infty} s_j^{k_2} t_j^{k_2} \right| \geq \left| \sum_{j=1}^{\infty} s_j^{k_2} t_j^{k_2} \right| - \sum_{j=1}^{n_1} \left| s_j^{k_2} t_j^{k_2} \right| > k_2^2.$$

Pick $n_2 > n_1$ such that $\left| \sum_{j=n_1+1}^{n_2} s_j^{k_2} t_j^{k_2} \right| > k_2^2$. Set $I_2 = [n_1 + 1, n_2]$ so

$$\left| s^{k_2} \cdot \chi_{I_2} t^{k_2} \right| > k_2^2.$$

Now just continue and relabel.

Theorem 8.18. *Let λ have weak μ-GHP. Assume*

(▲) $\{e^k : k \in \mathbb{N}\}$ *is $\beta(\lambda, \lambda^{\beta})$ bounded in λ.*

If $A \subset \lambda$ is bounded and $B \subset \lambda^\beta$ is $\sigma(\lambda^\beta, \lambda)$ bounded, then $|B \cdot A| < \infty$.

Proof: Suppose the conclusion fails and let the notation be as in Lemma 8.17. Let $\{n_j\}$ be as in Definition 8.16. Define a linear map $T : \mu \to \lambda$ by $Ts = \sum_{j=1}^\infty s_j \chi_{I_{n_j}} t^{n_j}$ [coordinate sum]. We claim that T is $\sigma(\mu, \mu^\beta) - \sigma(\lambda, \lambda^\beta)$ continuous. Let $s \in \mu, t \in \lambda^\beta$. Then

$$(*) \quad t \cdot Ts = \sum_{j=1}^\infty s_j (t \cdot \chi_{I_{n_j}} t^{n_j}).$$

Equation $(*)$ implies that $\{t \cdot \chi_{I_{n_j}} t^{n_j}\} \in \lambda^\beta$ and $t \cdot Ts = s \cdot \{t \cdot \chi_{I_{n_j}} t^{n_j}\}$ so T is $\sigma(\mu, \mu^\beta) - \sigma(\lambda, \lambda^\beta)$ continuous. Hence, T is $\beta(\mu, \mu^\beta) - \beta(\lambda, \lambda^\beta)$ continuous ([Wi] 11.2.6, [Sw2] 26.15). Thus, by ($\blacktriangle$), $\{Te^j\} = \{\chi_{I_{n_j}} t^{n_j}\}$ is $\beta(\lambda, \lambda^\beta)$ bounded. But, this contradicts the conclusion of Lemma 8.17.

Corollary 8.19. *Under the hypothesis of Theorem 8.18, if $\lambda' \subset \lambda^\beta$, then λ is a Banach-Mackey space.*

Conditions for $\lambda' \subset \lambda^\beta$ are given in Proposition 2.5.

We next consider some results of the Spanish school in Cadiz which use series to characterize completeness and barrelledness of normed linear spaces ([AP2], [PBA]).

For the remainder of this chapter let X be a normed space. Let $\{x_j\} \subset X$ and let $M^\infty(\sum x_j)$ be the space of bounded multipliers for the series $\sum x_j$:

$$M^\infty\left(\sum x_j\right) = \left\{ \{t_j\} \in l^\infty : \sum_{j=1}^\infty t_j x_j \text{ converges} \right\}.$$

We equip $M^\infty(\sum x_j)$ with the sup-norm from l^∞.

When X is a Banach space, we give necessary and sufficient conditions for the space $M^\infty(\sum x_j)$ to be complete. Recall a series $\sum_j x_j$ in a normed space is wuc iff $\sum_{j=1}^\infty |\langle x', x_j \rangle| < \infty$ for all $x' \in X'$ (see Definition 3.7).

Theorem 8.20. *Let X be a Banach space and $\sum_j x_j$ be a series in X. Then $M^\infty(\sum x_j)$ is complete iff $\sum_j x_j$ is wuc.*

Proof: Suppose $\sum_j x_j$ is wuc. Then $E = \{\sum_{j=1}^n t_j x_j : |t_j| \leq 1, n \in \mathbb{N}\}$ is bounded (Proposition 3.8) so let $M > 0$ be such that $\|x\| \leq M$ for all $x \in E$. Let $\{t^k\}$ be a Cauchy sequence in $M^\infty(\sum x_j)$ and let $t \in l^\infty$ be such that $\|t^k - t\|_\infty \to 0$. Let $\epsilon > 0$ and pick n such that $\|t^n - t\|_\infty < \epsilon/2M$. Since $\sum_{j=1}^\infty t_j^n x_j$ is convergent, there exists $N > n$ such that $q > p \geq N$ implies

$\left\| \sum_{j=p}^{q} t_j^n x_j \right\| < \epsilon/2$. Since $\frac{2M}{\epsilon} \sum_{j=p}^{q} (t_j^n - t_j) x_j \in E$, $\left\| \sum_{j=p}^{q} (t_j^n - t_j) x_j \right\| < \epsilon/2$ so $\left\| \sum_{j=p}^{q} t_j x_j \right\| < \epsilon$ for $q > p \geq N$. Hence, $\sum_j t_j x_j$ converges since X is a Banach space and $t \in M^\infty(\sum x_j)$.

Assume that $M^\infty(\sum x_j)$ is complete. Let $t \in c_0$. It suffices to show that $t \in M^\infty(\sum x_j)$ (Proposition 3.8). For each n, $t^n = \sum_{j=1}^{n} t_j e^j \in M^\infty(\sum x_j)$ and $\|t^n - t\|_\infty \to 0$ so $t \in M^\infty(\sum x_j)$.

We can use Theorem 8.20 and the multiplier space $M^\infty(\sum x_j)$ to characterize completeness of X.

Theorem 8.21. *The normed space X is complete iff for every wuc series $\sum_j x_j$ in X the space $M^\infty(\sum x_j)$ is complete.*

Proof: Suppose that X is not complete. Then there exists a non-convergent series $\sum_j x_j$ with $\sum_{j=1}^{\infty} j \|x_j\| < \infty$. Set $z_j = j x_j$. Then $\sum_j z_j$ is wuc since $|\langle x', z_j \rangle| \leq \|x'\| \|z_j\|$ for $x' \in X'$ but $\sum_j (1/j) z_j$ does not converge. That is, $\{i/j\} \notin M^\infty(\sum x_j)$. Since $\{i/j\} \in c_0$, $M^\infty(\sum x_j)$ is not complete.

The converse follows from Theorem 8.20.

We can also characterize wuc series in terms of the "summing operator" $T : M^\infty(\sum x_j) \to X$ defined by $Tt = T\{t_j\} = \sum_{j=1}^{\infty} t_j x_j$, $t \in M^\infty(\sum x_j)$ (Theorem 2.2).

Theorem 8.22. *The summing operator T is continuous iff $\sum_j x_j$ is wuc. In this case,*

$$\|T\| = \sup \left\{ \left\| \sum_{j=1}^{n} t_j x_j \right\| : |t_j| \leq 1, n \in \mathbb{N} \right\}.$$

Proof: Suppose that T is continuous. Now $c_{00} \subset M^\infty(\sum x_j)$ and if $t \in c_{00}, \|t\|_\infty \leq 1$ and $t_i = 0$ for $i \geq n$, then $\|Tt\| = \left\| \sum_{j=1}^{n} t_j x_j \right\| \leq \|T\|$. Hence, $\{ \left\| \sum_{j=1}^{n} t_j x_j \right\| : |t_j| \leq 1, n \in \mathbb{N} \}$ is bounded so $\sum_j x_j$ is wuc by Proposition 3.8.

Suppose that $\sum_j x_j$ is wuc and set $M = \sup\{ \left\| \sum_{j=1}^{n} t_j x_j \right\| : |t_j| \leq 1, n \in \mathbb{N}\}$. Let $t \in M^\infty(\sum x_j), \|t\|_\infty \leq 1$. For every n,

$$\left\| T(\sum_{j=1}^{n} t_j e^j) \right\| = \left\| \sum_{j=1}^{n} t_j x_j \right\| \leq M$$

so $\left\| \sum_{j=1}^{\infty} t_j x_j \right\| = \|Tt\| \leq M$ and T is continuous with $\|T\| \leq M$.
The last statement follows from the computations above.

We next consider weakly convergent series and the associated multiplier spaces. Let $\sum_j x_j$ be a series in X. Define

$$M_w^{\infty}\left(\sum x_j\right) = \left\{ \{t_j\} \in l^{\infty} : \sum_{j=1}^{\infty} t_j x_j \text{ is weakly convergent} \right\}.$$

Again we supply $M_w^{\infty}(\sum x_j)$ with the sup-norm topology from l^{∞}.
From Proposition 3.8, we have

Lemma 8.23. *Let X be a Banach space and $\sum_j x_j$ a series in X. Then $\sum_j x_j$ is wuc iff $c_0 \subset M_w^{\infty}(\sum x_j)$.*

We have the analogue of Theorem 8.20.

Theorem 8.24. *Let X be a Banach space and $\sum_j x_j$ a series in X. The space $M_w^{\infty}(\sum x_j)$ is complete iff $\sum_j x_j$ is wuc.*

Proof: Suppose that $\sum_j x_j$ is wuc. Let $\{t^k\}$ be a sequence in $M_w^{\infty}(\sum x_j)$ which converges to $t \in l^{\infty}$. Let $z_k = \sum_{j=1}^{\infty} t_j^k x_j$, where this is the $\sigma(X, X')$ sum of the series. Let $E = \{\sum_{j=1}^{n} a_j x_j : |a_j| \leq 1, n \in \mathbb{N}\}$ and $\sup\{\|x\| : x \in E\} < M$. Let $\epsilon > 0$. There exists n such that $\|t^k - t\|_{\infty} < \epsilon/3M$ for $k \geq n$. Then

$$(*) \qquad \left\| \sum_{j=1}^{m} (t_j^k - t_j) x_j \right\| \leq \epsilon/3 \text{ for } k \geq n, m \in \mathbb{N}$$

so

$$\left\| \sum_{j=1}^{m} (t_j^k - t_j^l) x_j \right\| \leq 2\epsilon/3 \text{ for } k, l \geq n, m \in \mathbb{N}.$$

Thus, $\|z_k - z_l\| \leq \epsilon$ for $k, l \geq n$. Let $z = \|\cdot\| - \lim z_k$.

We claim that $\sum_j t_j x_j$ is weakly convergent to z. There exists $N > n$ such that $\|z_k - z\| < \epsilon/3$ for $k \geq N$. If $x' \in X', \|x'\| \leq 1$, then by $(*)$ $\left| \sum_{j=1}^{m} (t_j^k - t_j) \langle x', x_j \rangle \right| \leq \epsilon/3$ for $k \geq N, m \in \mathbb{N}$. If $m \in \mathbb{N}$, then

$$\left| \left\langle x', \sum_{j=1}^{m} t_j x_j - z \right\rangle \right| \leq 2\epsilon/3 + \left| \left\langle x', \sum_{j=1}^{m} t_j^N x_j - z_N \right\rangle \right|.$$

There exists N_1 such that $\left|\left\langle x', \sum_{j=1}^{m} t_j^N - z_N \right\rangle\right| < \epsilon/3$ for $m \geq N_1$. Hence, $m \geq N_1$ implies that $\left|\left\langle x', \sum_{j=1}^{m} t_j x_j - z \right\rangle\right| \leq \epsilon$ and the claim is established.

Suppose that $M_w^\infty(\sum x_j)$ is complete. Let $t \in c_0$. For every n, $t^n = \sum_{j=1}^{n} t_j e^j \in M_w^\infty(\sum x_j)$ so $t = \|\cdot\| - \lim t^n \in M_w^\infty(\sum x_j)$ by Lemma 8.23.

The analogue of Theorem 8.22 also holds.

Theorem 8.25. *Define* $T : M_w^\infty(\sum x_j) \to X$ *by* $Tt = \sigma(X, X') - \lim \sum_{j=1}^{n} t_j x_j$. *Then* T *is continuous iff* $\sum_j x_j$ *is wuc. In this case,*

$$\|T\| = \sup\left\{\left\|\sum_{j=1}^{n} t_j x_j\right\| : |t_j| \leq 1, n \in \mathbb{N}\right\}.$$

Proof: Suppose that T is continuous. Now $c_{00} \subset M_w^\infty(\sum x_j)$ and if $t \in c_{00}, \|t\|_\infty \leq 1$ and $t_i = 0$ for $i \geq n$, then $\|Tt\| = \left\|\sum_{j=1}^{n} t_j x_j\right\| \leq \|T\|$. Hence, $\{\left\|\sum_{j=1}^{n} t_j x_j\right\| : |t_j| \leq 1, n \in \mathbb{N}\}$ is bounded so $\sum_j x_j$ is wuc by Proposition 3.8.

Suppose that $\sum_j x_j$ is wuc and set $M = \sup\{\left\|\sum_{j=1}^{n} t_j x_j\right\| : |t_j| \leq 1, n \in \mathbb{N}\}$. If $t \in M_w^\infty(\sum x_j), \|t\|_\infty \leq 1$, then for every n we have $\left\|T(\sum_{j=1}^{n} t_j e^j)\right\| = \left\|\sum_{j=1}^{n} t_j x_j\right\| \leq M$. Therefore, if $x' \in X', \|x'\| \leq 1$, then

$$|\langle x', Tt \rangle| = \left|\left\langle x', \sum_{j=1}^{\infty} t_j x_j \right\rangle\right| = \left|\sum_{j=1}^{\infty} \langle x', t_j x_j \rangle\right| \leq M.$$

Hence, $\|Tt\| \leq M$ and T is continuous with $\|T\| \leq M$.

The last statement follows from the computations above.

We next consider the analogue of Theorem 8.21 for $M_w^\infty(\sum x_j)$. We say that a series $\sum_j x_j$ is bounded multiplier Cauchy if the series $\sum_j t_j x_j$ is Cauchy for every $t \in l^\infty$.

Lemma 8.26. *Suppose* $\sum_j x_j$ *is bounded multiplier Cauchy. Then* $M_w^\infty(\sum x_j) = M^\infty(\sum x_j)$.

Proof: Let $t \in M_w^\infty(\sum x_j)$ and let $x = \sum_{j=1}^{\infty} t_j x_j$ [weak sum]. Since the partial sums of the series $\sum_{j=1}^{\infty} t_j x_j$ form a Cauchy sequence in X, there exists $x'' \in X''$ such that $x'' = \sum_{j=1}^{\infty} t_j x_j$ [norm limit in X'']. If $x' \in X'$, then $\left\langle x', \sum_{j=1}^{\infty} t_j x_j \right\rangle = \langle x', x \rangle = \langle x'', x' \rangle$ so $x'' = x$ and $t \in M^\infty(\sum x_j)$.

Theorem 8.27. *X is a Banach space iff for every wuc series $\sum_j x_j$ the space $M_w^\infty(\sum x_j)$ is complete.*

Proof: Suppose that X is not complete. By the proof of Theorem 8.21, there is an absolutely convergent series $\sum_j x_j$ which is wuc but for which $M^\infty(\sum x_j)$ is not complete. The series $\sum_j x_j$ is bounded multiplier Cauchy since it is absolutely convergent so $M_w^\infty(\sum x_j) = M^\infty(\sum x_j)$ by Lemma 8.26 and $M_w^\infty(\sum x_j)$ is not complete.

The converse follows from Theorem 8.24.

We next consider series with values in the dual space X' and use the series to characterize barrelled spaces. If $\sum_j x_j'$ is a series in X', define

$$M_{w*}^\infty\left(\sum x_j'\right)$$

$$= \{\{t_j\} \in l^\infty : \sum_{j=1}^\infty t_j x_j' \text{ converges in } X' \text{ with respect to } \sigma(X', X)\}.$$

We supply $M_{w*}^\infty(\sum x_j')$ with the sup-norm topology from l^∞.

Theorem 8.28. *Let $\sum_j x_j'$ be a series in X'. Consider the following conditions:*

(i) $\sum_j x_j'$ is wuc
(ii) $M_{w*}^\infty(\sum x_j') = l^\infty$
(iii) $\sum_{j=1}^\infty |\langle x_j', x \rangle| < \infty$ for every $x \in X$.

Then (i) implies (ii) implies (iii). The conditions (i), (ii) and (iii) are equivalent iff X is barrelled.

Proof: (i) implies (ii): If $\sum_j x_j'$ is wuc and $t \in l^\infty$, then $\sum_j t_j x_j'$ is also wuc. Hence, $\{\sum_{j=1}^n t_j x_j' : n \in \mathbb{N}\}$ is a bounded (equicontinuous) sequence in X' that is also $\sigma(X', X)$ Cauchy and, therefore, $\sigma(X', X)$ convergent. Hence, $t \in M_{w*}^\infty(\sum x_j)$.

(ii) implies (iii): For every $t \in l^\infty$ and $x \in X$, $\sum_{j=1}^\infty t_j \langle x_j', x \rangle$ converges. Hence, $\{\langle x_j', x \rangle\} \in l^1$ and (iii) holds.

Suppose that X is barrelled and (iii) is satisfied for the series $\sum_j x_j'$. The set $\{\sum_{j \in \sigma} x_j' : \sigma \text{ finite}\}$ is pointwise bounded on X and is, therefore, norm bounded since X is barrelled. Thus, $\sum_j x_j'$ is wuc by Proposition 3.8.

Suppose (i),(ii) and (iii) are equivalent and X is not barrelled. Then there exists a subset $F \subset X'$ which is weak* bounded but not norm bounded. Pick $y_j' \in F$ such that $\|y_j'\| > 2^{2j}$ and set $x_j' = y_j'/2^j$. Then $\sum_{j=1}^\infty |\langle x_j', x \rangle| < \infty$ for every $x \in X$ so (iii) holds. But, $\|x_j'\| > 2^j$ so $\sum_{j=1}^\infty x_j'/2^j$ does not converge in norm and $\sum_j x_j'$ is not wuc and (i) fails to hold.

Chapter 9

The Antosik Interchange Theorem

A problem often encountered in analysis is the interchange of two limiting processes. For example, the Lebesgue Dominated Convergence Theorem gives sufficient conditions to interchange the pointwise limit of a sequence of integrable functions with the Lebesgue integral, i.e., to take the "limit under the integral sign". In this chapter we consider sufficient conditions for the equality of two iterated series. For real valued series one of the most useful criterion for interchanging the limit of an iterated series $\sum_{i=1}^{\infty} \sum_{j=1}^{\infty} t_{ij}$ is the absolute convergence of the iterated series. However, absolute convergence for series with values in an LCTVS is a very strong condition and is, therefore, not appropriate. Antosik has given a sufficient condition involving subseries convergence of an iterated series with values in a topological group which has proven to be useful in a number of applications ([A]). We begin this chapter with a presentation of Antosik's result for series with values in a TVS. We then give generalizations of Antosik's result to multiplier convergent series.

Throughout this chapter let X be a TVS and let λ be a sequence space containing c_{00}. Let $x_{ij} \in X$ for $i, j \in \mathbb{N}$. The double series $\sum_{i,j} x_{ij}$ converges to $x \in X$ if for every neighborhood, U, of 0 in X, there exists N such that $\sum_{i=1}^{p} \sum_{j=1}^{q} x_{ij} - x \in U$ for $p, q \geq N$. We have the following familiar properties of double series.

Proposition 9.1. *Let $\sum_{i,j} x_{ij}$ be a double series.*

(i) *If the double series $\sum_{i,j} x_{ij}$ converges to $x \in X$ and if the series $\sum_{j=1}^{\infty} x_{ij}$ converge for each i, then the iterated series $\sum_{i=1}^{\infty} \sum_{j=1}^{\infty} x_{ij}$ converges to x.*

(ii) *If the series $\{\sum_{i=1}^{m} \sum_{j=1}^{\infty} x_{ij} : m \in \mathbb{N}\}$ converge uniformly and if the* *iterated series* $\sum_{i=1}^{\infty} \sum_{j=1}^{\infty} x_{ij}$ *converges to* x, *then the double series* $\sum_{i,j} x_{ij}$ *converges to* x.

Proof: (i): Let U be a neighborhood of 0 in X and let V be a symmetric neighborhood such that $V + V \subset U$. There exists N_1 such that $p, q \geq N_1$ implies that $\sum_{i=1}^{p} \sum_{j=1}^{q} x_{ij} - x \in V$. For each p there exists $N_2(p)$ such that $\sum_{i=1}^{p} \sum_{j=1}^{\infty} x_{ij} - \sum_{i=1}^{p} \sum_{j=1}^{q} x_{ij} \in V$ for $q \geq N_2(p)$. Let $p \geq N_1$ and fix $q \geq \max\{N_1, N_2(p)\}$. Then

$$\sum_{i=1}^{p} \sum_{j=1}^{\infty} x_{ij} - x = \sum_{i=1}^{p} \sum_{j=1}^{\infty} x_{ij} - \sum_{i=1}^{p} \sum_{j=1}^{q} x_{ij} + \sum_{i=1}^{p} \sum_{j=1}^{q} x_{ij} - x \in V + V \subset U.$$

(ii): There exists N such that $\sum_{i=1}^{p} \sum_{j=q+1}^{\infty} x_{ij} \in V$ for $q > N$ and for every $p \in \mathbb{N}$. There exists $M > N$ such that $\sum_{i=1}^{p} \sum_{j=1}^{\infty} x_{ij} - x \in V$ for $p \geq M$. If $p, q \geq M$, then

$$\sum_{i=1}^{p} \sum_{j=1}^{q} x_{ij} - x = \sum_{i=1}^{p} \sum_{j=1}^{\infty} x_{ij} - x - \sum_{i=1}^{p} \sum_{j=q+1}^{\infty} x_{ij} \in V + V \subset U.$$

Theorem 9.2. *(Antosik) Let* $\{x_{ij}\} \subset X$. *Suppose the series* $\sum_{i=1}^{\infty} \sum_{j=1}^{\infty} x_{im_j}$ *converges for every increasing sequence* $\{m_j\}$. *Then the double series* $\sum_{i,j} x_{ij}$ *converges and*

$$(*) \quad \sum_{i,j} x_{ij} = \sum_{i=1}^{\infty} \sum_{j=1}^{\infty} x_{ij} = \sum_{j=1}^{\infty} \sum_{i=1}^{\infty} x_{ij}.$$

Proof: Note that the series $\sum_{i=1}^{\infty} x_{ik}$ converges for every k [consider the difference between the two series $\sum_{i=1}^{\infty} \sum_{j=1}^{\infty} x_{in_j}$ and $\sum_{i=1}^{\infty} \sum_{j=1}^{\infty} x_{im_j}$, where $n_j = j$ for every j and $\{m_j\}$ is the sequence $\{1, ..., k-1, k+1, ...\}$]. Set $z_{mj} = \sum_{i=1}^{m} x_{ij}$. Then for $\sigma \subset \mathbb{N}$, $\sum_{j \in \sigma} z_{mj} = \sum_{i=1}^{m} \sum_{j \in \sigma} x_{ij}$ converges to $\sum_{i=1}^{\infty} \sum_{j \in \sigma} x_{ij}$ as $m \to \infty$ by hypothesis. By the Hahn-Schur Theorem 7.17, the series $\sum_{j=1}^{\infty} (\sum_{i=1}^{\infty} x_{ij})$ is subseries convergent and

$$\lim_m \sum_{i=1}^{m} \sum_{j \in \sigma} x_{ij} = \sum_{j \in \sigma} \sum_{i=1}^{\infty} x_{ij}$$

uniformly for $\sigma \subset \mathbb{N}$. In particular, $\sum_{i=1}^{\infty} \sum_{j=1}^{\infty} x_{ij} = \sum_{j=1}^{\infty} \sum_{i=1}^{\infty} x_{ij}$. By Proposition 9.1 the uniform convergence implies that the double series $\sum_{i,j} x_{ij}$ converges and $(*)$ holds.

We give applications of Theorem 9.2 later in the chapter.

Although Antosik's Theorem is easy to apply in many concrete situations, it is only a necessary condition for the equality of the two iterated series. For example, suppose that $a_j, b_j \in \mathbb{R}$ and $x_{ij} = a_i b_j$. If both series $\sum_j a_j$ and $\sum_j b_j$ converge, then

$$\sum_{i=1}^{\infty}\sum_{j=1}^{\infty} x_{ij} = \sum_{j=1}^{\infty}\sum_{i=1}^{\infty} x_{ij} = \sum_{i=1}^{\infty} a_i \sum_{j=1}^{\infty} b_j.$$

However, if the "inner" series, $\sum_{j=1}^{\infty} b_j$ is conditionally convergent, the hypothesis in Theorem 9.2 is not satisfied. Stuart has given a result which covers this case ([St3]). The result uses Stuart's weaker form of the Hahn-Schur Theorem 7.24.

Theorem 9.3. *Let $\mathcal{F}$ be an $IQ\sigma$ family which contains $\mathbb{N}$ (Appendix B.23). If the series $\sum_{i=1}^{\infty}\sum_{j\in\sigma} x_{ij}$ converges for every $\sigma \in \mathcal{F}$, then*

$$\sum_{i=1}^{\infty}\sum_{j=1}^{\infty} x_{ij} = \sum_{j=1}^{\infty}\sum_{i=1}^{\infty} x_{ij}.$$

Proof: As before in Theorem 9.2, the series $\sum_{i=1}^{\infty} x_{ij}$ converges for every j. As in the proof of Theorem 9.2, set $z_{mj} = \sum_{i=1}^{m} x_{ij}$. For any $\sigma \in \mathcal{F}$, $\lim_m \sum_{j\in\sigma} z_{mj}$ exists and equals $\sum_{i=1}^{\infty}\sum_{j\in\sigma} x_{ij}$. By the version of the Hahn-Schur Theorem for $IQ\sigma$ families given in Theorem 7.24, $\sum_{j=1}^{\infty}(\sum_{i=1}^{\infty} x_{ij})$ converges and equals $\sum_{i=1}^{\infty}\sum_{j=1}^{\infty} x_{ij}$.

Note that by employing the weaker version of the Hahn-Schur Theorem in Theorem 7.24, we cannot assert the uniform convergence of the limit, $\lim_m \sum_{j\in\sigma} z_{mj}$, and the existence of the double series. However, Theorem 9.3 is sufficiently strong to cover the case mentioned prior to Theorem 9.3 since if $\sum_j b_j$ is a conditionally convergent series, the set $\mathcal{F} = \{\sigma \subset \mathbb{N} : \sum_{j\in\sigma} b_j$ converges$\}$ is an $IQ\sigma$ family containing $\mathbb{N}$ (Appendix B.24).

As has been done before, Antosik's Interchange Theorem can be viewed as a result concerning m_0 multiplier convergent series and this suggests generalizations to more general λ multiplier convergent series. The hypothesis in Theorem 9.2 that the series $\sum_{i=1}^{\infty}\sum_{j=1}^{\infty} x_{im_j}$ converges for every increasing sequence $\{m_j\}$ can be restated to assert that the series $\sum_{i=1}^{\infty}\sum_{j=1}^{\infty} t_j x_{ij}$ converges for every $t = \{t_j\} \in m_0$. This suggests that we might generalize Antosik's theorem by replacing m_0 by other sequence spaces λ. We now give such a generalization. Recall that λ has the signed weak gliding hump property (signed-WGHP) if whenever $t \in \lambda$ and $\{I_j\}$ is an increasing sequence of intervals, there exist a sequence of signs $\{s_j\}$ and a subsequence

$\{n_j\}$ such that the coordinate sum of the series $\sum_{j=1}^{\infty} s_j \chi_{I_{n_j}} t$ belongs to λ; if the signs can all be chosen equal to 1, then λ has the weak gliding hump property (WGHP) [see Appendix B for examples].

Theorem 9.4. *Let λ have signed-WGHP. Let $\{x_{ij}\} \subset X$. Suppose that the series $\sum_j x_{ij}$ is λ multiplier convergent for every i and the iterated series $\sum_{i=1}^{\infty} \sum_{j=1}^{\infty} t_j x_{ij}$ converges for every $t = \{t_j\} \in \lambda$. Then for every $t \in \lambda$, the sequence of iterated series $\{\sum_{i=1}^{m} \sum_{j=1}^{\infty} t_j x_{ij}\}$ converge uniformly for $m \in \mathbb{N}$.*

Proof: If the conclusion fails to hold, then there exists a neighborhood of 0, U, in X such that for every k there exist $j_k > k$ and m_k such that

$$\sum_{i=1}^{m_k} \sum_{j=j_k}^{\infty} t_j x_{ij} \notin U.$$

Pick a balanced neighborhood of 0, V, such that $V + V \subset U$. There exists $l_k > j_k$ such that

$$\sum_{i=1}^{m_k} \sum_{j=l_k+1}^{\infty} t_j x_{ij} \in V$$

so

$$(*) \quad \sum_{i=1}^{m_k} \sum_{j=j_k}^{l_k} t_j x_{ij} \notin V.$$

By the condition in $(*)$ for $k = 1$, there exist $j_1 < l_1$ and m_1 such that $\sum_{i=1}^{m_1} \sum_{j=j_1}^{l_1} t_j x_{ij} \notin V$. There exists $J_1 > j_1$ such that

$$(**) \quad \sum_{i=1}^{m} \sum_{j=n}^{n+p} t_j x_{ij} \in V \quad \text{for } 1 \le m \le m_1, n > J_1 \text{ and } p > 0.$$

By $(*)$ there exist $l_2 > j_2 > J_1$ and m_2 such that $\sum_{i=1}^{m_2} \sum_{j=j_2}^{l_2} t_j x_{ij} \notin V$. By $(**)$, $m_2 > m_1$. We can continue this construction to produce increasing sequences m_k, l_k, j_k with $l_{k-1} < j_k < l_k$ and

$$(***) \quad \sum_{i=1}^{m_k} \sum_{j=j_k}^{l_k} t_j x_{ij} \notin V.$$

Put $I_k = [j_k, l_k]$ so $\{I_k\}$ is an increasing sequence of intervals.

Define a matrix

$$M = [m_{pq}] = [\sum_{i=1}^{m_p} \sum_{j \in I_q} t_j x_{ij}].$$

We claim that M is a signed $\mathcal{K}$-matrix (Appendix D.3). First, note that the series $\sum_{i=1}^{\infty} t_j x_{ij}$ converges for every j by setting $t = e^j$ in the hypothesis. Thus, each column of M converges. Next, given an increasing sequence of positive integers there is a subsequence $\{n_q\}$ and a sequence of signs $\{s_q\}$ such that the coordinate sum $u = \sum_{q=1}^{\infty} s_q \sum_{j \in I_{n_q}} t_j \in \lambda$. Then

$$\sum_{q=1}^{\infty} s_q m_{pn_q} = \sum_{q=1}^{\infty} s_q \sum_{i=1}^{m_p} \sum_{j \in I_{n_q}} t_j x_{ij} = \sum_{i=1}^{m_p} \sum_{q=1}^{\infty} s_q \sum_{j \in I_{n_q}} t_j x_{ij} = \sum_{i=1}^{m_p} \sum_{j=1}^{\infty} u_j x_{ij}$$

so $\lim_p \sum_{q=1}^{\infty} s_q m_{pn_q} = \sum_{i=1}^{\infty} \sum_{j=1}^{\infty} u_j x_{ij}$ exists. Hence, M is a signed $\mathcal{K}$-matrix and by the signed version of the Antosik-Mikusinski Matrix Theorem, the diagonal of M converges to 0 (Appendix D.3). But, this contradicts $(* * *)$.

From Proposition 9.1 and Theorem 9.4, we obtain

Corollary 9.5. *Under the hypothesis of Theorem 9.4, for every $t \in \lambda$ the double series $\sum_{i,j} t_j x_{ij}$ converges and*

$$\sum_{i,j} t_j x_{ij} = \sum_{i=1}^{\infty} \sum_{j=1}^{\infty} t_j x_{ij} = \sum_{j=1}^{\infty} \sum_{i=1}^{\infty} t_j x_{ij}.$$

By strengthening the hypothesis on the multiplier space λ, we can also strengthen the conclusions of Theorem 9.4 and Corollary 9.5 to uniform convergence over bounded sets in the multiplier space. Recall that the K-space λ has the signed strong gliding hump property (signed-SGHP) if whenever $\{t^j\}$ is a bounded sequence in λ and $\{I_j\}$ is an increasing sequence of intervals, there exist a sequence of signs $\{s_j\}$ and a subsequence $\{n_j\}$ such that the coordinate sum of the series $\sum_{j=1}^{\infty} s_j \chi_{I_{n_j}} t^{n_j}$ belongs to λ; if all of the signs can be chosen equal to 1, then λ has the strong gliding hump property (SGHP) [see Appendix B for examples].

Theorem 9.6. *Let λ have signed-SGHP. Let $\{x_{ij}\} \subset X$. Suppose that the series $\sum_j x_{ij}$ is λ multiplier convergent for every i and the iterated series $\sum_{i=1}^{\infty} \sum_{j=1}^{\infty} t_j x_{ij}$ converges for every $t = \{t_j\} \in \lambda$. Then the family of iterated series*

$$\left\{ \sum_{i=1}^{m} \sum_{j=1}^{\infty} t_j x_{ij} : m \in \mathbb{N}, t \in B \right\}$$

converge uniformly for every bounded subset $B \subset \lambda$.

Proof: The proof is similar to the proof of Theorem 9.4 which we now sketch. If the conclusion fails, there exists a neighborhood U of 0 in X such that for every k there exist $j_k > k, t^k \in B$ and m_k such that

$$\sum_{i=1}^{m_k} \sum_{j=j_k}^{\infty} t_j^k x_{ij} \notin U.$$

Pick a balanced neighborhood V such that $V + V \subset U$. Then there exists $l_k > j_k$ such that

$$\sum_{i=1}^{m_k} \sum_{j=l_k+1}^{\infty} t_j^k x_{ij} \in V$$

so

$$(*) \quad \sum_{i=1}^{m_k} \sum_{j=j_k}^{l_k} t_j^k x_{ij} \notin V.$$

By the condition in $(*)$ for $k = 1$, there exist $j_1 < l_1, t^1 \in B$ and m_1 such that

$$(**) \quad \sum_{i=1}^{m_1} \sum_{j=j_1}^{l_1} t_j^1 x_{ij} \notin V.$$

By Theorem 2.16, there exists $J_1 > j_1$ such that

$$(***) \quad \sum_{i=1}^{m} \sum_{j=n}^{n+p} t_j x_{ij} \in V \quad \text{when } 1 \le m \le m_1, n > J_1, p > 0 \text{ and } t \in B$$

(this is where signed-SGHP is used to guarantee the uniform convergence over B). By $(**)$ there exist $l_2 > j_2 > J_1, m_2$ and $t^2 \in B$ such that $\sum_{i=1}^{m_2} \sum_{j=j_2}^{l_2} t_j^2 x_{ij} \notin V$. By $(***)$ $m_2 > m_1$. Continuing this construction produces increasing sequences m_k, l_k, j_k with $l_{k-1} < j_k < l_k$ and $t^k \in B$ such that

$$(****) \quad \sum_{i=1}^{m_k} \sum_{j=j_k}^{l_k} t_j^k x_{ij} \notin V.$$

Put $I_k = [j_k, l_k]$ so $\{I_k\}$ is an increasing sequence of intervals.

Define a matrix

$$M = [m_{pq}] = [\sum_{i=1}^{m_p} \sum_{j \in I_q} t_j^q x_{ij}].$$

Using the signed-SGHP, the proof of Theorem 9.4 can now be employed to show that M is a signed $\mathcal{K}$-matrix. By the signed version of the Antosik-Mikusinski Matrix Theorem, the diagonal of M converges to 0 (Appendix D.3). But, this contradicts $(\ast\ast\ast\ast)$.

We can now use Theorem 9.6 to give a generalization of Corollary 9.5. Let $x_{ij}(a) \in X$ for $i, j \in \mathbb{N}$ and $a \in A$. The double series $\sum_{i,j} x_{ij}(a)$ converge uniformly for $a \in A$ if for every neighborhood, U, of 0 in X, there exists N such that $\sum_{i=p}^{\infty} \sum_{j=q}^{\infty} x_{ij}(a) \in U$ for $p, q \geq N$ and $a \in A$. The proof of Proposition 9.1 shows that if the series $\sum_{i=1}^{m} \sum_{j=1}^{\infty} x_{ij}(a)$ converge uniformly for $m \in \mathbb{N}$ and $a \in A$, then the double series $\sum_{i,j} x_{ij}(a)$ converge uniformly for $a \in A$. Thus, from Theorem 9.6, we have

Corollary 9.7. *Under the hypothesis of Theorem 9.6, the double series $\sum_{i,j} t_j x_{ij}$ converge uniformly for $t \in B$.*

Using the proof of Theorem 9.6 and Theorem 2.22 on the uniform convergence over null sequences for multiplier spaces with 0-GHP in place of Theorem 2.16 on the uniform convergence of multiplier convergent series over bounded sets with signed-SGHP, we can obtain the following results.

Theorem 9.8. *Let λ have 0-GHP. Let $\{x_{ij}\} \subset X$. Suppose that the series $\sum_j x_{ij}$ is λ multiplier convergent for every i and the iterated series $\sum_{i=1}^{\infty} \sum_{j=1}^{\infty} t_j x_{ij}$ converges for every $t \in \lambda$. Then the family of iterated series*

$$\left\{ \sum_{i=1}^{m} \sum_{j=1}^{\infty} t_j^k x_{ij} : m, k \in \mathbb{N} \right\}$$

converge uniformly whenever $\{t^k\}$ is a null sequence in λ.

Corollary 9.9. *Under the hypothesis of Theorem 9.8, the double series $\sum_{i,j} t_j^k x_{ij}$ converge uniformly for any null sequence $\{t^k\}$ in λ.*

As an application of Theorem 9.4, we establish a multiplier convergent version of an Orlicz-Pettis Theorem for non-locally convex TVS with a Schauder basis due to Stiles ([Sti]) which was considered for LCTVS in Chapter 4.74. Stiles' result seems to be the first version of an Orlicz-Pettis Theorem for non-locally convex spaces. Indeed, Kalton remarks that Stiles' result motivated his far-reaching generalization of the Orlicz-Pettis Theorem for series with values in a topological group ([Ka2]). Stiles' version of the Orlicz-Pettis Theorem is for subseries convergent series with values in an F-space with a Schauder basis and his proof uses the metric properties of the space. Other proofs of Stiles' result have been given in [Bs] and [Sw5]. We will establish a version of Stiles' result for multiplier convergent series which requires no metrizability assumption.

Let X be a TVS with a Schauder basis $\{b_j\}$ and associated coordinate functionals $\{f_j\}$. That is, every $x \in X$ has a unique series representation $x = \sum_{j=1}^{\infty} t_j b_j$ and $f_j : X \to \mathbb{R}$ is defined by $\langle f_j, x \rangle = t_j$. We do not assume that the coordinate functionals $\{f_j\}$ are continuous although this is the case if X is a complete metric linear space ([Sw2] 10.1.13). Let $F = \{f_j : j \in \mathbb{N}\}$. We consider the weak topology $\sigma(X, F)$ on X and use the interchange theorem to establish a multiplier convergent version of the Orlicz-Pettis Theorem for $\sigma(X, F)$ and the original topology of X.

Theorem 9.10. *Let λ have signed-WGHP. If $\sum_j x_j$ is λ multiplier convergent with respect to $\sigma(X, F)$, then $\sum_j x_j$ is λ multiplier convergent with respect to the original topology of X.*

Proof: Let $t \in \lambda$. Consider the iterated series $\sum_{i=1}^{\infty} \sum_{j=1}^{\infty} t_j \langle f_i, x_j \rangle b_i$. Let $x = \sum_{j=1}^{\infty} t_j x_j$ be the $\sigma(X, F)$ sum of the series. Then for each i, $\sum_{j=1}^{\infty} t_j \langle f_i, x \rangle = \langle f_i, x \rangle$ so $\sum_{j=1}^{\infty} t_j \langle f_i, x \rangle b_i = \langle f_i, x \rangle b_i$, where the convergence is in the original topology of X. But,

$$\sum_{i=1}^{\infty} \sum_{j=1}^{\infty} t_j \langle f_i, x_j \rangle b_i = \sum_{i=1}^{\infty} \langle f_i, x \rangle b_i = x$$

with convergence in the original topology of X since $\{b_j\}$ is a Schauder basis. By Corollary 9.5,

$$\sum_{i=1}^{\infty} \sum_{j=1}^{\infty} t_j \langle f_i, x_j \rangle b_i = \sum_{j=1}^{\infty} \sum_{i=1}^{\infty} t_j \langle f_i, x_j \rangle b_i = \sum_{j=1}^{\infty} t_j x_j$$

with convergence in the original topology of X.

The use of Corollary 9.5 in the proof of Theorem 9.10 removes the metrizability and completeness assumptions in Stiles' Theorem. Note that we did not use the continuity of the coordinate functional $\{f_j\}$ in the proof so the topology of X and the topology $\sigma(X, F)$ may not even be compatible!

Remark 9.11. Theorem 9.10 covers the non-locally convex case when $\lambda = l^p$, $0 < p < 1$. The case when $\lambda = bs$ gives a generalization of Stiles' result for subseries convergent series. If X is an AK-space, then $\{e^j\}$

is a Schauder basis and the weak topology $\sigma(X, \{e^j\})$ is just the topology of coordinatewise convergence on X. Thus, from Theorem 9.10 a series in X which is λ multiplier convergent with respect to the topology of coordinatewise convergence is λ multiplier convergent with respect to the original topology of X. In particular, these remarks apply to the classical sequence spaces $X = l^p, 0 < p < \infty, c_0, c_{00}$.

We next consider a generalization of Theorem 9.10 for multiplier convergent series; see also Theorem 4.75 for a locally convex version. Let (X, τ) be a TVS and assume that there exists a sequence of linear operators $P_j : X \to X$ such that $x = \sum_{j=1}^{\infty} P_j x$ [convergence in X] for every $x \in X$. When each P_j is a continuous projection, $\{P_j\}$ is called a Schauder decomposition ([LT]). If X has a Schauder basis $\{b_j\}$ with continuous coordinate functionals $\{f_j\}$ and $P_j x = \langle f_j, x \rangle b_j$ for $x \in X$, then $\{P_j\}$ is a Schauder decomposition for X.

Theorem 9.12. *Let λ have signed-WGHP. Let (X, τ) be a TVS and σ a Hausdorff topology on X. Assume that each $P_j : X \to X$ is $\sigma - \tau$ continuous. If $\sum_j x_j$ is λ multiplier convergent with respect to σ, then $\sum_j x_j$ is λ multiplier convergent with respect to τ.*

Proof: Let $t \in \lambda$ and $\sum_{j=1}^{\infty} t_j x_j$ be the σ sum of the series. For each i, $\sum_{j=1}^{\infty} t_j P_i x_j = P_i(\sum_{j=1}^{\infty} t_j x_j)$, where the series is τ convergent by the continuity of P_i. Hence,

$$\sum_{i=1}^{\infty} \sum_{j=1}^{\infty} P_i(t_j x_j) = \sum_{i=1}^{\infty} P_i(\sum_{j=1}^{\infty} t_j x_j)$$

converges with respect to τ. By Antosik's Interchange Theorem 9.4 (Corollary 9.5),

$$\sum_{i=1}^{\infty} \sum_{j=1}^{\infty} P_i(t_j x_j) = \sum_{j=1}^{\infty} t_j x_j,$$

where the convergence is with respect to τ.

Note that Theorem 9.10 is a corollary of Theorem 9.12. For suppose that $\{b_j\}$ is a Schauder basis for X with coordinate functionals $\{f_j\}$ and define $P_j : X \to X$ by $P_j x = \langle f_j, x \rangle b_j$. If $F = \{f_j : j \in \mathbb{N}\}$ and $\sigma = \sigma(X, F)$, then each P_j is $\sigma - \tau$ continuous and Theorem 9.12 applies.

We indicate applications to vector valued sequence spaces; see also Example 4.76.

Example 9.13. Let E be a vector space of X valued sequences equipped with a vector topology τ. E is a K-space if the coordinate maps $f_j : E \to X$, $f_j(\{x_j\}) = x_j$ are continuous for every j. If E is a K-space and every $x = \{x_j\} \in E$ has a representation $x = \sum_{j=1}^{\infty} e^j \otimes x_j$, then E is an AK-space [Appendix C; here $e^j \otimes x$ is the sequence with x in the jth coordinate and 0 in the other coordinates]. If E is an AK-space, define $P_j : E \to E$ by $P_j(\{x_j\}) = e^j \otimes x_j$. If E is an AK-space and σ is the topology of coordinatewise convergence on E, then each P_j is $\sigma - \tau$ continuous, where τ is the original topology of E. Thus, Theorem 9.12 is applicable and any series $\sum_j x_j$ in E which is coordinatewise λ multiplier convergent is λ multiplier convergent in E.

For situations where Example 9.13 is applicable, let X be a metric linear space whose topology is generated by the quasi-norm $|\cdot|$. For $0 < p < \infty$, let $l^p(X)$ be all X-valued sequences such that $\sum_{j=1}^{\infty} |x_j|^p < \infty$. If $1 \le p < \infty$ $(0 < p < 1)$, equip $l^p(X)$ with the quasi-norm $|\{x_j\}|_p = (\sum_{j=1}^{\infty} |x_j|^p)^{1/p}$ $(|\{x_j\}|_p = \sum_{j=1}^{\infty} |x_j|^p)$. Then $l^p(X)$ is an AK-space so Example 9.13 applies. Similarly, let $c_0(X)$ be the space of X valued sequences which converge to 0 and equip $c_0(X)$ with the sup-norm, $|\{x_k\}|_{\infty} = \sup_k |x_k|$. Then $c_0(X)$ is an AK-space so Example 9.13 applies. If X is an LCTVS, the spaces $l^p(X)$ and $c_0(X)$ can be defined similarly and Example 9.13 is also applicable in this case.

Let X be a metric linear space whose topology is generated by the quasi-norm $|\cdot|$. Let $c(X)$ be the space of all X valued sequences which converge and equip $c(X)$ with the sup-norm, $|\{x_k\}|_{\infty} = \sup_k |x_k|$. Although $c(X)$ is not an AK-space, we can use Theorem 9.12 to obtain a result similar to those for $l^p(X)$ and $c_0(X)$.

If $x = \{x_k\} \in c(X)$, define $x_0 = \lim_k x_k$. Now define $P_0 : c(X) \to c(X)$ by $P_0 x = \{x_0, x_0, ...\}$ and $P_i : c(X) \to c(X)$ by $P_i x = e^i \otimes (x_i - x_0)$ for $i \in \mathbb{N}$. Then $x = \sum_{i=0}^{\infty} P_i x$ in $c(X)$ for every $x \in c(X)$. If σ is the topology of coordinatewise convergence on $c(X)$, then each P_i is $\sigma - |\cdot|_{\infty}$ continuous so Theorem 9.12 applies and any series in $c(X)$ which is coordinatewise λ multiplier convergent is λ multiplier convergent in the sup-norm. We indicate how this result for $c(X)$ can be used to treat pointwise convergence for subseries convergent series in spaces of continuous functions; see Example 4.68 for a multiplier convergent version.

Example 9.14. Let S be a compact Hausdorff space and X a metric linear space whose topology is generated by the quasi-norm $|\cdot|$. Let $C_X(S)$ be the space of all X valued continuous functions defined on S equipped with the sup-norm $|f|_\infty = \sup\{|f(s)| : s \in S\}$. Suppose that $\sum_j f_j$ is subseries convergent with respect to the topology of pointwise convergence on S. We show that $\sum_j f_j$ is subseries convergent with respect to the sup-norm. First, consider the case when S is metrizable. For each j pick $t_j \in S$ such that $|f_j(t_j)| = |f_j|_\infty$. There is a subsequence $\{t_{n_j}\}$ such that $\{t_{n_j}\}$ converges to some $t \in S$. Define a continuous linear operator $U : C_X(S) \to c(X)$ by $Uf = \{f(t_{n_j})\}$. The series $\sum_j Uf_j$ is subseries convergent in $c(X)$ with respect to the topology of coordinatewise convergence so by the result above is subseries convergent with respect to the sup-norm. In particular, $|f_{n_j}|_\infty = |f_{n_j}(t_{n_j})| \leq \sup\{|f_{n_j}(t_{n_k})| : k \in \mathbb{N}\} = |Uf_{n_j}|_\infty$ so $\sum_j f_j$ is subseries convergent in the sup-norm by Lemma A.6 of Appendix A. Now consider the general case when S is not metrizable. Define an equivalence relation $\sim$ on S by $t \sim s$ iff $f_j(t) = f_j(s)$ for all j. Let S_0 be the equivalence classes of S under this equivalence relation and let $\pi : S \to S_0$ be the quotient map which associates $s \in S$ with the equivalence class determined by s. Now S_0 is compact and metrizable under the metric $d(\pi(s), \pi(t)) = \sum_{j=1}^\infty |f_j(s) - f_j(t)|/2^j$ ([DS] VI.7.6). Define continuous functions $F_j : S_0 \to X$ by $F_j(\pi(t)) = f_j(t)$. Now $\sum_j F_j$ is subseries convergent with respect to the topology of pointwise convergence on S_0 so by the part above $\sum_j F_j$ is subseries convergent with respect to the sup-norm. Hence. $\sum_j f_j$ is subseries convergent with respect to the sup-norm.

The proof of the first part of the result in Example 9.14 also yields the following result.

Proposition 9.15. *Let S be compact and metrizable and let D be a dense subset of S. If $\sum_j f_j$ is subseries convergent with respect to the topology of pointwise convergence on D, then $\sum_j f_j$ is subseries convergent with respect to the sup-norm.*

These Orlicz-Pettis Theorems for the space of continuous functions are due to Thomas ([Th]). Thomas also makes the interesting observation that the scalar case of Example 9.14 can be used to derive the version of the Orlicz-Pettis Theorem for normed spaces. Indeed, let X be a normed space and let S be the closed unit ball of X' equipped with the weak* topology. If $\sum_j x_j$ is subseries convergent in the weak topology of X , then $\sum_j x_j$ is subseries convergent in $C(S)$ with respect to the topology of pointwise convergence on S. By the result in Example 9.14 the series $\sum_j x_j$ is subseries convergent with respect to the sup-norm in $C(S)$ so $\sum_j x_j$ is subseries convergent in the norm topology of X.

The analogue of the result in Example 9.14 is false for the space of bounded measurable functions. For example, consider the series $\sum_j e^j$ in l^∞. This example, also shows that the AK-assumption in Example 9.13 is important.

Chapter 10

Automatic Continuity of Matrix Mappings

In this chapter we will consider the automatic continuity and boundedness of matrix mappings between sequence spaces which are K-spaces. Let λ and μ be sequence spaces which contain c_{00}, and let $A = [a_{ij}]$ be an infinite matrix with real entries. The matrix A maps λ into μ, denoted by $A : \lambda \to \mu$, if for each $t \in \lambda$, the series $\sum_{j=1}^{\infty} a_{ij} t_j$ converge for every i and the sequence $\{\sum_{j=1}^{\infty} a_{ij} t_j\}_i$ belongs to μ; we write $At = \{\sum_{j=1}^{\infty} a_{ij} t_j\}_i$. Thus, for a matrix A to map λ into μ, it is necessary for the rows of A to be λ multiplier convergent. We will consider the continuity and boundedness of a matrix $A : \lambda \to \mu$ with respect to various topologies on λ and μ. Such results are often referred to as automatic continuity (boundedness) results since there are no apriori assumptions on the matrix A. The first such automatic continuity result for matrix mappings is due to Hellinger and Toeplitz who showed that a matrix A which maps l^2 into itself is automatically continuous ([HT]). Another general automatic continuity which follows from the Closed Graph Theorem is that any matrix A mapping an FK-space λ into another FK-space μ is continuous [recall that a sequence space is an FK-space if it is a K-space under a complete metrizable vector topology (Appendix B)]. We now give a proof of a generalization of this result.

Theorem 10.1. *Let λ be a barrelled K-space and μ an FK-space. If $A : \lambda \to \mu$, then A is continuous.*

Proof: Let $a^i = \{a_{ij}\}_{j=1}^{\infty}$ be the ith row of $A = [a_{ij}]$. Since $A : \lambda \to \mu$, each $a^i \in \lambda^\beta$ and since λ is barrelled, $a^i \in \lambda'$ (Proposition 2.5). We show that A is closed so A is continuous by the Closed Graph Theorem ([Sw2] 26.3, [Wi] 12.5.7). Suppose that $t^k \to t$ in λ and $At^k \to s$ in μ. We show that $At = s$. By the continuity of each a^i noted above, $\lim_k a^i \cdot t^k = a^i \cdot t$.

Since μ is a K-space, for each i, $\lim_k e^i \cdot At^k = \lim_k a^i \cdot t^k = s_i$. Hence, $a^i \cdot t = s_i$ or $At = s$ as desired.

Theorem 10.1 yields the original result of Hellinger and Toeplitz on the continuity of a matrix which maps l^2 into itself since l^2 is an FK-space.

Theorem 10.1 applies to FK-spaces as well as spaces such as $\lambda = m_0$ which are barrelled but not complete. Another similar result is given by

Theorem 10.2. *Let λ be a barrelled K-space and μ an AK-space. If $A : \lambda \to \mu$, then A is continuous.*

Proof: As in Theorem 10.1, let a^i be the ith row of A so $a^i \in \lambda'$. For any $t \in \lambda$, $At = \{a^i \cdot t\} = \sum_{i=1}^{\infty}(a^i \cdot t)e^i$ since μ is an AK-space. For each n ,the linear map $t \to \sum_{i=1}^{n}(a^i \cdot t)e^i$ is continuous from λ into μ so $A : \lambda \to \mu$ is continuous since λ is barrelled and this linear map is the limit of a sequence of continuous linear maps ([Sw2] 24.12, [Wi] 9.3.7).

Theorem 10.2 covers the case where $\mu = c_{00}$ which is not covered by Theorem 10.1; $\mu = l^{\infty}$ is covered by Theorem 10.1 but not by Theorem 10.2.

We next consider continuity for matrix maps with respect to weak topologies. If A is an infinite matrix, we denote the transpose of A by A^T.

Theorem 10.3. *Let λ and μ be K-spaces and assume that $\sigma(\lambda^{\beta}, \lambda)$ is sequentially complete. If $A : \lambda \to \mu$, then $A^T : \mu^{\beta} \to \lambda^{\beta}$ and*
$$(\#) \quad s \cdot At = A^T s \cdot t \ \ for \ t \in \lambda, s \in \mu^{\beta}.$$

Proof: Let $t \in \lambda, s \in \mu^{\beta}$. Then
$$s \cdot At = \lim_m \sum_{i=1}^{m} s_i \sum_{j=1}^{\infty} a_{ij}t_j = \lim_m \sum_{j=1}^{\infty} t_j \sum_{i=1}^{m} a_{ij}s_i$$
which implies that the sequence $s^m = \{\sum_{i=1}^{m} a_{ij}s_i\}_{j=1}^{\infty} \in \lambda^{\beta}$ is $\sigma(\lambda^{\beta}, \lambda)$ Cauchy. Therefore, there exists $u \in \lambda^{\beta}$ such that $\sigma(\lambda^{\beta}, \lambda) - \lim_m s^m = u$ so $u = \{\sum_{i=1}^{\infty} a_{ij}s_i\} = A^T s \in \lambda^{\beta}$ and $s \cdot At = \lim_m t \cdot s^m = A^T s \cdot t$.

Remark 10.4. From Corollary 2.28, it follows that if λ has signed-WGHP and $A : \lambda \to \mu$, then $A^T : \mu^{\beta} \to \lambda^{\beta}$ and (#) holds.

From condition (#), we obtain

Proposition 10.5. *If $A : \lambda \to \mu, A^T : \mu^{\beta} \to \lambda^{\beta}$ and (#) holds, then A is $\sigma(\lambda, \lambda^{\beta}) - \sigma(\mu, \mu^{\beta})$ continuous and A^T is $\sigma(\mu^{\beta}, \mu) - \sigma(\lambda^{\beta}, \lambda)$ continuous.*

If w is a Hellinger-Toeplitz topology (Appendix A.1), then A is $w(\lambda, \lambda^\beta) - w(\mu, \mu^\beta)$ continuous and A^T is $w(\mu^\beta, \mu) - w(\lambda^\beta, \lambda)$ continuous.

Without some condition on λ even if $A : \lambda \to \mu$, the transpose matrix may not map μ^β into λ^β.

Example 10.6. Let A be the matrix which maps $t = \{t_j\}$ into the sequence $\{t_j - t_{j-1}\}$, where $t_0 = 0$. Thus, if $A = [a_{ij}]$, then $a_{ii} = 1, a_{i,i-1} = -1$ and $a_{ij} = 0$ otherwise. Then $A : bv \to l^1$. The transpose matrix $A^T = [b_{ij}]$ of A is the matrix $b_{ii} = 1, b_{i,i+1} = -1$ for $i \geq 2$ and $b_{ij} = 0$ otherwise. Thus, $A^T s = (s_1, s_2 - s_3, s_3 - s_4, ...)$. Thus, if $s^0 = \{(-1)^j\} \in l^\infty = (l^1)^\beta$, then $A^T s^0 = (-1, 2, -2, 2, -2, ...) \notin cs = (bv)^\beta$ so A^T doesn't map $l^\infty = (l^1)^\beta$ into $cs = (bv)^\beta$.

Condition $(\#)$ above is a sufficient condition for the weak continuity of both the matrix map A and its transpose A^T. We now consider this condition and several weak continuity results. For this and results to follow, we make several elementary observations. If $t \in \lambda$, we let $P_n t = t^{[n]} = (t_1, ..., t_n, 0, 0, ...)$ be the nth section of t.

(a) For every $s \in \lambda^\beta$, $P_n s \to s$ in $\sigma(\lambda^\beta, \lambda)$ and $t^{[n]} \to t$ in $\sigma(\lambda, \lambda^\beta)$ for every $t \in \lambda$.

(b) If $A : \lambda \to \mu$, then $\sum_{i=1}^{\infty} a_{ij} s_i$ converges for every $j \in \mathbb{N}$ and $s \in \mu^\beta$, i.e., $A^T : \mu^\beta \to s$, the space of all sequences. [Take $t = e^j$ in $s \cdot At$.]

(c) For every $n \in \mathbb{N}, t \in \lambda$ and $s \in \mu^\beta$, we have $s \cdot At^{[n]} = A^T s \cdot t^{[n]}$ and $s^{[n]} \cdot At = A^T s^{[n]} \cdot t$.

(d) $\{P_n\}$ is pointwise bounded on λ with respect to $\sigma(\lambda, \lambda^\beta)$, i.e., $\{s \cdot P_n t\}$ is bounded for every $t \in \lambda, s \in \mu^\beta$.

Theorem 10.7. *Let $A : \lambda \to \mu$ and assume that $A^T : \mu^\beta \to \lambda^\beta$ is $\sigma(\mu^\beta, \mu) - \sigma(\lambda^\beta, \lambda)$ sequentially continuous. Then*

$$(\#) \quad s \cdot At = A^T s \cdot t \quad for \ t \in \lambda, s \in \mu^\beta$$

and A is $\sigma(\lambda, \lambda^\beta) - \sigma(\mu, \mu^\beta)$ continuous and A^T is actually $\sigma(\mu^\beta, \mu) - \sigma(\lambda^\beta, \lambda)$ continuous.

Proof: Let $t \in \lambda, s \in \mu^\beta$. By (a), $s^{[n]} \to s$ in $\sigma(\mu^\beta, \mu)$. Therefore, $A^T s^{[n]} \to A^T s$ in $\sigma(\lambda^\beta, \lambda)$ or from (c), $\lim_n A^T s^{[n]} \cdot t = \lim_n s^{[n]} \cdot At = A^T s \cdot t = s \cdot At$ and $(\#)$ holds. The remaining statements follow from $(\#)$.

Theorem 10.8. *Suppose that $A : \lambda \to \mu$ is $\sigma(\lambda, \lambda^\beta) - \sigma(\mu, \mu^\beta)$ sequentially continuous. Then $A^T : \mu^\beta \to \lambda^\beta$ and*

$$(\#) \quad s \cdot At = A^T s \cdot t \quad for \ t \in \lambda, s \in \mu^\beta$$

holds. Thus, A^T is $\sigma(\mu^\beta, \mu) - \sigma(\lambda^\beta, \lambda)$ continuous and A is actually $\sigma(\lambda, \lambda^\beta) - \sigma(\mu, \mu^\beta)$ continuous.

Proof: Let $t \in \lambda, s \in \mu^\beta$. Then $t^{[n]} \to t$ in $\sigma(\lambda, \lambda^\beta)$ (condition (a)) so $At^{[n]} \to At$ in $\sigma(\mu, \mu^\beta)$ and using (b), $s \cdot At = \lim_n s \cdot At^{[n]} = \lim_n \sum_{j=1}^n t_j \sum_{i=1}^\infty a_{ij} s_i$. Hence, $A^T s \in \lambda^\beta$ and (#) holds. The remaining statements follow from (#).

The matrix A may map λ into μ and the transpose matrix A^T may map μ^β into λ^β but (#) may fail to hold.

Example 10.9. Let $a_{ij} = -1$ if $i = j$, $a_{ij} = 1$ if $j = i + 1$ and $a_{ij} = 0$ otherwise. Then $At = \{t_{i+1} - t_i\}$ and if $s_0 = 0$, then $A^T s = \{s_{i-1} - s_i\}$. Thus, $A : c \to cs$ and $A^T : bv = (cs)^\beta \to l^1 = c^\beta$. If e is the constant sequence with 1 in each coordinate, then $e \in c$ and $e \in bv$ and $Ae \cdot e - e \cdot A^T e = -1$ so (#) fails.

We now establish several automatic continuity/boundedness results which require gliding hump assumptions. Recall λ has the 0-GHP if whenever $\{t^j\}$ converges to 0 in λ and $\{I_j\}$ is an increasing sequence of intervals, there is a subsequence $\{n_j\}$ such that the coordinate sum of the series $\sum_{j=1}^\infty \chi_{I_{n_j}} t^{n_j} \in \lambda$ (Appendix B). Examples are given in Appendix B.

Theorem 10.10. Let (λ, τ) be a K-space with 0-GHP. If $A : \lambda \to \mu$, then A is $\tau - \sigma(\mu, \mu^\beta)$ sequentially continuous.

Proof: If the conclusion fails, there exist $\{t^k\} \subset \lambda$, $t^k \to 0$ in λ, $s \in \mu^\beta$ and $\delta > 0$ such that

$$(*) \quad \left| s \cdot At^k \right| > \delta \text{ for every } k.$$

Set $k_1 = 1$. Pick m_1, n_1 such that

$$(**) \quad \left| \sum_{i=1}^{m_1} s_i \sum_{j=1}^{n_1} a_{ij} t_j^{k_1} \right| = \left| P_{m_1} s \cdot A P_{n_1} t^{k_1} \right| > \delta.$$

From (b) and (c),

$$(***) \quad s \cdot A P_{n_1} t^k = A^T s \cdot P_{n_1} t^k = \sum_{j=1}^{n_1} t_j^k (A^T s)_j \text{ for every } k.$$

Since λ is a K-space, $\lim_k t_j^k = 0$ for every j so it follows from $(***)$ that $s \cdot A P_{n_1} t^k \to 0$ as $k \to \infty$. Therefore, there exists $k_2 > k_1$ such that

$$(\clubsuit) \quad \left| s \cdot A P_{n_1} t^{k_2} \right| < \delta/2.$$

From (∗) and (♣),

$$\left|s \cdot A(t^{k_2} - P_{n_1} t^{k_2})\right| > \delta/2.$$

Pick $m_2 > m_1, n_2 > n_1$ such that

$$\left|\sum_{i=1}^{m_2} s_i \sum_{j=n_1+1}^{n_2} a_{ij} t_j^{k_2}\right| = \left|P_{m_2} s \cdot A(P_{n_2} t^{k_2} - P_{n_1} t^{k_2})\right| > \delta/2.$$

Continuing this construction produces increasing sequences $\{k_p\}, \{m_p\}$, and $\{n_p\}$ such that

$$\left|P_{m_p} s \cdot A(P_{n_p} t^{k_p} - P_{n_{p-1}} t^{k_p})\right| > \delta/2.$$

Let $I_p = \{j \in \mathbb{N} : n_{p-1} < j \leq n_p\}$. So $\{I_p\}$ is an increasing sequence of intervals with

$$(\spadesuit) \quad \left|P_{m_p} s \cdot A\chi_{I_p} t^{k_p}\right| > \delta/2.$$

Define a matrix

$$M = [m_{pq}] = [P_{m_p} s \cdot A\chi_{I_q} t^{k_q}] = [A^T P_{m_p} s \cdot \chi_{I_q} t^{k_q}].$$

We claim that M is a $\mathcal{K}$-matrix (Appendix D.2). First, by (a), the columns of M converge to $s \cdot A\chi_{I_q} t^{k_q}$. Next, given any increasing sequence $\{r_q\}$, by 0-GHP there is a further subsequence, still denoted by $\{r_q\}$, such that $t = \sum_{q=1}^{\infty} \chi_{I_{r_q}} t^{k_{r_q}} \in \lambda$. Therefore, the sequence

$$\{\sum_{q=1}^{\infty} m_{pr_q}\} = \{A^T P_{m_p} s \cdot t\} = \{P_{m_p} s \cdot At\}$$

converges with limit $s \cdot At$ by (a) and (c). Hence, M is a $\mathcal{K}$-matrix and the diagonal of M converges to 0 by the Antosik-Mikusinski Matrix Theorem (Appendix D.2). But, this contradicts ($\spadesuit$).

We give several corollaries of Theorem 10.10.

Corollary 10.11. *Let (λ, τ) be a K-space with 0-GHP. Let η be a vector topology on μ such that $(\mu, \eta)' = \mu' \subset \mu^{\beta}$. If $A : \lambda \to \mu$, then A is $\tau - \eta$ bounded.*

Proof: The hypothesis implies that $\sigma(\mu, \mu^{\beta})$ bounded sets are η bounded so the result follows from Theorem 10.10.

For example, the hypothesis in Corollary 10.11 is satisfied if μ is an AK-space (Proposition 2.5). Corollary 10.11 is not applicable to l^{∞} or its subspace m_0. We give a result which is applicable to these spaces.

Corollary 10.12. *Let (λ, τ) be a K-space with 0-GHP. Let η be a vector topology on μ such that η and $\sigma(\mu, \mu^\beta)$ have the same bounded sets. If $A : \lambda \to \mu$, then A is $\tau - \eta$ bounded.*

Since $(l^\infty)^\beta = (m_0)^\beta = c^\beta = l^1$, Corollary 10.12 is applicable if $\mu = c, m_0$ or l^∞ with the sup-norm. Even though bs, the space of bounded series, is not an AK-space, Corollary 10.12 is applicable to bs with its natural topology since $bs = (bv_0)'$ and $(bs)^\beta = bv_0$ (Appendix B or [KG] p. 69).

We next establish another automatic continuity result which requires assumptions on the range space μ.

Theorem 10.13. *Let (λ, τ) be a K-space with 0-GHP. Assume that (μ, η) is a separable K-space such that $\mu' = (\mu, \eta)' \subset \mu^\beta$ and*

(1) *the sectional projections $\{P_n\}$ are η equicontinuous.*

If $A : \lambda \to \mu$, then A is $\tau - \eta$ sequentially continuous.

Proof: Let $x^k \to 0$ in τ. It suffices to show that $y^k \cdot Ax^k \to 0$ when $\{y^k\}$ is an equicontinuous subset of μ'. If this fails we may assume that there exists $\delta > 0$ such that

$$\left| y^k \cdot Ax^k \right| > \delta$$

for every k. Set $k_1 = 1$ and pick m_1, n_1 such that

$$\left| P_{m_1} y^{k_1} \cdot P_{n_1} x^{k_1} \right| > \delta.$$

By (b), $A^T y^k \in s$, the space of all sequences, and since

$$A^T y^k \cdot e^j = y^k \cdot Ae^j = (A^T y^k)_j$$

and $\{y^k\} \subset \mu'$ is $\sigma(\mu', \mu)$ bounded, $\{(A^T y^k)_j : k \in \mathbb{N}\}$ is bounded for every j. Since $\lim_k x_j^k = 0$ for every j, by Lemma 7.7, $\lim_k (A^T y^k)_j x_j^k = 0$ for every j. Therefore, there exists $k_2 > k_1$ such that

$$\left| \sum_{j=1}^{n_1} (A^T y^{k_2})_j x_j^{k_2} \right| = \left| y^{k_2} \cdot A P_{n_1} x^{k_2} \right| < \delta/2.$$

Hence,

$$\left| y^{k_2} \cdot A(x^{k_2} - P_{n_1} x^{k_2}) \right| > \delta/2.$$

Pick $m_2 > m_1, n_2 > n_1$ such that

$$\left| P_{m_2} y^{k_2} \cdot A(P_{n_2} x^{k_2} - P_{n_1} x^{k_2}) \right| > \delta/2.$$

Continuing this construction produces increasing sequences $\{k_p\}, \{m_p\}$ and $\{n_p\}$ such that

$$\text{(✠)} \quad |P_{m_p} y^{k_p} \cdot A\chi_{I_p} x^{k_p}| > \delta/2,$$

where $I_p = \{j \in \mathbb{N} : n_{p-1} < j \leq n_p\}$.

Since $\{P_n\}$ is equicontinuous, $\{P'_n = P_n\}$ will carry equicontinuous subsets of μ' into equicontinuous subsets of μ'. Therefore, $\{P_n y^k : n, k \in \mathbb{N}\} \subset \mu'$ is equicontinuous. Since η is separable, $\{P_{m_p} y^{k_p}\}$ has a subsequence, still denoted by $\{P_{m_p} y^{k_p}\}$, which is $\sigma(\mu', \mu)$ convergent to some $y \in \mu' \subset \mu^\beta$ ([Wi] 9.5.3, [Sw] 18.9).

Define a matrix

$$M = [m_{pq}] = [P_{m_p} y^{k_p} \cdot A\chi_{I_q} x^{k_q}].$$

We claim that M is a $\mathcal{K}$-matrix (Appendix D.2). First, the columns of M converge to $y \cdot A\chi_{I_q} x^{k_q}$. Next, given any subsequence $\{r_q\}$ there is a further subsequence, still denoted by $\{r_q\}$, such that $t = \sum_{q=1}^\infty \chi_{I_{r_q}} x^{k_{r_q}} \in \lambda$. Then

$$\sum_{q=1}^\infty m_{pr_q} = P_{m_p} y^{k_p} \cdot At \to y \cdot At.$$

Hence, M is a $\mathcal{K}$-matrix and by the Antosik-Mikusinski Matrix Theorem (Appendix D.2), the diagonal of M converges to 0. But, this contradicts (✠).

Proposition 2.5 gives sufficient conditions for $\mu' = (\mu, \eta)' = \mu^\beta$ to be satisfied. If μ is a barrelled AB-space (Appendix B.3), then condition (1) is satisfied. Thus, Theorem 10.13 in particular applies to the spaces $l^p, 1 \leq p < \infty$.

Without some assumptions on the domain space λ the continuity conclusions in Theorems 10.10 and 10.13 may fail.

Example 10.14. Let $A = [a_{ij}]$ be the matrix $a_{ij} = 1$ if $i \leq j$ and $a_{ij} = 0$ otherwise. Then $A : c_{00} \to c$ is not norm continuous with respect to the sup-norm and c is separable ($\|A(\sum_{j=1}^n e^j)\|_\infty = n$).

The method of proof of Theorem 10.10 can also be used to establish another boundedness result with respect to the strong topology $\beta(\mu, \mu^\beta)$. In contrast to Theorem 10.10 this result requires an assumption on the range space μ. Actually we are able to establish a Uniform Boundedness result for families of pointwise bounded matrix mappings.

Theorem 10.15. *Let (λ, τ) be a K-space with 0-GHP. Let $A^k = [a_{ij}^k]:\lambda \to \mu$ for every $k \in \mathbb{N}$. Assume that μ satisfies*

(2) *The sectional projections* $P_n : \mu^\beta \to \mu^\beta$ *are uniformly bounded on* $\sigma(\mu^\beta, \mu)$ *bounded subsets with respect to* $\sigma(\mu^\beta, \mu)$ *(i.e., if* $B \subset \mu^\beta$ *is* $\sigma(\mu^\beta, \mu)$ *bounded and* $t \in \mu$, *then* $\{P_n s \cdot t : s \in B, n \in \mathbb{N}\}$ *is bounded). If* $\{A^k\}$ *is pointwise bounded on* λ *with respect to* $\beta(\mu, \mu^\beta)$, *then* $\{A^k\}$ *is uniformly bounded on bounded subsets of* λ *with respect to* $\beta(\mu, \mu^\beta)$.

Proof: If the conclusion fails, we may assume, by passing to a subsequence if necessary, that there exist $x^k \to 0$ in λ, $\{y^k\} \subset \mu^\beta$ which is $\sigma(\mu^\beta, \mu)$ bounded, $t_k \to 0$ and $\delta > 0$ such that

$$(*) \quad |t_k y^k \cdot A^k x^k| > \delta \text{ for all } k.$$

Set $k_1 = 1$. Pick m_1, n_1 such that $\left|\sum_{i=1}^{m_1} t_{k_1} y_i^{k_1} \sum_{j=1}^{n_1} A^{k_1} x_j^{k_1}\right| > \delta$. Now

$$t_k y^k \cdot A^k P_{n_1} x^k = t_k (A^k)^T y^k \cdot P_{n_1} x^k = \sum_{j=1}^{n_1} t_k x_j^k ((A^k)^T y^k)_j.$$

Since λ is a K-space, $\lim_k t_k x_j^k = 0$ for each j, and the pointwise boundedness assumption implies that $\{((A^k)^T y^k)_j : k \in \mathbb{N}\}$ is bounded for each j. Hence, $\lim_k t_k y^k \cdot A^k P_{n_1} x^k = 0$ so there exist $k_2 > k_1$ such that

$$|t_{k_2} y^{k_2} \cdot A^{k_2} P_{n_1} x^{k_2}| < \delta/2.$$

Therefore, from $(*)$,

$$|t_{k_2} y^{k_2} \cdot A^{k_2} (x^{k_2} - P_{n_1} x^{k_2})| > \delta/2.$$

Pick $m_2 > m_1, n_2 > n_1$ such that

$$|t_{k_2} P_{m_2} y^{k_2} \cdot A^{k_2} (P_{n_2} x^{k_2} - P_{n_1} x^{k_2})| > \delta/2.$$

Continuing this construction produces increasing sequences $\{k_p\}, \{m_p\}$ and $\{n_p\}$ such that

$$(\heartsuit) \quad |t_{k_p} P_{m_p} y^{k_p} \cdot A^{k_p} \chi_{I_p} x^{k_p}| > \delta/2,$$

where $I_p = \{j \in \mathbb{N} : n_{p-1} < j \le n_p\}$.

Define a matrix

$$M = [m_{pq}] = [t_{k_p} P_{m_p} y^{k_p} \cdot A^{k_p} \chi_{I_q} x^{k_q}].$$

We claim that M is a $\mathcal{K}$-matrix (Appendix D.2). Since $\{P_n y^k : n, k \in \mathbb{N}\}$ is $\sigma(\mu^\beta, \mu)$ bounded by (2) and $\{A^{k_p} \chi_{I_q} x^{k_q} : p \in \mathbb{N}\}$ is $\beta(\mu, \mu^\beta)$ bounded, the columns of M converge to 0. Since $x^k \to 0$ in λ, by the 0-GHP, given a subsequence $\{r_q\}$ there is a further subsequence, still denoted by $\{r_q\}$, such that $x = \sum_{q=1}^\infty \chi_{I_{r_q}} x^{k_{r_q}} \in \lambda$. Therefore,

$$\sum_{q=1}^\infty m_{pr_q} = t_{k_p} (A^{k_p})^T P_{m_p} y^{k_p} \cdot x = t_{k_p} P_{m_p} y^{k_p} \cdot A^{k_p} x \to 0$$

as above. Hence, M is a $\mathcal{K}$-matrix and by the Antosik-Mikusinski Matrix Theorem the diagonal of M converges to 0 (Appendix D.2). But, this contradicts ($\heartsuit$).

Of course, Theorem 10.15 is applicable to a single matrix A and gives a boundedness result. We now give sufficient conditions for (2) to hold.

Proposition 10.16. *Consider the following conditions:*

(α) *If* (μ, μ^β) *is a Banach-Mackey pair [i.e., if* $\sigma(\mu^\beta, \mu)$ *bounded sets are* $\beta(\mu^\beta, \mu)$ *bounded], then (2) holds.*

(β) *If* $(\mu, \beta(\mu, \mu^\beta))$ *is an AB-space (Appendix B.3), then (2) holds.*

Proof: Let $B \subset \mu^\beta$ be $\sigma(\mu^\beta, \mu)$ bounded and $x \in \mu$.

(α) : By (d), $\{P_n x\}$ is $\sigma(\mu, \mu^\beta)$ bounded so

$$\sup\{|P_n y \cdot x| : y \in B, n \in \mathbb{N}\} = \sup\{|y \cdot P_n x| : y \in B, n \in \mathbb{N}\} < \infty$$

and (2) holds.

(β) : $\{P_n x\}$ is $\beta(\mu, \mu^\beta)$ so (2) holds by the computation in part (α).

We can use the methods of Theorem 10.15 to establish a Banach-Steinhaus type result.

Theorem 10.17. *Let* (λ, τ) *be a K-space with 0-GHP. Let* $A^k = [a_{ij}^k]:\lambda \to \mu$ *for every* $k \in \mathbb{N}$. *Assume that* μ *satisfies*

(2) The sectional projections $P_n : \mu^\beta \to \mu^\beta$ *are uniformly bounded on* $\sigma(\mu^\beta, \mu)$ *bounded subsets with respect to* $\sigma(\mu^\beta, \mu)$ *(i.e., if* $B \subset \mu^\beta$ *is* $\sigma(\mu^\beta, \mu)$ *bounded and* $t \in \mu$, *then* $\{P_n s \cdot t : s \in B, n \in \mathbb{N}\}$ *is bounded). If* $\lim_k A^k x$ *exists with respect to the strong topology* $\beta(\mu, \mu^\beta)$, *then* $\{A^k\}$ *is* $\tau - \beta(\mu, \mu^\beta)$ *sequentially equicontinuous.*

Proof: If the conclusion fails to hold, we may assume, by passing to a subsequence if necessary, that there exist $\delta > 0$, $x^j \to 0$ in λ and $\{y^j\} \subset \mu^\beta$ which is $\sigma(\mu^\beta, \mu)$ bounded such that

$$|y^j \cdot A^j x^j| > \delta.$$

Set $k_1 = 1$ and pick m_1, n_1 such that $\left|P_{m_1} y^{k_1} \cdot A^{k_1} P_{n_1} x^{k_1}\right| > \delta$. Since $\{A^k e^j : k \in \mathbb{N}\}$ is strong bounded for every j,

$$\{(A^k)^T y^k \cdot e^j = y^k \cdot A^k e^j = ((A_k)^T y^k)_j : k \in \mathbb{N}\}$$

is bounded for every j. Since $\lim_k x_j^k = 0$ for every j, by Lemma 7.7,

$$\lim_k ((A_k)^T y^k)_j x_j^k = 0$$

for every j. Therefore, there exist $k_2 > k_1$ such that

$$\left| y^{k_2} \cdot A^{k_2} P_{n_1} x^{k_2} \right| < \delta/2.$$

Therefore,

$$\left| y^{k_2} \cdot A^{k_2} (x^{k_2} - P_{n_1} x^{k_2}) \right| > \delta/2.$$

Pick $m_2 > m_1, n_2 > n_1$ such that

$$\left| P_{m_2} y^{k_2} \cdot A^{k_2} (P_{n_2} x^{k_2} - P_{n_1} x^{k_2}) \right| > \delta/2.$$

Continuing this construction produces increasing sequences $\{k_p\}, \{m_p\}$ and $\{n_p\}$ such that

$$(*) \quad \left| P_{m_p} y^{k_p} \cdot A^{k_p} \chi_{I_p} x^{k_p} \right| > \delta/2,$$

where $I_p = \{ j \in \mathbb{N} : n_{p-1} < j \leq n_p \}$.

Define a matrix

$$M = [m_{pq}] = [P_{m_p} y^{k_p} \cdot A^{k_p} \chi_{I_q} x^{k_q}].$$

We claim that M is a $\mathcal{K}$-matrix (Appendix D.2). First, the columns of M converge since $\{A^k\}$ converges pointwise with respect to the strong topology and (2) implies that $\{P_{m_p} y^{k_p}\}$ is $\sigma(\mu^\beta, \mu)$ bounded. Next, given an increasing sequence $\{r_q\}$ there exists a further subsequence, still denoted by $\{r_q\}$, such that $t = \sum_{q=1}^\infty \chi_{I_{r_q}} x^{k_{r_q}} \in \lambda$. Then

$$\sum_{q=1}^\infty m_{pr_q} = P_{m_p} y^{k_p} \cdot A^{k_p} t$$

and the sequence $\{P_{m_p} y^{k_p} \cdot A^{k_p} t\}$ converges by the same argument that the columns converge. Hence, M is a $\mathcal{K}$-matrix so by the Antosik-Mikusinski Matrix Theorem (Appendix D.2), the diagonal of M converges to 0. But, this contradicts $(*)$.

The original version of the Hellinger-Toeplitz Theorem asserts that if the series $a(x, y) = \sum_{i=1}^\infty \sum_{j=1}^\infty a_{ij} x_j y_i$ converges for every $x, y \in l^2$, then there exists M such that $|a(x, y)| \leq M$ for $\|x\|_2 \leq 1, \|y\|_2 \leq 1$ ([HT]). That is, the bilinear form a induced by the matrix A is a continuous bilinear form $a : l^2 \times l^2 \to \mathbb{R}$. We can use the method of proof in Theorem 10.10 to establish the sequential continuity of bilinear forms between products of sequence spaces which are induced by matrices. If the series $a(x, y) = \sum_{i=1}^\infty \sum_{j=1}^\infty a_{ij} x_j y_i = y \cdot Ax$ converges for every $x \in \lambda, y \in \mu$, where λ and μ are sequence spaces, then a is a bilinear form on $\lambda \times \mu$. We use the method of proof in Theorem 10.10 to establish a sequential continuity result for such bilinear forms.

Theorem 10.18. *Let (λ, τ) be a K-space with 0-GHP and let (μ, η) be a K-space such that $\mu^\beta \subset \mu' = (\mu, \eta)'$. Assume*

(3) the sectional projections $P_n : \mu \to \mu$ are sequentially equicontinuous with respect to η.

If $a(x, y) = \sum_{i=1}^{\infty} \sum_{j=1}^{\infty} a_{ij} x_j y_i = y \cdot Ax$ converges for every $x \in \lambda, y \in \mu$, then $a : \lambda \times \mu \to \mathbb{R}$ is sequentially $\tau \times \eta$ continuous.

Proof: If the conclusion fails, there exist sequences $x^k \to 0$ in τ, $y^k \to 0$ in η and $\delta > 0$ such that

$$\left| a(x^k, y^k) \right| = \left| y^k \cdot Ax^k \right| > \delta$$

for all k. Set $k_1 = 1$ and pick m_1, n_1 such that

$$\left| \sum_{i=1}^{m_1} \sum_{j=1}^{n_1} a_{ij} x_j^{k_1} y_j^{k_1} \right| > \delta.$$

Note that $Ax \in \mu^\beta \subset \mu'$ for each $x \in \lambda$ by the convergence of the series $\sum_{i=1}^{\infty} \sum_{j=1}^{\infty} a_{ij} x_j y_i$ for each $y \in \mu$. For each j the series

$$a(e^j, y^k) = \sum_{i=1}^{\infty} a_{ij} y_i^k = y^k \cdot Ae^j = (A^T y^k)_j$$

converges, and since $\{y^k\}$ is η convergent to 0 and $Ae^j \in \mu'$ by the observation above, for each j, $\{(A^T y^k)_j : k \in \mathbb{N}\}$ is bounded. Since λ is a K-space, $\lim_k x_j^k = 0$ for for each j. Therefore, by Lemma 7.7,

$$\lim_k \sum_{j=1}^{n_1} x_j^k (A^T y^k)_j = \lim_k \sum_{j=1}^{n_1} \sum_{i=1}^{\infty} a_{ij} x_j^k y_i^k = 0$$

so there exists $k_2 > k_1$ such that

$$\left| \sum_{j=1}^{n_1} \sum_{i=1}^{\infty} a_{ij} x_j^{k_2} y_i^{k_2} \right| = \left| y^{k_2} \cdot AP_{n_1} x^{k_2} \right| < \delta/2.$$

Hence,

$$\left| y^{k_2} \cdot A(x^{k_2} - P_{n_1} x^{k_2}) \right| > \delta/2.$$

Pick $m_2 > m_1, n_2 > n_1$ such that

$$\left| \sum_{i=1}^{m_2} \sum_{j=n_1+1}^{n_2} a_{ij} x_j^{k_2} y_j^{k_2} \right| > \delta/2.$$

Continuing this construction produces increasing sequences $\{k_p\}, \{m_p\}$ and $\{n_p\}$ such that

$$\left|P_{m_p}y^{k_p} \cdot A(P_{n_p} - P_{n_{p-1}})x^{k_p}\right| > \delta/2.$$

Set $I_p = \{j \in \mathbb{N} : n_{p-1} < j \le n_p\}$ so

$$(\Diamond) \quad \left|P_{m_p}y^{k_p} \cdot A\chi_{I_p}x^{k_p}\right| > \delta/2.$$

Define a matrix

$$M = [m_{pq}] = [P_{m_p}y^{k_p} \cdot A\chi_{I_q}x^{k_q}].$$

We claim that M is a $\mathcal{K}$-matrix (Appendix D.2). First, since $y^k \to 0$ in η, from (3) $P_k y^k \to 0$ in η and $A\chi_{I_q}x^{k_q} \in \mu'$, the columns of M converge to 0. Next, given any subsequence $\{r_q\}$, there is a further subsequence, still denoted by $\{r_q\}$, such that $t = \sum_{q=1}^{\infty} \chi_{I_{r_q}}x^{k_{r_q}} \in \lambda$ by the 0-GHP assumption. Then

$$\sum_{q=1}^{\infty} m_{pr_q} = P_{m_p}y^{k_p} \cdot At \to 0$$

since $At \in \mu'$ by the observation above. Hence, M is a $\mathcal{K}$-matrix and by the Antosik-Mikusinski Matrix Theorem (Appendix D.2), the diagonal of M converges to 0. But, this contradicts $(\Diamond)$.

Proposition 2.5 gives sufficient conditions for the condition $\mu^{\beta} \subset \mu' = (\mu, \eta)'$ to be satisfied. Note that this condition is satisfied by the topology $\sigma(\mu, \mu^{\beta})$. If μ is a barrelled AB-space, then condition (3) is satisfied. Thus, Theorem 10.18 is applicable to a wide range of sequence spaces including $l^p, 0 < p < \infty$, and includes the original Hellinger-Toeplitz result.

Chapter 11

Operator Valued Series and Vector Valued Multipliers

In this chapter we will consider operator valued series but allow the space of multipliers to be vector valued with values in the domain space of the operators. Throughout this chapter let X, Y be LCTVS and $L(X, Y)$ the space of all continuous linear operators from X into Y. Let E be a vector valued sequence space with values in X which contains $c_{00}(X)$, the space of all X valued sequences which are eventually 0.

Definition 11.1. A series $\sum_j T_j$ in $L(X, Y)$ is E multiplier convergent if the series $\sum_{j=1}^{\infty} T_j x_j$ converges in Y for every $x = \{x_j\} \in E$. The series $\sum_j T_j$ is E multiplier Cauchy if the series $\sum_{j=1}^{\infty} T_j x_j$ is Cauchy for every $x = \{x_j\} \in E$. The elements of E are called multipliers.

If $E = l^{\infty}(X)$, a series $\sum_j T_j$ which is $l^{\infty}(X)$ multiplier convergent is said to be bounded multiplier convergent. If $E = m_0(X)$, then a series $\sum_j T_j$ which is $m_0(X)$ multiplier convergent is subseries convergent in the strong operator topology.

We now consider the basic properties of E multiplier convergent series. Many of these properties are the same as those for scalar valued multiplier convergent series and the proofs are essentially identical. When this phenomena occurs we will make references to the appropriate scalar results. However, due to the presence of continuous linear operators, the results for vector valued multipliers and operator valued series often require additional technical assumptions which we will indicate.

Let $\sum_j T_j$ be E multiplier convergent. The *summing operator* $S : E \to Y$ (with respect to $\sum_j T_j$ and E) is defined by

$$Sx = \sum_{j=1}^{\infty} T_j x_j, x = \{x_j\} \in E.$$

Recall the β-dual of E (with respect to the scalar field) is defined to be

$$E^\beta = \{\{x'_j\} : x'_j \in X', \sum_{j=1}^\infty \langle x'_j, x_j \rangle \text{ converges for every } \{x_j\} \in E\}$$

and E, E^β form a dual pair under the pairing $x' \cdot x = \sum_{j=1}^\infty \langle x'_j, x_j \rangle$ (Appendix C). Also, recall that a locally convex topology w defined for dual pairs is a Hellinger-Toeplitz topology if whenever X, X' and Y, Y' are dual pairs and

$$T : (X, \sigma(X, X')) \to (Y, \sigma(Y, Y'))$$

is a continuous linear operator, then

$$T : (X, w(X, X')) \to (Y, w(Y, Y'))$$

is continuous (Appendix A.1). As in Theorem 2.2, we have

Theorem 11.2. *The summing operator* $S : E \to Y$ *is* $\sigma(E, E^\beta) - \sigma(Y, Y')$ *continuous. Therefore, S is $w(E, E^\beta) - w(Y, Y')$ continuous for any Hellinger-Toeplitz topology w.*

As in Corollaries 2.3 and 2.4, we have

Corollary 11.3. *If B is $\sigma(E, E^\beta)$ bounded, then $SB = \{\sum_{j=1}^\infty T_j x_j : x \in B\}$ is bounded in Y.*

Corollary 11.4. *Let E be a K-space. If $E^\beta \subset E'$ and $B \subset E$ is bounded, then $SB = \{\sum_{j=1}^\infty T_j x_j : x \in B\}$ is bounded in Y.*

For conditions which guarantee that $E^\beta \subset E'$, we have the analogue of Proposition 2.5. For this we need the following property of vector valued sequence spaces. If $z \in X$ and $j \in \mathbb{N}$, recall that $e^j \otimes z$ is the sequence with z in the jth coordinate and 0 in the other coordinates. The space E has the property (I) if the maps $z \to e^j \otimes z$ are continuous from X into E for every j.

Proposition 11.5. *We have the following conditions:*

(i) If E is a barrelled K-space, then $E^\beta \subset E'$.
(ii) If E is an AK-space with property (I), then $E' \subset E^\beta$.
(iii) If E is a barrelled AK-space with property (I), then $E' = E^\beta$.

Proof: (i): Let $y \in E^\beta$. For each n define $F_n : E \to \mathbb{R}$ by $F_n(x) = \sum_{j=1}^{n} \langle y_j, x_j \rangle$. Since E is a K-space, each F_n is continuous and linear. Since $F_n(x) \to y \cdot x$ for every $x \in E$, the map $x \to y \cdot x$ is continuous by the barrelledness assumption. Hence, $y \in E'$.

(ii): Let $F \in E'$. Define $y_j : X \to \mathbb{R}$ by $\langle y_j, x \rangle = \langle F, e^j \otimes x \rangle$. Since E has property (I), $y_j \in X'$. Set $y = \{y_j\}$. If $x \in E$, then

$$\langle F, x \rangle = \left\langle F, \sum_{j=1}^{\infty} e^j \otimes x_j \right\rangle = \sum_{j=1}^{\infty} \langle F, e^j \otimes x_j \rangle = \sum_{j=1}^{\infty} \langle y_j, x_j \rangle.$$

Therefore, $y \in E^\beta$ and $\langle F, x \rangle = y \cdot x$.

(iii) follows from (i) and (ii).

Corollary 11.6. *Assume that E is a barrelled AB-space (Appendix C.3) and $\sum_j T_j$ is E multiplier convergent. If $B \subset E$ is bounded, then $\{\sum_{j=1}^{n} T_j x_j : n \in \mathbb{N}, x \in B\}$ is $\beta(Y, Y')$ bounded.*

Proof: Let $P_n : E \to E$ be the section operator $P_n(x) = \sum_{j=1}^{n} e^j \otimes x_j$. By the AB assumption $\{P_n : n\}$ is pointwise bounded on E and, therefore, equicontinuous since E is barrelled. Since E is barrelled, E has the strong topology $\beta(E, E')$ so $\{P_n x : n \in \mathbb{N}, x \in B\}$ is $\beta(E, E')$ bounded. By Proposition 11.5 , $E^\beta \subset E'$ so $\{P_n x : n \in \mathbb{N}, x \in B\}$ is $\beta(E, E^\beta)$ bounded. The result now follows from Theorem 11.2 since the strong topology is a Hellinger-Toeplitz topology.

Recall the condition that $E^\beta \subset E'$ is important even in the scalar case (Example 2.9). From Corollary C.7 in Appendix C if X is a Frechet space, then $l^\infty(X)$ is a Frechet space with property AB so we have

Corollary 11.7. *Let $\sum_j T_j$ be bounded multiplier convergent. If X is a Frechet space and $B \subset l^\infty(X)$ is bounded, then $\{\sum_{j=1}^{\infty} T_j x_j : x \in B\}$ is $\beta(Y, Y')$ bounded.*

We have the analogue of Theorem 2.11. Recall that E has 0-GHP if whenever $x^j \to 0$ in E and $\{I_j\}$ is an increasing sequence of intervals, there is a subsequence $\{n_j\}$ such that the coordinatewise sum of the series $\sum_{j=1}^{\infty} \chi_{I_{n_j}} x^{n_j} \in E$ (Appendix C).

Theorem 11.8. *Let E be a K-space with 0-GHP. If $\sum_j T_j$ is E multiplier convergent, then the summing operator $S : E \to Y$ is sequentially continuous and, therefore, bounded.*

We next consider the uniform convergence of operator valued series over bounded subsets of the multiplier space E. Recall that E has the signed-SGHP if whenever $\{x^j\}$ is a bounded sequence in E and $\{I_j\}$ is an increasing sequence of intervals, there exist a sequence of signs $\{s_j\}$ and a subsequence $\{n_j\}$ such that the coordinatewise sum of the series $\sum_{j=1}^{\infty} s_j \chi_{I_{n_j}} x^{n_j} \in E$ (Appendix C). We have the analogue of Theorem 2.16.

Theorem 11.9. *Let E have signed-SGHP. If $\sum_j T_j$ is E multiplier convergent, then the series $\sum_{j=1}^{\infty} T_j x_j$ converge uniformly for $x = \{x_j\}$ belonging to bounded subsets of E.*

Without the signed-SGHP assumption, the conclusion of Theorem 11.9 may fail.

Example 11.10. Let $1 \leq p < \infty$ and define $Q_k : l^p \to l^p$ by $Q_k t = t_k e^k$. For $t \in l^p$ and $\sigma \subset \mathbb{N}$, $\sum_{k \in \sigma} Q_k t = \sum_{k \in \sigma} t_k e^k$ converges in l^p so $\sum_k Q_k$ is $m_0(l^p)$ multiplier convergent since $m_0(l^p) = span\{\chi_\sigma t : \sigma \subset \mathbb{N}, t \in l^p\}$. However, the series $\sum_k Q_k t_k$ do not converge uniformly for $t = \{t_k\}$ belonging to bounded subsets of $m_0(l^p)$. For let t^k be the constant sequence in $m_0(l^p)$ with e^k in each coordinate. Then $\sum_{j=n}^{\infty} Q_j t_j^k = \sum_{j=n}^{\infty} Q_j e^k = e^k$ if $k \geq n$.

As a corollary of Theorem 11.9 we have an important property of bounded multiplier convergent series which was established for Banach spaces by Batt ([Bt]).

Corollary 11.11. *Let $\sum_j T_j$ be bounded multiplier convergent. Then the series $\sum_{j=1}^{\infty} T_j x_j$ converge uniformly for $x = \{x_j\}$ belonging to bounded subsets of $l^\infty(X)$.*

Remark 11.12. We can give another interesting proof of Corollary 11.11 above by employing the lemma of Li (Lemma 3.29). Suppose that $B \subset l^\infty(X)$ is bounded and let p be a continuous semi-norm on X. There exists $M > 0$ such that $\sup\{p(x_j) : x = \{x_j\} \in B\} \leq M$. Put $E_j = \{x \in X : p(x) \leq M\}$ and define $f_j : E_j \to X$ by $f_j(x) = T_j x$. Then $\sum_{j=1}^{\infty} f_j(x_j) = \sum_{j=1}^{\infty} T_j x_j$ converges for every $x = \{x_j\} \in \Pi_{j=1}^{\infty} E_j$. By Lemma 3.29, the series $\sum_{j=1}^{\infty} T_j x_j$ converge uniformly for $x = \{x_j\} \in \Pi_{j=1}^{\infty} E_j$ so the series converge uniformly for $\{x_j\} \in B$.

The analogue of Theorem 2.22 also holds.

Theorem 11.13. *Let E be a K-space with 0-GHP. If $\sum_j T_j$ is E multiplier convergent and $x^k \to 0$ in E, then the series $\sum_{j=1}^{\infty} T_j x_j^k$ converge uniformly for $k \in \mathbb{N}$.*

We next consider uniform convergence results for families of E multiplier convergent series. The β-dual of E with respect to Y is

$$E^{\beta Y} = \{\{T_j\} : T_j \in L(X, Y), \sum_{j=1}^{\infty} T_j x_j \text{ converges for every } \{x_j\} \in E\}.$$

The topology $w(E^{\beta Y}, E)$ on $E^{\beta Y}$ is defined to be the weakest topology such that the mappings $T = \{T_j\} \to T \cdot x = \sum_{j=1}^{\infty} T_j x_j$ from $E^{\beta Y}$ into Y are continuous for every $x \in E$. If X is the scalar field, these notations agree with those employed in Chapter 2. The analogue of Theorem 2.26 holds. Recall E has signed-WGHP if for every $x \in E$ and every increasing sequence $\{I_j\}$, there exist a sequence of signs $\{s_j\}$ and a subsequence $\{n_j\}$ such that the coordinatewise sum of the series $\sum_{j=1}^{\infty} s_j \chi_{I_{n_j}} x \in E$ (Appendix C).

Theorem 11.14. *Assume that E has signed-WGHP. If $\{T^k\} \subset E^{\beta Y}$ is such that $\lim_k T^k \cdot x$ exists for every $x \in E$, then for every $x \in E$ the series $\sum_{j=1}^{\infty} T_j^k x_j$ converge uniformly for $k \in \mathbb{N}$.*

We next consider the analogue of Stuart's weak completeness theorem (Corollary 2.28). This is the point where the significant differences between the scalar and vector cases appear.

Definition 11.15. *The pair (X, Y) has the Banach-Steinhaus property if whenever $\{T_j\} \subset L(X, Y)$ is pointwise convergent, $\lim T_j x = Tx$ exists for every $x \in X$, then $T \in L(X, Y)$.*

For example, if X is barrelled, then (X, Y) has the Banach-Steinhaus property for every LCTVS Y ([Sw2] 24.12, [Wi] 9.3.7).

Lemma 11.16. *Let $\{T^k\} \subset E^{\beta Y}$ be such that there for every j, there exists $T_j \in L(X, Y)$ with $\lim_k T_j^k x = T_j x$ for every $x \in X$. If for every $x \in E$, $\lim_k T^k \cdot x$ exists and the series $\sum_{j=1}^{\infty} T_j^k x_j$ converge uniformly for $k \in \mathbb{N}$, then $T = \{T_j\} \in E^{\beta Y}$ and $T^k \to T$ in $w(E^{\beta Y}, E)$.*

Proof: Let $x \in E$ and set $u = \lim_k T^k \cdot x$. It suffices to show that $u = \sum_{j=1}^{\infty} T_j x_j$. Let U be a balanced neighborhood of 0 in Y and pick a balanced neighborhood of 0, V, such that $V + V + V \subset U$. There exists p

such that $\sum_{j=n}^{\infty} T_j^k x_j \in V$ for $n \geq p, k \in \mathbb{N}$. Fix $n \geq p$. There exists $k = k_n$ such that $\sum_{j=1}^{\infty} T_j^k x_j - u \in V$ and $\sum_{j=1}^{n} (T_j^k - T_j)x_j \in V$. Then

$$\sum_{j=1}^{n} T_j x_j - u = \left(\sum_{j=1}^{\infty} T_j^k x_j - u \right) - \sum_{j=1}^{n} (T_j^k - T_j)x_j - \sum_{j=n+1}^{\infty} T_j^k x_j \in V + V + V \subset U$$

and the result follows.

Lemma 11.17. *Let $\{T^k\} \subset E^{\beta Y}$ and let (X, Y) have the Banach-Steinhaus property. If for every $x \in E$, $\lim_k T^k \cdot x$ exists and the series $\sum_{j=1}^{\infty} T_j^k x_j$ converge uniformly for $k \in \mathbb{N}$, then there exists $T \in E^{\beta Y}$ such that $T^k \to T$ in $w(E^{\beta Y}, E)$.*

Proof: For each j define a linear map $T_j : X \to Y$ by

$$T_j z = \lim_k T^k \cdot (e^j \otimes z) = \lim_k T_j^k z.$$

By the Banach-Steinhaus assumption, $T_j \in L(X, Y)$. The result now follows from Lemma 11.16.

From Theorem 11.14 and Lemma 11.17, we can now obtain Stuart's completeness result for vector valued sequence spaces.

Corollary 11.18. *(Stuart) Let E have signed-WGHP, Y be sequentially complete and (X, Y) have the Banach-Steinhaus property. If $\{T^k\}$ is Cauchy in $w(E^{\beta Y}, E)$, then there exists $T \in E^{\beta Y}$ such that $T^k \to T$ in $w(E^{\beta Y}, E)$. That is, $w(E^{\beta Y}, E)$ is sequentially complete.*

Proof: For each $z \in X$ and j, the sequence $\{T_j^k z\}_k = \{T^k \cdot (e^j \otimes z)\}_k$ is Cauchy in Y. By the sequential completeness assumption, $\lim_k T_j^k z = T_j z$ exists. The result now follows from Theorem 11.14 and Lemma 11.17.

The assumption that the pair (X, Y) has the Banach-Steinhaus property is necessary for the conclusion of Corollary 11.18 to hold. For suppose that $T_k \in L(X, Y)$ and $\lim T_k x = T x$ exists for every $x \in X$. Define $T^k \in E^{\beta Y}$ by $T^k = (T_k, 0, 0, ...)$. Then $\lim T^k \cdot x = \lim T_k x_1$ exists for every $\{x_j\} \in E$ so $\{T^k\}$ is $w(E^{\beta Y}, E)$ Cauchy. If $w(E^{\beta Y}, E)$ is sequentially complete and $T^0 = w(E^{\beta Y}, E) - \lim T^k$, then $T_1^0 = T \in L(X, Y)$ so (X, Y) has the Banach-Steinhaus property.

We state another corollary of Theorem 11.14. A subset F of $E^{\beta Y}$ is conditionally $w(E^{\beta Y}, E)$ sequentially compact if every sequence $\{T^k\} \subset F$ has a subsequence $\{T^{n_k}\}$ which is such that $\lim T^{n_k} \cdot x$ exists for every $x \in E$. From Theorem 11.14, we have

Corollary 11.19. *Let E have signed-WGHP and (X, Y) have the Banach-Steinhaus property. If $F \subset E^{\beta Y}$ is conditionally $w(E^{\beta Y}, E)$ sequentially compact and $x \in E$, then the series $\sum_{j=1}^{\infty} T_j x_j$ converge uniformly for $T \in F$.*

We next consider a uniform convergence result for the strong topology of Y. This requires the ∞-GHP assumption. The space E has the ∞-GHP if whenever $x \in E$ and $\{I_j\}$ is an increasing sequence of intervals there exist a subsequence $\{n_j\}$ and $a_{n_j} > 0, a_{n_j} \to \infty$ such that every subsequence of $\{n_j\}$ has a further subsequence $\{p_j\}$ such that the coordinatewise sum of the series $\sum_{j=1}^{\infty} a_{p_j} \chi_{I_{p_j}} x \in E$ (Appendix C).

Theorem 11.20. *Assume that E has ∞-GHP. If $F \subset E^{\beta Y}$ is pointwise bounded on E with respect to $\beta(Y, Y')$, then for every $x \in E$ the series $\sum_{j=1}^{\infty} T_j x_j$ converge uniformly in $\beta(Y, Y')$ for $T \in F$.*

Proof: If the conclusion fails, there exist $\epsilon > 0, \{T^k\} \subset F, \{y_k'\} \subset Y'$ $\sigma(Y', Y)$ bounded and an increasing sequence of intervals $\{I_k\}$ such that

$$(*) \quad \left| \left\langle y_k', \sum_{l \in I_k} T_l^k x_l \right\rangle \right| > \epsilon \text{ for all } k.$$

By ∞-GHP, there exist $\{n_k\}, a_{p_k} > 0, a_{p_k} \to 0$ such that every subsequence of $\{n_k\}$ has a further subsequence $\{p_k\}$ such that the coordinatewise sum of the series $\sum_{j=1}^{\infty} a_{p_j} \chi_{I_{p_j}} x \in E$. Define a matrix

$$M = [m_{ij}] = \left[\left\langle y_{n_i}'/a_{n_i}, \sum_{l \in I_j} T_l^{n_i}(a_{p_j} x_l) \right\rangle \right].$$

We claim that M is a $\mathcal{K}$-matrix (Appendix D.2). First, since F is pointwise $\beta(Y', Y)$ bounded on E, $\{y_i'\}$ is $\sigma(Y', Y)$ bounded and $1/a_{p_i} \to 0$, the columns of M converge to 0. Next, given any subsequence there is a further subsequence $\{p_j\}$ such that $u = \sum_{j=1}^{\infty} a_{p_j} \chi_{I_{p_j}} x \in E$. Therefore,

$$\sum_{j=1}^{\infty} m_{ip_j} = \left\langle y_{n_i}'/a_{n_i}, \sum_{j=1}^{\infty} T^{p_i} u_j \right\rangle \to 0$$

by the same argument as above. Hence, M is a $\mathcal{K}$-matrix and the diagonal of M converges to 0 by the Antosik-Mikusinski Matrix Theorem (Appendix D.2). But, this contradicts $(*)$.

From Theorem 11.20 and Lemma 11.17, we have the following weak sequential completeness result.

Corollary 11.21. *Let E have ∞-GHP, Y be a sequentially complete, barrelled space and (X,Y) have the Banach-Steinhaus property. Then $w(E^{\beta Y}, E)$ is sequentially complete.*

Proof: Since Y is barrelled the original topology of Y is just $\beta(Y, Y')$. Thus, the result follows from Theorem 11.20 and Lemma 11.17 since any $w(E^{\beta Y}, E)$ Cauchy sequence $\{T^k\}$ is pointwise bounded on E and $\lim T^k \cdot x$ exists for every $x \in E$ by the sequential completeness of Y.

As noted in Remark 2.34, the ∞-GHP and signed-WGHP are independent so the results in Corollaries 11.18 and 11.21 cover different spaces.

We have the vector analogues of Theorems 2.35 and 2.39.

Theorem 11.22. *Assume that E has signed-SGHP. If $\{T^k\} \subset E^{\beta Y}$ is such that $\lim T^k \cdot x$ exists for every $x \in E$ and $B \subset E$ is bounded, then the series $\sum_{j=1}^{\infty} T_j^k x_j$ converge uniformly for $k \in \mathbb{N}, x \in B$.*

Theorem 11.23. *Assume that E has 0-GHP. If $\{T^k\} \subset E^{\beta Y}$ is such that $\lim T^k \cdot x$ exists for every $x \in E$ and $x^k \to 0$ in E, then the series $\sum_{j=1}^{\infty} T_j^k x_j^l$ converge uniformly for $k, l \in \mathbb{N}$.*

Examples 2.30 and 2.40 show the gliding hump assumptions above are important.

We next discuss the relationships between absolute convergence, bounded multiplier convergence and subseries convergence for operator valued series. For the topologies on the space $L(X,Y)$ which are employed, see Appendix C.

Theorem 11.24. *Let Y be sequentially complete. If the series $\sum_j T_j$ is absolutely convergent in $L_b(X,Y)$, then the series is bounded multiplier convergent.*

Proof: Let $\{x_j\} \subset X$ be bounded and set $A = \{x_j : j \in \mathbb{N}\}$. Let p be a continuous semi-norm on Y. If $n > m$, then

$$p(\sum_{j=m}^{n} T_j x_j) \leq \sum_{j=m}^{n} p(T_j x_j) \leq \sum_{j=m}^{n} p_A(T_j)$$

so the result follows from the completeness of Y.

The converse of the result above does not hold.

Example 11.25. Define $T_j : \mathbb{R} \to c_0$ by $T_j t = (t/j)e^j$. Then T_j is continuous, linear with $\|T_j\| = 1/j$. If $\{t_j\}$ is bounded in $\mathbb{R}$, then the series

$\sum_{j=1}^{\infty} t_j T_j = \sum_{j=1}^{\infty} (t_j/j) e^j$ converges in c_0 so the series $\sum_j T_j$ is bounded multiplier convergent but not absolutely convergent.

Theorem 11.26. *If the series $\sum_j T_j$ is bounded multiplier convergent, then the series is subseries convergent in $L_b(X, Y)$.*

Proof: We first claim that the series $\sum_j T_j$ is convergent in $L_b(X, Y)$. If this fails to hold, there exist a continuous semi-norm p on Y, a bounded subset $A \subset X$, an increasing sequence of intervals $\{I_j\}$ and $\epsilon > 0$ such that $p_A(\sum_{i \in I_j} T_i) > \epsilon$ for all j. Pick $x_j \in A$ such that $p((\sum_{i \in I_j} T_i) x_j) > \epsilon$ for all j. Define $z \in l^{\infty}(X)$ by $z_i = x_j$ if $i \in I_j$ and $z_i = 0$ otherwise. Then the series $\sum_{j=1}^{\infty} T_j z_j$ is not convergent since the series fails the Cauchy criterion.

Since the same argument can be applied to any subseries of the series $\sum_j T_j$, the result follows.

The converse of the result in Theorem 11.26 fails to hold.

Example 11.27. Let $X = Y = l^1$ and define $T_j : l^1 \to l^1$ by $T_j t = (t_j/j) e^j$. If $\sigma \subset \mathbb{N}$, then $\left\| \sum_{j \in \sigma} T_j \right\| \leq \sup_{j \in \sigma} |1/j|$ so the series $\sum_j T_j$ is subseries convergent in the uniform operator topology. However, the series $\sum_j T_j$ is not bounded multiplier convergent since the series $\sum_{j=1}^{\infty} T_j e^j = \sum_{j=1}^{\infty} (1/j) e^j$ is not convergent in l^1.

Obviously, a series which is subseries convergent in $L_b(X, Y)$ is subseries convergent in the strong operator topology, $L_s(X, Y)$, but the converse is false.

Example 11.28. Let $X = Y = l^1$ and define $T_j : l^1 \to l^1$ by $T_j t = t_j e^j$ so T_j is continuous, linear with $\|T_j\| = 1$. If $\sigma \subset \mathbb{N}$ and $t \in l^1$, then

$$\left\| \sum_{j \in \sigma} T_j t_j \right\| = \sum_{j \in \sigma} |t_j|$$

so the series $\sum_j T_j$ is subseries convergent in the strong operator topology. However, since $\|T_j\| = 1$ the series is not subseries convergent in the uniform operator topology.

We establish partial converses to the statements above.

Proposition 11.29. *Let X be a normed space and Y a Banach space. Every $l^{\infty}(X)$ multiplier convergent series is absolutely convergent iff Y is finite dimensional.*

Proof: Suppose that Y is infinite dimensional. By the Dvoretsky-Rogers Theorem ([Sw2] 30.1.1, [Day]), there is a subseries convergent series $\sum_j y_j$ in Y which is not absolutely convergent. Pick $x' \in X', x' \neq 0$. Define $T_j \in L(X,Y)$ by $T_j x = \langle x', x \rangle y_j$. Then $\sum_j T_j$ is $l^\infty(X)$ multiplier convergent since the series $\sum_j y_j$ is also bounded multiplier convergent (2.54). However, $\sum_j T_j$ is not absolutely convergent since $\|T_j\| = \|x'\| \|y_j\|$.

Suppose $Y = \mathbb{R}^n$ and $\sum_j T_j$ is $l^\infty(X)$ multiplier convergent in $L(X,Y)$. Since $l^\infty(X)$ is monotone, the series $\sum_{j=1}^\infty T_j x_j$ is subseries convergent in Y for every $\{x_j\} \in l^\infty(X)$. Therefore, $\sum_{j=1}^\infty \|T_j x_j\| < \infty$. For each j, pick $x_j \in X, \|x_j\| \leq 1$, such that $\|T_j\| \leq \|T_j x_j\| + 1/2^j$. Then $\sum_{j=1}^\infty \|T_j\| < \infty$.

Proposition 11.30. *If $X = \mathbb{R}^n$ and Y is a Banach space, then every series $\sum_j T_j$ which is subseries convergent in the strong operator topology of $L(X,Y)$ is $l^\infty(X)$ multiplier convergent. In particular, the series $\sum_j T_j$ is subseries convergent in $L_b(X,Y)$.*

Proof: Let $\{e^i\}_{i=1}^n$ be the canonical base in $\mathbb{R}^n$. For $i = 1,...,n$ and $x \in l^\infty(X)$, the series $\sum_{j=1}^\infty (e^i \cdot x_j) T_j e^i$ is bounded multiplier convergent since Y is complete (2.54). Therefore, the series

$$\sum_{i=1}^n \sum_{j=1}^\infty (e^i \cdot x_j) T_j e^i = \sum_{j=1}^\infty \sum_{i=1}^n (e^i \cdot x_j) T_j e^i = \sum_{j=1}^\infty T_j x_j$$

converges in Y.

Proposition 11.31. *Suppose X is a normed space and Y is a Banach space. Then every series in $L(X,Y)$ which is subseries convergent in the uniform operator topology is $l^\infty(X)$ multiplier convergent iff X is finite dimensional.*

Proof: Suppose X is infinite dimensional. Since X' is infinite dimensional, there exists a series $\sum_j x_j'$ in X' which is subseries convergent in X' but

$$\sum_{j=1}^\infty \|x_j'\| = \infty$$

(Dvoretsky-Rogers Theorem ([Day], [Sw2] 30.1.1)). Pick $y \in Y, y \neq 0$. Define $T_j : X \to Y$ by $T_j x = \langle x_j', x \rangle y$. Then $\sum_j T_j$ is subseries convergent in the uniform operator topology but not $l^\infty(X)$ multiplier convergent (pick $x_j \in X, \|x_j\| \leq 1$, such that $\|x_j'\| \leq \langle x_j', x_j \rangle + 1/2^j$).

The converse follows from Theorem 11.30.

We next establish several results relating vector and scalar multiplier convergent series. For the definition of the spaces $\nu\{X\}$, see Appendix C.

Proposition 11.32. *If the series $\sum_j T_j$ is $\nu\{X\}$ multiplier convergent, then the series is ν multiplier convergent in the strong operator topology.*

Proof: Let $t \in \nu$ and $x \in X$ so $tx \in \nu\{X\}$ and $\sum_{j=1}^{\infty} T_j(t_j x) = \sum_{j=1}^{\infty} t_j T_j x$ converges.

The converse of the result above fails to hold.

Example 11.33. Define $T_j \in L(l^2, l^2)$ by $T_j t = t_j e^j$. The series $\sum_j T_j$ is bounded multiplier convergent in the strong operator topology since if $s \in l^{\infty}$ and $t \in l^2$, the series $\sum_{j=1}^{\infty} s_j T_j t = \sum_{j=1}^{\infty} s_j t_j e^j$ converges in l^2. However, $\{e^j\} \in l^{\infty}(l^2)$ and $\sum_{j=1}^{\infty} T_j e^j = \sum_{j=1}^{\infty} e^j$ does not converge in l^2 so the series $\sum_j T_j$ is not $l^{\infty}(l^2)$ multiplier convergent.

We do have a partial converse to Proposition 11.32.

Proposition 11.34. *Let $X = \mathbb{R}^n$ be finite dimensional. If the series $\sum_j T_j$ is ν multiplier convergent in the strong operator topology of $L(X,Y)$, then the series is $\nu\{X\}$ multiplier convergent.*

Proof: Let $\{e^i\}_{i=1}^n$ be the canonical basis for $\mathbb{R}^n$. If $x \in \nu\{X\}$, then $\{\|x_j\|\} \in \nu$ and since ν is normal and $\|x_j\| \geq |e^i \cdot x_j|$ for every i, $\{e^i \cdot x_j\}_j \in \nu$. Therefore,

$$\lim_m \sum_{j=1}^m T_j x_j = \lim_m \sum_{j=1}^m T_j (\sum_{i=1}^n (e^i \cdot x_j) e^i)$$

$$= \lim_m \sum_{i=1}^n \sum_{j=1}^m (e^i \cdot x_j) T_j e^i = \sum_{i=1}^n \sum_{j=1}^{\infty} (e^i \cdot x_j) T_j e^i$$

so the series $\sum_{j=1}^{\infty} T_j x_j$ converges.

The results in Propositions 11.32 and 11.34 hold if $\nu = c_0$ or $\nu = l^p$ ($0 < p \leq \infty$).

We now consider some applications to operator valued set functions. For this we assume that X and Y are normed spaces. Let $\mathcal{A}$ be an algebra of subsets of a set S and $\mu : \mathcal{A} \to L(X,Y)$ be finitely additive. The (operator) semi-variation of μ is defined to be

$$\hat{\mu}(A) = \sup\left\{ \left\| \sum_{i=1}^{n} \mu(A_i) x_i \right\| : \{A_i\} \subset \mathcal{A} \text{ a partition of } A, \|x_i\| \le 1 \right\}.$$

It is clear that the semi-variation is subadditive and monotone. The operator semi-variation is employed when integrating X valued measurable functions with respect to operator valued measures with values in $L(X, Y)$ ([Bar]). Since $\hat{\mu}(A) \ge \|\mu\|(A)$ for any $A \in \mathcal{A}$ (as computed with respect to the norm in $L(X, Y)$), if $\hat{\mu}(S) < \infty$, then μ has bounded semi-variation and is, therefore, bounded (Proposition 3.45). However, μ can be bounded and have infinite operator semi-variation.

Example 11.35. Define $\mu : 2^{\mathbb{N}} \to L(l^2, \mathbb{R}) = l^2$ by $\mu(A) = \sum_{j \in A} T_j$ where $T_j = (1/j)e^j \in L(l^2, \mathbb{R}) = l^2$. Then μ is countably additive and bounded. However, $T_j e^j = (1/j)e^j \cdot e^j = 1/j$ so $\left| \sum_{j=1}^{n} T_j e^j \right| = \sum_{j=1}^{n} 1/j$ and $\hat{\mu}(\{1, ...n\}) = \sum_{j=1}^{n} 1/j \to \infty$ so the operator semi-variation of μ is infinite.

We have a multiplier convergent characterization of operator valued set functions with finite semi-variation.

Theorem 11.36. *The following conditions are equivalent:*

(i) μ has finite (operator) semi-variation.

(ii) For every pairwise disjoint sequence $\{A_j\} \subset \mathcal{A}$, the series $\sum_j \mu(A_j)$ is $c_0(X)$ multiplier Cauchy.

Proof: Assume (i). Let $\{A_j\} \subset \mathcal{A}$ be pairwise disjoint and let $x \in c_0(X)$. Put $z_j = x_j / \|x_j\|$. For any finite $\sigma \subset \mathbb{N}$, $\left\| \sum_{j \in \sigma} \mu(A_j) z_j \right\| \le \hat{\mu}(S)$. Hence, $\sum_j \mu(A_j) z_j$ is c_0 multiplier Cauchy (Proposition 3.8). Therefore,

$$\sum_j \|x_j\| \mu(A_j) z_j = \sum_j \mu(A_j) x_j$$

is Cauchy so (ii) holds.

Assume (ii). Suppose (i) fails so $\hat{\mu}(S) = \infty$. First, note that μ is bounded. For, if $t_j \to 0$ and $\{x_j\} \subset X$ is bounded, then $\{t_j x_j\} \in c_0(X)$ so $\sum_j \mu(A_j) t_j x_j$ is Cauchy and $\mu(A_j) t_j x_j \to 0$. Since $\{t_j\} \in c_0$ is arbitrary, $\{\mu(A_j) x_j\}$ is bounded for any bounded sequence $\{x_j\}$. Hence, $\{\|\mu(A_j)\|\}$ is bounded, and μ is bounded by Theorem 3.28. Put $M = \sup\{\|\mu(A)\| : A \in \mathcal{A}\}$. There exist a partition $\{A_1^1, ..., A_n^1, A_{n+1}^1\}$ and $x_1^1, ..., x_{n+1}^1, \|x_j^1\| \le 1$, with $\left\| \sum_{j=1}^{n+1} \mu(A_j^1) x_j^1 \right\| > 1 + M$, where some $\{A_j^1\}$, say A_{n+1}^1, satisfies $\hat{\mu}(A_{n+1}^1) = \infty$. Then

$$\left\| \sum_{j=1}^{n} \mu(A_j^1) x_j^1 \right\| \geq 1 + M - \left\| \mu(A_{n+1}^1) x_{n+1}^1 \right\| \geq 1.$$

Now treat A_{n+1}^1 as S above to obtain a partition $\{A_1^2, ..., A_m^2, A_{m+1}^2\}$ of A_{n+1}^1 and $x_1^2, ..., x_{m+1}^2$, $\left\| x_j^2 \right\| \leq 1$, with $\left\| \sum_{j=1}^{m} \mu(A_j^2) x_j^2 \right\| \geq 2$ and $\hat{\mu}(A_{m+1}^2) = \infty$. Continuing this construction produces a pairwise disjoint sequence

$$\{B_j\} = \{A_1^1, ..., A_n^1, A_1^2, ..., A_m^2, ...\}$$

and a null sequence

$$\{z_j\} = \{x_1^1, ..., x_n^1, x_1^2/2, ..., x_m^2/2, ...\}$$

such that the series $\sum_j \mu(B_j)$ corresponding to the $\{B_j\}$ is not $c_0(X)$ multiplier Cauchy.

For sequentially complete spaces, we can use Li's Lemma 3.29 to give an improvement to Theorem 11.36.

Corollary 11.37. *Let Y be sequentially complete. The following are equivalent:*

(i) *μ has finite (operator) semi-variation.*
(ii) *For every pairwise disjoint sequence $\{A_j\} \subset \mathcal{A}$, the series $\sum_j \mu(A_j)$ is $c_0(X)$ multiplier convergent.*
(iii) *For every pairwise disjoint sequence $\{A_j\} \subset \mathcal{A}$ and $x = \{x_j\} \in c_0(X)$, the series $\sum_j \mu(B_j) z_j$ converge uniformly for $B_j \in \mathcal{A}$, $B_j \subset A_j$, $\|z_j\| \leq \|x_j\|$.*

Proof: (i) and (ii) are equivalent by Theorem 11.36. Clearly, (iii) implies (ii).

Assume (ii). We use Li's Lemma 3.29. For this, set

$$E_j = \{(B, z) : B \in \mathcal{A}, B \subset A_j, z \in X, \|z\| \leq \|x_j\|\}$$

and define $f_j : E_j \to Y$ by $f_j(B, z) = \mu(B) z$. The condition (iii) now follows from Li's Lemma.

Another important property of operator valued set functions is that the semi-variation is continuous from above in the sense that if $\{A_j\} \subset \mathcal{A}$ and $A_j \downarrow \varnothing$, then $\hat{\mu}(A_j) \to 0$. For example, this property was utilized by Bartle in developing properties of bilinear vector integrals ([Bar]).

Note that if $\hat{\mu}$ is continuous from above, then μ is countably additive ($\|\mu(E)\| \leq \hat{\mu}(E)$ for $E \in \Sigma$). However, μ may be countably additive and $\hat{\mu}$ may fail to be continuous from above.

Example 11.38. Let X be a Banach space and $\sum_j T_j$ a series in $L(X)$ which is subseries convergent in $L(X)$ with respect to the operator norm but not bounded multiplier convergent (Example 11.27). Define $\mu : 2^{\mathbb{N}} \to L(X)$ by $\mu(E) = \sum_{j \in E} T_j$. Then μ is countably additive with respect to the operator norm. Since the series $\sum_j T_j$ is not bounded multiplier convergent, there exist $\epsilon > 0$, an increasing sequence of intervals $\{I_j\}$ and $x_j \in X, \|x_j\| \leq 1$ such that $\left\|\sum_{k \in I_j} T_k x_k\right\| > \epsilon$. Put $E_j = \cup_{k \geq j} I_k$. Then $E_j \downarrow \emptyset$ and $\hat{\mu}(E_j) \geq \hat{\mu}(I_j) \geq \epsilon$ so $\hat{\mu}$ is not continuous from above.

We give a series characterization for countably additive measures whose operator semi-variation is continuous from above.

Theorem 11.39. *Let Σ be a σ-algebra of subsets of a set S and $\mu : \Sigma \to L(X, Y)$ be countably additive in $L_s(X, Y)$. The following are equivalent:*

(i) *$\hat{\mu}$ is continuous from above.*
(ii) *For every pairwise disjoint sequence $\{A_j\} \subset \Sigma$, the series $\sum_j \mu(A_j)$ is $l^\infty(X)$ multiplier Cauchy.*

Proof: Assume (i). Let $\{A_j\} \subset \Sigma$ be pairwise disjoint and $\|x_j\| \leq 1$. Then

$$\left\|\sum_{j=n}^{n+p} \mu(A_j)x_j\right\| \leq \hat{\mu}(\cup_{j=n}^{n+p}A_j) \leq \hat{\mu}(\cup_{j=n}^{\infty}A_j) \to 0$$

since $\cup_{j=n}^{\infty}A_j \downarrow \varnothing$. Therefore, $\sum_j \mu(A_j)x_j$ is Cauchy and (ii) holds.

Assume (ii). If (i) fails, there exist $\epsilon > 0, A_j \downarrow \varnothing, A_j \in \Sigma$, such that $\hat{\mu}(A_j) > \epsilon$. There exist $n_1, \{E_j : j = 1, ..., n_1\} \subset \Sigma$ pairwise disjoint and $\{x_j : j = 1, ..., n_1\}, \|x_j\| \leq 1$, such that $\left\|\sum_{j=1}^{n_1} \mu(A_1 \cap E_j)x_j\right\| > \epsilon$. Since μ is countably additive in $L_s(X, Y)$, $\lim_i \mu(A_i \cap E_j)x_j = 0$ for $j = 1, ..., n_1$.

Put $k_1 = 1$. There exists $k_2 > k_1$ such that

$$\left\| \sum_{j=1}^{n_1} (\mu(A_{k_1} \cap E_j)x_j - \mu(A_{k_2} \cap E_j)x_j) \right\| = \left\| \sum_{j=1}^{n_1} \mu((A_{k_1} \backslash A_{k_2}) \cap E_j)x_j \right\| > \epsilon.$$

Treat A_{k_2} as A_1 above to obtain a partition $\{E_j : j = n_1+1, ..., n_2\}$ of A_{k_2}, $\|x_j\| \leq 1$ for $j = n_1 + 1, ..., n_2$, such that $\left\| \sum_{j=n_1+1}^{n_2} \mu(A_{k_2} \cap E_j)x_j \right\| > \epsilon$ and $k_3 > k_2$ such that $\left\| \sum_{j=n_1+1}^{n_2} \mu((A_{k_2} \backslash A_{k_3}) \cap E_j)x_j \right\| > \epsilon$. Continuing this construction produces a disjoint sequence $\{E_j\} \subset \Sigma, \|x_j\| \leq 1$, and increasing sequences $\{n_j\}, \{k_j\}$ with

$$\left\| \sum_{i=n_j+1}^{n_{j+1}} \mu((A_{k_{j+1}} \backslash A_{k_j}) \cap E_i)x_i \right\| > \epsilon$$

for every j. The series

$$\sum_{j=1}^{\infty} \sum_{i=n_j+1}^{n_{j+1}} \mu((A_{k_{j+1}} \backslash A_{k_j}) \cap E_i)x_i$$

is not Cauchy so (ii) fails.

Again if Y is sequentially complete, we can obtain an improvement of Theorem 11.39.

Corollary 11.40. *Let Y be sequentially complete and μ as in the theorem above. The following are equivalent:*

(i) $\hat{\mu}$ is continuous from above.

(ii) For every pairwise disjoint sequence $\{A_j\} \subset \Sigma$, the series $\sum_j \mu(A_j)$ is $l^{\infty}(X)$ multiplier convergent.

(iii) For every pairwise disjoint sequence $\{A_j\} \subset \Sigma$, the series $\sum_{j=1}^{\infty} \mu(B_j)x_j$ converge uniformly for $B_j \subset A_j, B_j \in \Sigma$ and $x_j \in X, \|x_j\| \leq 1$.

Proof: (i) and (ii) are equivalent by Theorem 11.39. Clearly, (iii) implies (ii). Assume (ii). We use Li's Lemma 3.29. Set

$$E_j = \{(B, x) : B \in \Sigma, B \subset A_j, x \in X, \|x\| \leq 1\}$$

and define $f_j : E_j \to Y$ by $f_j(B, x) = \mu(B)x$. Then (iii) follows from Li's Lemma.

A sequence of finitely additive set functions $\mu_i : \Sigma \to L(X,Y)$ is uniformly continuous from above if $E_j \downarrow \emptyset$ implies that $\lim_j \hat{\mu}_i(E_j) = 0$ uniformly for $i \in \mathbb{N}$. We have a characterization of uniform continuity from above in terms of multiplier convergent series.

Theorem 11.41. *Let $\mu_i : \Sigma \to L(X,Y)$ be countably additive in $L_s(X,Y)$ for every $i \in \mathbb{N}$. The following are equivalent:*

(i) *$\{\mu_i\}$ is uniformly continuous from above.*
(ii) *For every pairwise disjoint sequence $\{A_j\} \subset \Sigma$ the series $\{\sum_{j=1}^{\infty} \mu_i(A_j) : i \in \mathbb{N}\}$ are uniformly $l^{\infty}(X)$ multiplier Cauchy.*

Proof: That (i) implies (ii) follows as in Theorem 11.39.

Suppose that (i) fails to hold. Then there exist $\epsilon > 0, A_j \downarrow \emptyset$ such that for every j there exist $n_j > j$ and m_j with $\hat{\mu}_{m_j}(A_{n_j}) > \epsilon$. For $j = 1$ there exist n_1, m_1 such that $\hat{\mu}_{m_1}(A_{n_1}) > \epsilon$. There exist pairwise disjoint $\{E_l : 1 \leq l \leq N_1\}$ and $x_l \in X, \|x_l\| \leq 1$ such that

$$\left\| \sum_{l=1}^{N_1} \mu_{m_1}(A_{n_1} \cap E_l)x_l \right\| > \epsilon.$$

Since each μ_i is countably additive in $L_s(X,Y)$, $\lim_j \mu_{m_1}(A_j \cap E_l)x_l = 0$ for $1 \leq l \leq N_1$. There exist $k_1 > n_1$ such that

$$\left\| \sum_{l=1}^{N_1} (\mu_{m_1}(A_{n_1} \cap E_l)x_l - \mu_{m_1}(A_{k_1} \cap E_l)x_l) \right\|$$

$$= \left\| \sum_{l=1}^{N_1} \mu_{m_1}((A_{n_1} \setminus A_{k_1}) \cap E_l)x_l \right\| > \epsilon.$$

There exists $n_2 > k_1, m_2$ such that $\hat{\mu}_{m_2}(A_{n_2}) > \epsilon$. By the construction above there exist N_2, pairwise disjoint $\{E_l : N_1 + 1 \leq l \leq N_2\} \subset \Sigma, x_l \in X, \|x_l\| \leq 1, N_1 + 1 \leq l \leq N_2$ and $k_2 > n_2$ such that

$$\left\| \sum_{l=N_1+1}^{N_2} \mu_{m_2}(A_{n_2} \cap E_l)x_l \right\| > \epsilon$$

and

$$\left\| \sum_{l=N_1+1}^{N_2} \mu_{m_2}((A_{n_2} \setminus A_{k_2}) \cap E_l)x_l \right\| > \epsilon.$$

Continuing this construction produces a disjoint sequence $\{E_l\}, x_l \in X, \|x_l\| \leq 1$, a sequence $\{m_i\}$ and increasing sequences $\{n_j\}, \{k_j\}, k_j > n_j$ satisfying

$$\left\| \sum_{l=N_j+1}^{N_{j+1}} \mu_{m_i}((A_{n_j} \setminus A_{k_j}) \cap E_l)x_l \right\| > \epsilon.$$

The series

$$\left\{ \sum_{j=1}^{\infty} \sum_{l=N_j+1}^{N_{j+1}} \mu_{m_i}((A_{n_j} \setminus A_{k_j}) \cap E_l)x_l : i \in \mathbb{N} \right\}$$

are not uniformly $l^{\infty}(X)$ multiplier Cauchy so (ii) fails to hold.

Chapter 12

Orlicz-Pettis Theorems for Operator Valued Series

In this chapter we consider Orlicz-Pettis type theorems for operator valued series and vector valued multipliers. Throughout this chapter let X and Y be LCTVS and $L(X, Y)$ the space of all continuous linear operators from X into Y. Let E be a vector space of X valued sequences which contains the subspace $c_{00}(X)$ of all sequences which are eventually 0. If $\sum_j T_j$ is a series in $L(X, Y)$ and $x = \{x_j\}$ is a multiplier, then the series $\sum_{j=1}^{\infty} T_j x_j$ has values in Y so any Orlicz-Pettis Theorem must focus on the topology of Y. We first have a straightforward result.

Theorem 12.1. *Assume that E is monotone. If the series $\sum_j T_j$ is E multiplier convergent with respect to the weak topology $\sigma(Y, Y')$ of Y, then the series $\sum_j T_j$ is E multiplier convergent with respect to the topologies $\gamma(Y, Y')$ and $\lambda(Y, Y')$ (see Appendix A for these topologies). In particular, the series $\sum_j T_j$ is E multiplier convergent in the original topology of Y.*

Proof: Let $x \in E$. Since the space E is monotone, the series $\sum_{j=1}^{\infty} T_j x_j$ is subseries convergent in the weak topology $\sigma(Y, Y')$. By the Orlicz-Pettis result in Corollary 4.11, the result follows immediately.

Examples of vector valued sequence spaces which are monotone are given in Appendix C.

We also have the vector analogue of Theorem 4.5. Recall the β-dual of E (with respect to the scalar field) is

$$E^{\beta} = \left\{ \{x_j'\} \subset X' : \sum_{j=1}^{\infty} \langle x_j', x_j \rangle \text{ converges for every } x = \{x_j\} \in E \right\}$$

and that E, E^β form a dual pair under the bilinear pairing

$$x' \cdot x = \{x'_j\} \cdot \{x_j\} = \sum_{j=1}^{\infty} \langle x'_j, x_j \rangle$$

(Appendix C). Also, a locally convex topology w defined for dual pairs X, X' is a Hellinger-Toeplitz topology if whenever X, X' and Y, Y' are dual pairs and

$$T : (X, \sigma(X, X')) \to (Y, \sigma(Y, Y'))$$

is linear and continuous, then

$$T : (X, w(X, X')) \to (Y, w(Y, Y'))$$

is continuous (Appendix A.1).

Theorem 12.2. *Let w be a Hellinger-Toeplitz topology for dual pairs. If $(E, w(E, E^\beta))$ is an AK-space and $\sum_j T_j$ is E multiplier convergent with respect to $\sigma(Y, Y')$, then $\sum_j T_j$ is E multiplier convergent with respect to $w(Y, Y')$.*

Proof: Let $S : E \to Y$ be the summing operator with respect to the series $\sum_j T_j$ and the topology $\sigma(Y, Y')$, $Sx = \sum_{j=1}^{\infty} T_j x_j$ ($\sigma(Y, Y')$ sum; Chapter 11). By Theorem 11.2, S is $w(E, E^\beta) - w(Y, Y')$ continuous. If $x \in E$, $x = w(E, E^\beta) - \lim_n \sum_{j=1}^{n} e^j \otimes x_j$ so

$$Sx = w(Y, Y') - \lim_n \sum_{j=1}^{n} S(e^j \otimes x_j) = w(Y, Y') - \lim_n \sum_{j=1}^{n} T_j x_j = \sum_{j=1}^{\infty} T_j x_j.$$

Examples of vector valued AK-spaces are given in Appendix C.

We next consider the strong topology on Y. The following example shows that without some condition on the multiplier space E the series $\sum_j T_j x_j$ will not, in general, converge in the strong topology of Y.

Example 12.3. Equip $X = l^\infty$ with the weak topology $\sigma(l^\infty, l^1)$ and let $E = l^\infty(X)$. Define $Q_k : l^\infty \to l^\infty$ by $Q_k t = t_k e^k$. If $x = \{x^k\} \in E$, then $\sum_{k=1}^{\infty} Q_k x^k = \sum_{k=1}^{\infty} x_k^k e^k$ is $\sigma(l^\infty, l^1)$ convergent, but if $x = \{e^k\} \in E$, then $\sum_{k=1}^{\infty} Q_k e^k = \sum_{k=1}^{\infty} e^k$ is not $\beta(l^\infty, l^1) = \|\cdot\|_\infty$ convergent.

We now establish the analogue of Theorem 5.7. The space E has ∞-GHP if whenever $x \in E$ and $\{I_j\}$ is an increasing sequence of intervals, there exist a subsequence $\{n_j\}$ and $a_{n_j} > 0, a_{n_j} \to \infty$ such that every subsequence of $\{n_j\}$ has a further subsequence $\{p_j\}$ such that the coordinatewise sum of the series $\sum_{j=1}^{\infty} a_{p_j} \chi_{I_{p_j}} x \in E$ (Appendix C).

Theorem 12.4. *Let E have ∞-GHP. If $\sum_j T_j$ is E multiplier convergent with respect to $\sigma(Y, Y')$, then $\sum_j T_j$ is E multiplier convergent with respect to $\beta(Y, Y')$.*

Proof: If the conclusion fails to hold, there exist $x \in E, \{y_k'\}$ $\sigma(Y', Y)$ bounded, $\epsilon > 0$ and an increasing sequence of intervals $\{I_k\}$ such that

$$(*) \qquad \left| \sum_{j \in I_k} \langle y_k', T_j x_j \rangle \right| > \epsilon \text{ for all } k.$$

Since E has ∞-GHP, there exist $\{p_k\}, a_{p_k} > 0, a_{p_k} \to \infty$ such that every subsequence of $\{p_k\}$ has a further subsequence $\{q_k\}$ such that the coordinatewise sum of the series $\sum_{k=1}^{\infty} a_{q_k} \chi_{I_{q_k}} x \in E$.

Define a matrix

$$M = [m_{ij}] = [\sum_{l \in I_{p_j}} \langle y_{p_i}'/a_{p_i}, T_l(a_{p_j} x_l) \rangle].$$

We claim that M is a $\mathcal{K}$-matrix (Appendix D.2). First, the columns of M converge to 0 since $\{y_i'\}$ is $\sigma(Y', Y)$ bounded and $a_{p_i} \to \infty$. Next, given a subsequence there is a further subsequence $\{q_j\}$ such that $y = \sum_{j=1}^{\infty} a_{q_j} \chi_{I_{q_j}} x \in E$. Let

$$z = \sum_{l=1}^{\infty} T_l y_l = \sum_{j=1}^{\infty} \sum_{l \in I_{q_j}} T_l(a_{q_j} x_l)$$

be the $\sigma(Y, Y')$ sum of this series. Then

$$\sum_{j=1}^{\infty} m_{iq_j} = \langle y_{p_i}'/a_{p_i}, z \rangle \to 0$$

so M is a $\mathcal{K}$-matrix. By the Antosik-Mikusinski Matrix Theorem (Appendix D.2), the diagonal of M converges to 0. But, this contradicts $(*)$.

Examples of vector valued sequence spaces with ∞-GHP are given in Appendix C.

Chapter 13

Hahn-Schur Theorems for Operator Valued Series

In this chapter we consider versions of the Hahn-Schur Theorem for series with values in the space of continuous linear operators and with multipliers which are vector valued with values in the domain of the operators. Throughout this chapter let X and Y be LCTVS, $L(X, Y)$ the space of all continuous linear operators from X into Y, and E a vector space of X valued sequences which contains $c_{00}(X)$, the space of all X valued sequences which are eventually 0. The analogue of the hypothesis (H) for the Hahn-Schur Theorem in Chapter 7 for E multiplier convergent series is straightforward.

(H) Let $\sum_j T_{ij}$ be E multiplier convergent in $L(X, Y)$ for every i. Assume that $\lim_i \sum_{j=1}^{\infty} T_{ij} x_j$ exists for every $x = \{x_j\} \in E$ and that $T_j x = \lim_i T_{ij} x$ exists for every $x \in X$.

We, of course, need to have that $T_j \in L(X, Y)$ for every j for reasonable conclusions to hold. We will either assume this or impose conditions which guarantee that this holds.

Similarly, the analogues of the conclusions of the Hahn-Schur Theorems of Chapter 7 are also straightforward.

(C1) The series $\sum_j T_j$ is E multiplier convergent and $\lim_i \sum_{j=1}^{\infty} T_{ij} x_j = \sum_{j=1}^{\infty} T_j x_j$ for every $\{x_j\} \in E$.

(C2) $\lim_i \sum_{j=1}^{\infty} T_{ij} x_j = \sum_{j=1}^{\infty} T_j x_j$ uniformly for $x = \{x_j\}$ belonging to bounded subsets of E.

(C3) The series $\sum_{j=1}^{\infty} T_{ij} x_j$ converge uniformly for $x = \{x_j\}$ belonging to bounded subsets of E.

From Theorem 11.14 and Lemma 11.16, we have the following connection between (H) and conclusion (C1).

Theorem 13.1. *Assume (H), that E has signed-WGHP, and that there* *exist $T_j \in L(X,Y)$ such that $\lim_i T_{ij}x = T_j x$ for every $x \in E$. Then (C1) holds.*

Hypothesis (H) implies that $\lim_i T_{ij}x = T_j x$ exists for every $x \in X$ so if the pair (X,Y) has the Banach-Steinhaus property, it follows that $T_j \in L(X,Y)$ for every j and the hypothesis in Theorem 13.1 is satisfied.

The hypothesis in (H) implies in particular, that if $T_j \in L(X,Y)$, then

(i) For each $j \in \mathbb{N}$, $\lim_i T_{ij} = T_j$ in the strong operator topology of $L(X,Y)$.

To see this, take $x = e^j \otimes z$ for $j \in \mathbb{N}$ and $z \in X$.

The following example shows that even in the presence of the Banach-Steinhaus property, property SGHP, and condition (i), the hypothesis (H) does not imply conclusion (C2).

Example 13.2. Let $X = l^1$ and $Y = \mathbb{R}$ so $L(X,Y) = (l^1)' = l^\infty$. For $i,j \in \mathbb{N}$, let $T_{ij} = e^i/2^j$ so $T_{ij}t = (e^i/2^j) \cdot t = t_i/2^j$ for $t \in l^1$. Thus, for each j, $\lim_i T_{ij} = 0$ in the strong operator topology of $L(X,Y) = l^\infty$ which in this case is just $\sigma(l^\infty, l^1)$. Note, however, that $\{T_{ij}\}_i$ does not converge to 0 in the norm topology $\|\cdot\|_\infty$ of $L(X,Y) = l^\infty$. Let $E = l^\infty(l^1)$, where l^1 has the norm topology $\|\cdot\|_1$ and E has the sup norm. If $\{t^j\} \in E$ with $\|t^j\|_1 \leq 1$ for every j, we have

$$\sum_{j=1}^\infty T_{ij}t^j = \sum_{j=1}^\infty e^i \cdot t^j/2^j = \sum_{j=1}^\infty t_i^j/2^j$$

and since $\left| t_i^j \right| \leq 1$, the series $\sum_{j=1}^\infty T_{ij}t^j$ converges (absolutely). That is, for each i, the series $\sum_j T_{ij}$ is E multiplier convergent. Since

$$\sum_{i=1}^\infty \sum_{j=1}^\infty |T_{ij}t^j| = \sum_{i=1}^\infty \sum_{j=1}^\infty \left| t_i^j \right|/2^j = \sum_{j=1}^\infty \sum_{i=1}^\infty \left| t_i^j \right|/2^j \leq \sum_{j=1}^\infty 1/2^j < \infty,$$

$\lim_i \sum_{j=1}^\infty T_{ij}t^j = 0$. Thus, conditions (H) and (i) hold. However, condition (C2) does not hold. Indeed, for any i, let $t^i \in E$ be the constant sequence $t^i = \{e^i\}$. Then

$$\sum_{j=1}^\infty T_{ij}t^i = \sum_{j=1}^\infty e^i \cdot e^i/2^j = \sum_{j=1}^\infty 1/2^j = 1$$

while

$$\sum_{j=1}^\infty T_{kj}t^i = 0$$

for $k \neq i$. Thus, $\lim_i \sum_{j=1}^{\infty} T_{ij} t^k = 0$ does not hold uniformly for $\{t^k\}$ belonging to bounded subsets of E and conclusion (C2) fails.

Note that $E = l^{\infty}(l^1)$ has SGHP so a straightforward analogue of the Hahn-Schur Theorem given in Theorem 7.10 does not hold.

In order to obtain versions of the Hahn-Schur Theorem in which conclusions (C2) and (C3) hold, it is necessary to strengthen condition (i). In particular, we need to replace the assumption that the sequence $\{T_{ij}\}_i$ converges to T_j in the strong operator topology with convergence in $L_b(X, Y)$ in order to obtain the conclusion (C2). Actually, we consider the more general situation where the sequences $\{T_{ij}\}_i$ converge in $L_{\mathcal{A}}(X, Y)$ (see Appendix A).

We first consider the conclusion (C3).

Theorem 13.3. *Let $\sum_j T_{ij}$ be E multiplier convergent in $L(X, Y)$ for every i. Assume that E has signed-SGHP. If $\lim_i \sum_{j=1}^{\infty} T_{ij} x_j$ exists for every $x = \{x_j\} \in E$, then for every bounded set A the series $\sum_{j=1}^{\infty} T_{ij} x_j$ converge uniformly for $x \in A, i \in \mathbb{N}$, i.e., conclusion (C3) holds.*

Proof: If the conclusion fails to hold there exist a closed neighborhood of $0, U$, such that for every i there exist $k_i > i$, a finite interval I_i with $\min I_i > i$, $x^i \in A$ with

$$\sum_{k \in I_i} T_{k_i k} x_k^i \notin U.$$

Put $i_1 = 1$. By the condition above there exist $k_1 > 1$, an interval I_1 with $\min I_1 > i_1$, $x^1 \in A$ with

$$\sum_{k \in I_1} T_{k_1 k} x_k^1 \notin U.$$

By Theorem 11.9 (this uses signed-SGHP), there exists j_1 such that

$$\sum_{k=j}^{\infty} T_{ik} x_k \in U$$

for every $x \in A$, $j \geq j_1, 1 \leq i \leq k_1$. Set $i_2 = \max[I_1 + 1, j_1]$. By the condition above, there exist $k_2 > i_2$, an interval I_2 with $\min I_2 > i_2$, $x^2 \in A$ such that $\sum_{k \in I_2} T_{k_2 k} x_k^2 \notin U$. Note that $k_2 > k_1$. Continuing this construction produces an increasing sequence $\{k_i\}$, an increasing sequence of intervals $\{I_i\}$, $x^i \in A$ such that

$$(*) \quad \sum_{k \in I_i} T_{k_i} x_k^i \notin U.$$

Define an infinite matrix

$$M = [m_{ij}] = [\sum_{k \in I_j} T_{k_i k} x_k^j].$$

We claim that M is a signed $\mathcal{K}$-matrix (Appendix D). First the columns of M converge by hypothesis. Next, given any subsequence there exist a further subsequence $\{p_j\}$ and a sequence of signs $\{s_j\}$ such that $x = \sum_{j=1}^{\infty} s_j \chi_{I_{p_j}} x^{p_j} \in E$. Then

$$\sum_{j=1}^{\infty} s_j m_{ip_j} = \sum_{j=1}^{\infty} s_j \sum_{k \in I_{p_j}} T_{k_i k} x_k^{p_j} = \sum_{k=1}^{\infty} T_{k_i k} x_k$$

so $\lim_i \sum_{j=1}^{\infty} s_j m_{ip_j}$ exists by hypothesis. Hence, M is a signed $\mathcal{K}$-matrix so the diagonal of M converges to 0 by the signed version of the Antosik-Mikusinski Matrix Theorem (Appendix D.3). But, this contradicts $(*)$.

We next consider the converse of the Hahn-Schur result above.

In what follows $\mathcal{A}$ will denote a family of bounded subsets of X whose union is X(see Appendix A).

Theorem 13.4. Let $\sum_j T_{ij}$ be E multiplier convergent in $L(X,Y)$ for every i. Assume that there exist $T_j \in L(X,Y)$ such that $\lim_i T_{ij} = T_j$ in $L_{\mathcal{A}}(X,Y)$ and for every $A \in \mathcal{A}$ the series $\sum_{j=1}^{\infty} T_{ij} x_j$ converge uniformly for $x \in A, i \in \mathbb{N}$. Then the sequences $\{\sum_{j=1}^{\infty} T_{ij} x_j\}_i$ satisfy a Cauchy condition uniformly for $x \in A$.

Proof: Let U be a neighborhood of 0 in Y and pick a symmetric neighborhood of 0, V, such that $V + V + V \subset U$. There exists n such that $\sum_{j=n}^{\infty} T_{ij} x_j \in V$ for every $x \in A, i \in \mathbb{N}$. There exists m such that $\sum_{j=1}^{n-1} (T_{ij} - T_{kj}) x_j \in V$ for $x \in A$, and $i, k \geq m$. If $x \in A$ and $i, k \geq m$, then

$$\sum_{j=1}^{\infty} T_{ij} x_j - \sum_{j=1}^{\infty} T_{kj} x_j$$

$$= \sum_{j=1}^{n-1} (T_{ij} - T_{kj}) x_j + \sum_{j=n}^{\infty} T_{ij} x_j - \sum_{j=n}^{\infty} T_{kj} x_j \in V + V + V \subset U$$

and the conclusion holds.

From Theorem 13.4, we have condition (H).

Corollary 13.5. Let $\sum_j T_{ij}$ be E multiplier convergent in $L(X,Y)$ for every i. Assume that there exist $T_j \in L(X,Y)$ such that $\lim_i T_{ij} = T_j$ in

$L_s(X,Y)$. *If for each $x \in E$ the series $\sum_{j=1}^{\infty} T_{ij}x_j$ converge uniformly for $i \in \mathbb{N}$, then $\{\sum_{j=1}^{\infty} T_{ij}x_j\}_i$ is a Cauchy sequence. Hence, if Y is sequentially complete, hypothesis (H) is satisfied.*

We can also obtain a boundedness result from uniform convergence of series.

Proposition 13.6. *Let $\sum_j T_{ij}$ be E multiplier convergent in $L(X,Y)$ for every i. Assume that there exist $T_j \in L(X,Y)$ such that $\lim_i T_{ij} = T_j$ in $L_{\mathcal{A}}(X,Y)$ and for every $A \in \mathcal{A}$ the series $\sum_{j=1}^{\infty} T_{ij}x_j$ converge uniformly for $x \in A, i \in \mathbb{N}$. Then $B = \{\sum_{j=1}^{\infty} T_{ij}x_j : i \in \mathbb{N}, x \in A\}$ is bounded for every $A \in \mathcal{A}$.*

Proof: Let U be a neighborhood of 0 in Y and pick a balanced neighborhood, V, such that $V + V \subset U$. There exists n such that $\sum_{j=n}^{\infty} T_{ij}x_j \in V$ for $i \in \mathbb{N}, x \in A$. For every j, $\{T_{ij}x_j : i \in \mathbb{N}, x_j$ for $x = \{x_j\} \in A\}$ is bounded by the hypothesis so there exists $t > 1$ such that $\{\sum_{j=1}^{n-1} T_{ij}x_j : i \in \mathbb{N}, x \in A\} \subset tV$. Therefore, $\sum_{j=1}^{\infty} T_{ij}x_j = \sum_{j=1}^{n-1} T_{ij}x_j + \sum_{j=n}^{\infty} T_{ij}x_j \in tV + V \subset tU$ for $i \in \mathbb{N}, x \in A$.

Theorem 13.3 gives sufficient conditions for the uniform convergence hypothesis in Theorem 13.4, Corollary 13.5 and Proposition 13.6 to hold.

We next consider the conclusions (C1) and (C2).

Theorem 13.7. *Let $\sum_j T_{ij}$ be E multiplier convergent in $L(X,Y)$ for every i. Assume that there exist $T_j \in L(X,Y)$ such that $\lim_i T_{ij} = T_j$ in $L_{\mathcal{A}}(X,Y)$. If $\lim_i \sum_{j=1}^{\infty} T_{ij}x_j$ exists for every $x = \{x_j\} \in E$ and for every $A \in \mathcal{A}$ the series $\sum_{j=1}^{\infty} T_{ij}x_j$ converge uniformly for $x \in A, i \in \mathbb{N}$, then $\sum_j T_j$ is E multiplier convergent and $\lim_i \sum_{j=1}^{\infty} T_{ij}x_j = \sum_{j=1}^{\infty} T_j x_j$ uniformly for $x \in A$. That is, conclusions (C1) and (C2) hold for the family $\mathcal{A}$.*

Proof: That $\sum_j T_j$ is E multiplier convergent and

$$\lim_i \sum_{j=1}^{\infty} T_{ij}x_j = \sum_{j=1}^{\infty} T_j x_j$$

for $x \in E$ follows from Lemma 11.16. Let U be a neighborhood of 0 in Y and pick a closed, symmetric neighborhood of 0, V, such that $V + V + V \subset U$. There exists N such that $\sum_{j=m}^{n} T_{ij}x_j \in V$ for $n > m \geq N, i \in \mathbb{N}$ and $x \in A$. Hence, $\sum_{j=m}^{n} T_j x_j \in V$ for $n > m \geq N, i \in \mathbb{N}$ and $x \in A$ and, therefore, $\sum_{j=m}^{\infty} T_j x_j \in V$ for $m \geq N, x \in A$. There exists M such that $\sum_{j=1}^{N-1} (T_{ij} - T_j) x_j \in V$ for $i \geq M, x \in A$. If $x \in A$ and $i \geq M$, we have

$$\sum_{j=1}^{\infty} T_{ij} x_j - \sum_{j=1}^{\infty} T_j x_j = \sum_{j=N}^{\infty} T_{ij} x_j - \sum_{j=N}^{\infty} T_j x_j + \sum_{j=1}^{N-1} (T_{ij} - T_j) x_j \in V + V + V \subset U$$

and the result follows.

From Theorems 13.3 and 13.7, we have a Hahn-Schur Theorem for $L_{\mathcal{A}}(X, Y)$.

Corollary 13.8. *Assume that E has signed-SGHP. Let $\sum_j T_{ij}$ be E multiplier convergent in $L(X, Y)$ for every i. Assume that there exist $T_j \in L(X, Y)$ such that $\lim_i T_{ij} = T_j$ in $L_{\mathcal{A}}(X, Y)$. If $\lim_i \sum_{j=1}^{\infty} T_{ij} x_j$ exists for every $x = \{x_j\} \in E$, then $\sum_j T_j$ is E multiplier convergent, $\lim_i \sum_{j=1}^{\infty} T_{ij} x_j = \sum_{j=1}^{\infty} T_j x_j$ uniformly for $x \in A$ and the series $\sum_{j=1}^{\infty} T_{ij} x_j$ converge uniformly for $x \in A, i \in \mathbb{N}$. That is, conclusions (C2) and (C3) hold for the family $\mathcal{A}$.*

A sufficient condition for there to exist $T_j \in L(X, Y)$ with $\lim_i T_{ij} x = T_j x$ for every $x \in X$ is that the pair (X, Y) has the Banach-Steinhaus property. However, as Example 13.2 shows we must have convergence in $L_{\mathcal{A}}(X, Y)$ in order to obtain the conclusion in Theorem 13.7.

Corollary 13.9. *Assume that E has signed-SGHP and that Y is sequentially complete. Let $\sum_j T_{ij}$ be E multiplier convergent in $L(X, Y)$ for every i. Assume that there exist $T_j \in L(X, Y)$ such that $\lim_i T_{ij} = T_j$ in $L_{\mathcal{A}}(X, Y)$. The following are equivalent:*

(i) *(H).*
(ii) *For each $x \in E$ the series $\sum_{j=1}^{\infty} T_{ij} x_j$ converge uniformly for $i \in \mathbb{N}$.*
(iii) *For each $A \in \mathcal{A}$ the series $\sum_{j=1}^{\infty} T_{ij} x_j$ converge uniformly for $i \in \mathbb{N}, x \in A$.*
(iv) *For each $A \in \mathcal{A}$, the series $\sum_j T_j$ is E multiplier convergent and $\lim_i \sum_{j=1}^{\infty} T_{ij} x_j = \sum_{j=1}^{\infty} T_j x_j$ uniformly for $x \in A$.*

Proof: That (i) implies (iii) follows from Theorem 13.3; (iii) implies (iv) from Theorem 13.7; (ii) implies (i) from Corollary 13.5; clearly (iii) implies (ii) and (iv) implies (i).

We give a statement of the results above for the case of bounded ($l^{\infty}(X)$) multiplier convergent series. Note that $l^{\infty}(X)$ has SGHP so these results apply.

Corollary 13.10. *Let $\sum_j T_{ij}$ be bounded multiplier convergent for every i and assume that there exist $T_j \in L(X,Y)$ such that $\lim_i T_{ij} = T_j$ in $L_b(X,Y)$. If $\lim_i \sum_{j=1}^\infty T_{ij}x_j$ exists for every $x = \{x_j\} \in l^\infty(X)$, then the series $\sum_j T_j$ is bounded multiplier convergent, $\lim_i \sum_{j=1}^\infty T_{ij}x_j = \sum_{j=1}^\infty T_j x_j$ uniformly for $x = \{x_j\}$ belonging to bounded subsets $A \subset l^\infty(X)$ and the series $\sum_{j=1}^\infty T_{ij}x_j$ converge uniformly for $x = \{x_j\}$ belonging to bounded subsets $A \subset l^\infty(X)$.*

Corollary 13.11. *Let $\sum_j T_{ij}$ be bounded multiplier convergent for every i and assume that there exist $T_j \in L(X,Y)$ such that $\lim_i T_{ij} = T_j$ in $L_s(X,Y)$. If Y is sequentially complete and for every $x = \{x_j\} \in l^\infty(X)$, the series $\sum_{j=1}^\infty T_{ij}x_j$ converge uniformly for $i \in \mathbb{N}$, then $\lim_i \sum_{j=1}^\infty T_{ij}x_j$ exists for every $x = \{x_j\} \in l^\infty(X)$.*

From Proposition 13.6, we have a boundedness result.

Corollary 13.12. *Let $\sum_j T_{ij}$ be bounded multiplier convergent for every i and assume that there exist $T_j \in L(X,Y)$ such that $\lim_i T_{ij} = T_j$ in $L_b(X,Y)$. If for each bounded subset A of $l^\infty(X)$ the series $\sum_{j=1}^\infty T_{ij}x_j$ converge uniformly for $i \in \mathbb{N}, x \in A$, then $B = \{\sum_{j=1}^\infty T_{ij}x_j : i \in \mathbb{N}, x \in A\}$ is bounded.*

Unlike the case of scalar multipliers, the space $m_0(X)$ (or the subset $\{\chi_\sigma x : \sigma \subset \mathbb{N}, x \in X\}$) does not have SGHP so the results above do not apply to $m_0(X)$ multiplier convergent series. However, $m_0(X)$ does have WGHP so Theorem 13.1 applies and gives the following result.

Theorem 13.13. *Let $\sum_j T_{ij}$ be $m_0(X)$ multiplier convergent for every i and assume that there exist $T_j \in L(X,Y)$ such that $\lim_i T_{ij}x = T_j x$ for every $x \in X$. If $\lim_i \sum_{j=1}^\infty T_{ij}x_j$ exists for every $x = \{x_j\} \in m_0(X)$, then $\sum_j T_j$ is $m_0(X)$ multiplier convergent and $\lim_i \sum_{j=1}^\infty T_{ij}x_j = \sum_{j=1}^\infty T_j x_j$ for every $x = \{x_j\} \in m_0(X)$.*

The following example shows that the uniform convergence conclusions in (C2) and (C3) do not hold for $m_0(X)$ multiplier convergent series.

Example 13.14. Let $1 \le p < \infty$ and define $Q_j : l^p \to l^p$ by $Q_j t = t_j e^j$ as in Example 11.10. As noted in Example 11.10, the series $\sum_j Q_j$ is $m_0(l^p)$ multiplier convergent. Now define $T_{ij} = Q_j$ if $j \le i$ and $T_{ij} = 0$ if $j > i$. We have that

$$(*) \quad \lim_i \sum_{j=1}^\infty T_{ij}t_j = \sum_{j=1}^\infty Q_j t_j$$

for any $t = \{t_j\} \in m_0(l^p)$ and $\lim_i T_{ij} = Q_j$ in $L_b(l^p)$ for every j. However, the limit in (∗) is not uniform for t belonging to bounded subsets of $m_0(l^p)$ [take t to be the constant sequence $\{e^k\}$ in $m_0(l^p)$ so $\sum_{j=1}^\infty T_{ij}t_j = \sum_{j=1}^i Q_j e^k = e^k$ if $i \geq k$]. Similarly, the uniform convergence condition in (C3) does not hold.

From Theorems and 7.17 and Corollary 7.19, we do have the following results for $m_0(X)$ multiplier convergent series.

Theorem 13.15. *Let $\sum_j T_{ij}$ be $m_0(X)$ multiplier convergent for every i and assume that there exist $T_j \in L(X,Y)$ with $\lim_i T_{ij}x = T_j x$ for every $x \in X$. If $\lim_i \sum_{j=1}^\infty T_{ij}x_j$ exists for every $\{x_j\} \in m_0(X)$, then*

(i) *$\sum_j T_j$ is $m_0(X)$ multiplier convergent and $\lim_i \sum_{j=1}^\infty T_{ij}x_j = \sum_{j=1}^\infty T_j x_j$ for every $\{x_j\} \in m_0(X)$,*
(ii) *for every $x \in X$, $\lim_i \sum_{j\in\sigma} T_{ij}x = \sum_{j\in\sigma} T_j x$ uniformly for $\sigma \subset \mathbb{N}$,*
(iii) *for every $x \in X$, the series $\sum_{j\in\sigma} T_{ij}x$ converge uniformly for $i \in \mathbb{N}, \sigma \subset \mathbb{N}$.*

Theorem 13.16. *Let Y be sequentially complete. Let $\sum_j T_{ij}$ be $m_0(X)$ multiplier convergent for every i and assume that there exist $T_j \in L(X,Y)$ with $\lim_i T_{ij}x = T_j x$ for every $x \in X$. The following are equivalent:*

(1) *$\lim_i \sum_{j=1}^\infty T_{ij}x_j$ exists for every $\{x_j\} \in m_0(X)$.*
(2) *The series $\sum_j T_j$ is $m_0(X)$ multiplier convergent and for every $x \in X$, $\lim_i \sum_{j\in\sigma} T_{ij}x = \sum_{j\in\sigma} T_j x$ uniformly for $\sigma \subset \mathbb{N}$.*
(3) *For every $x \in X$ the series $\sum_{j\in\sigma} T_{ij}x$ converge uniformly for $i \in \mathbb{N}, \sigma \subset \mathbb{N}$.*
(4) *For every $x \in X$ and $\sigma \subset \mathbb{N}$ the series $\sum_{j\in\sigma} T_{ij}x$ converge uniformly for $i \in \mathbb{N}$.*

In the scalar case, if X is sequentially complete, $\sum_j x_{ij}$ is m_0 multiplier convergent for every i, $\lim_i x_{ij} = x_j$ exists for every j and $\lim_i \sum_{j\in\sigma} x_{ij}$ exists for every $\sigma \subset \mathbb{N}$, then the series $\sum_j x_j$ is l^∞ multiplier convergent and $\lim_i \sum_{j=1}^\infty t_j x_{ij} = \sum_{j=1}^\infty t_j x_j$ uniformly for $\|\{t_j\}\|_1 \leq 1$ (Theorem 7.29). As the following example shows, the analogue of this statement does not hold for $m_0(X)$ and $l^\infty(X)$ multiplier convergent series.

Example 13.17. Let $X = l^1$ and let $\sum_j T_j$ be a series in $L(X)$ which is subseries convergent but not $l^\infty(X)$ multiplier convergent (Example 11.27). Set $T_{ij} = T_j$ if $j \leq i$ and $T_{ij} = 0$ if $j > i$. If $\sigma \subset \mathbb{N}$ and $x \in X$, then $\lim_i \sum_{j \in \sigma} T_{ij} x = \sum_{j \in \sigma} T_j x$ so $\lim_i \sum_{j=1}^{\infty} T_{ij} x_j = \sum_{j=1}^{\infty} T_j x_j$ for any $\{x_j\} \in m_0(X)$. However, if $x = \{x_j\} \in l^\infty(X), \|\{x_j\}\| \le 1$, is such that $\sum_{j=1}^{\infty} T_j x_j$ does not converge, then $\lim_i \sum_{j=1}^{\infty} T_{ij} x_j$ does not exist.

In Chapter 7 it was noted that a scalar matrix $[a_{ij}]$ which maps m_0 into c also maps l^∞ into c (condition (S) following Theorem 7.2). It was also shown that a vector valued matrix $[x_{ij}]$ which maps m_0 into $c(X)$, the space of convergent X valued sequences, also maps l^∞ into $c(X)$ (see condition (S') following Theorem 7.29). The example above shows that an operator valued matrix $[T_{ij}]$ may map $m_0(X)$ into $c(Y)$ but fail to map $l^\infty(X)$ into $c(Y)$.

Finally, we give an application of the operator version of the Hahn-Schur Theorem to obtain a version of the Nikodym Convergence Theorem for operator valued measures. Let Σ be a σ-algebra of subsets of a set S and let $\mu_i : \Sigma \to L_A(X, Y)$ be countably additive. The Nikodym Convergence Theorem given in Theorem 7.47 implies that if $\lim_i \mu_i(E) = \mu(E)$ exists in $L_A(X, Y)$ for every $E \in \Sigma$, then $\mu : \Sigma \to L_A(X, Y)$ is countably additive and $\{\mu_i\}$ is uniformly countably additive in $L_A(X, Y)$. We seek to obtain a version of the Nikodym Convergence Theorem for operator valued measures whose operator semi-variation is continuous from above (see the definition following Corollary 11.37).

Let X, Y be normed spaces and $\mu : \Sigma \to L(X, Y)$ be finitely additive. Recall the operator semi-variation of μ, $\hat{\mu}$, is defined to be

$$\hat{\mu}(E) = \sup \left\{ \left\| \sum_{j=1}^{n} \mu(A_j) x_j \right\| : \{A_j\}_{j=1}^{n} \text{ a partition of } E, \|x_j\| \le 1 \right\}$$

(see the definition preceding Theorem 11.34). The semi-variation of μ, $\hat{\mu}$, is continuous from above if $E_j \downarrow \emptyset$ implies that $\hat{\mu}(E_j) \downarrow 0$. Theorem 11.39 gives necessary and sufficient conditions for $\hat{\mu}$ to be continuous from above. We derive a Nikodym Convergence Theorem for measures whose variation is continuous from above. First, the following example shows that a straightforward analogue of the Nikodym Convergence Theorem fails for such measures.

Example 13.18. Let X be a Banach space and $\sum_j T_j$ be a series in $L_b(X)$ which is subseries convergent in $L_b(X)$ with respect to the operator norm but not bounded multiplier convergent (Example 11.27). Define $\mu_i : 2^{\mathbb{N}} \to L_b(X)$ by $\mu_i(E) = \sum_{j \in E \cap [1, i]} T_j$. Then each μ_i is countably additive and its semi-variation is continuous from above. However, $\lim_i \mu_i(E) = \mu(E) =$

$\sum_{j \in E} T_j$ exists in $L_b(X)$ but the semi-variation of μ is not continuous from above (Example 11.38).

We obtain a version of the Nikodym Convergence Theorem for measures whose semi-variation is continuous from above by strengthening the condition that $\lim_i \mu_i(E) = \mu(E)$ exists in $L_b(X,Y)$ for every $E \in \Sigma$ to the condition that

> (#) $\lim_i \sum_{j=1}^{\infty} \mu_i(E_j) x_j$ exists in Y for every pairwise disjoint sequence $\{E_j\} \subset \Sigma$ and $\|x_j\| \leq 1$.

Recall that a sequence of finitely additive set functions $\{\mu_i\}$ with finite semi-variation is uniformly continuous from above if $E_j \downarrow \emptyset$ implies that $\lim_j \mu_i(E_j) = 0$ uniformly for $i \in \mathbb{N}$. Theorem 11.41 gives necessary and sufficient conditions for a sequence of set functions to be uniformly continuous from above.

Theorem 13.19. *Let Y be sequentially complete and let $\mu_i : \Sigma \to L_s(X,Y)$ be countably additive with $\hat{\mu}_i$ continuous from above. Assume that condition (#) above holds and for every $E \in \Sigma$ there exists $\mu(E) \in L(X,Y)$ with $\lim_i \mu_i(E) = \mu(E)$ in $L_b(X,Y)$. Then the semi-variation of μ, $\hat{\mu}$, is continuous from above and the semi-variations $\{\hat{\mu}_i\}$ are uniformly continuous from above.*

Proof: Let $\{E_j\} \subset \Sigma$ be pairwise disjoint and $x_j \in X, \|x_j\| \leq 1$. Then condition (#) and Corollary 13.10 imply that the series $\sum_j \mu(E_j)$ is bounded multiplier convergent and the series $\{\sum_j \mu_i(E_j)\}_{i \in \mathbb{N}}$ are uniformly bounded multiplier convergent. Theorems 11.39 and 11.41 now give the desired conclusion.

Chapter 14

Automatic Continuity for Operator Valued Matrices

In this chapter we will establish several automatic continuity results for operator valued matrices analogous to those for scalar valued matrices established in Chapter 10. We first consider the relationship between an operator matrix and its transpose which require some technical assumptions.

Let X, Y be TVS with $L(X, Y)$ the space of all continuous linear operators from X into Y. Let E $[F]$ be a vector space of X $[Y]$ valued sequences which contains $c_{00}(X)$ $[c_{00}(Y)]$, the space of X $[Y]$ valued sequences which are eventually 0 and let $A = [A_{ij}]$ be an infinite matrix with $A_{ij} \in L(X, Y)$. We say that A maps E into F if the series $\sum_{j=1}^{\infty} A_{ij} x_j$ converges for every $i \in \mathbb{N}$ and $x = \{x_j\} \in E$ and $Ax = \{\sum_{j=1}^{\infty} A_{ij} x_j\}_i \in F$ for every $x = \{x_j\} \in E$; we write $A : E \to F$ if A maps E into F. Note that if $A : E \to F$, then the rows of A must be E multiplier convergent. We begin by considering the analogue of Theorem 10.3.

Recall that the (scalar) β-dual of E is defined to be

$$E^{\beta} = \left\{ \{y_j\} : y_j \in X', \sum_{j=1}^{\infty} \langle y_j, x_j \rangle = y \cdot x \text{ converges for every } x = \{x_j\} \in E \right\}$$

and E, E^{β} form a dual pair under the pairing $y \cdot x$ (Appendix C).

Theorem 14.1. Let $A : E \to F$ and assume that $\sigma(E^{\beta}, E)$ is sequentially complete. Then A is $\sigma(E, E^{\beta}) - \sigma(F, F^{\beta})$ continuous and, therefore, $w(E, E^{\beta}) - w(F, F^{\beta})$ continuous for every Hellinger-Toeplitz topology w (Appendix A.1).

Proof: Let $y \in F^\beta, x \in E$. Then

$$y \cdot Ax = \sum_{i=1}^{\infty} \left\langle y_i, \sum_{j=1}^{\infty} A_{ij} x_j \right\rangle = \lim_m \sum_{i=1}^{m} \left\langle y_i, \sum_{j=1}^{\infty} A_{ij} x_j \right\rangle$$

$$= \lim_m \sum_{j=1}^{\infty} \sum_{i=1}^{m} \left\langle A'_{ij} y_i, x_j \right\rangle = \lim_m z^m \cdot x,$$

where $z_j^m = \sum_{i=1}^{m} A'_{ij} y_i, z^m = \{z_j^m\}_j \in E^\beta$. Then $\{z^m\}$ is $\sigma(E^\beta, E)$ Cauchy and, therefore, there exists $z \in E^\beta$ such that $z^m \to z$ with respect to $\sigma(E^\beta, E)$ with $y \cdot Ax = z \cdot x$ for every $x \in E$. This implies that A is $\sigma(E, E^\beta) - \sigma(F, F^\beta)$ continuous. The last statement follows from Appendix A.1.

A sufficient condition for $\sigma(E^\beta, E)$ to be sequentially complete is that E have the signed-WGHP and X' be $\sigma(X', X)$ sequentially complete (Corollary 11.18).

In the scalar case Theorem 14.1 also contains a statement concerning the transpose matrix of A; however, the operator case which we now discuss is more complicated. The transpose of the matrix A is defined to be $A^T = [A'_{ji}]$. In order for the transpose A^T to be defined on F^β it is necessary that the series $\sum_{i=1}^{\infty} A'_{ij} y_i$ converge in X' for each $y \in F^\beta$ with respect to some locally convex topology. In order for this to be the case we henceforth assume

(*) $(X', \sigma(X', X))$ is sequentially complete.

Under the assumption in (*) the transpose matrix A^T will map F^β into $s(X')$, the space of all X' valued sequences (Appendix C). Theorem 14.1 gives sufficient conditions for A^T to map F^β into E^β.

Corollary 14.2. *Let $A : E \to F$ and assume that $\sigma(E^\beta, E)$ is sequentially complete. Under assumption (*), $A^T : F^\beta \to E^\beta$ and*

$$y \cdot Ax = A^T y \cdot x$$

for $x \in E, y \in F^\beta$. Hence, A^T is $\sigma(F^\beta, F) - \sigma(E^\beta, E)$ continuous and, therefore, $w(F^\beta, F) - w(E^\beta, E)$ continuous with respect to any Hellinger-Toeplitz topology w (Appendix A.1).

Proof: Let the notation be as in the proof of Theorem 14.1. Then

$$y \cdot Ax = \lim_m z^m \cdot x = z \cdot x = A^T y \cdot x$$

since $\lim_m z_j^m = z_j = \sum_{i=1}^{\infty} A'_{ij} y_i$ for each j. The last statements are immediate from the equation $y \cdot Ax = A^T y \cdot x$.

Recall that even in the scalar case the matrix A may map E into F but the transpose matrix A^T may fail to map F^β into E^β (Example 10.6). It can also be the case that $A : E \to F$ and $A^T : F^\beta \to E^\beta$ but the condition $y \cdot Ax = A^T y \cdot x$ may fail to hold (Example 10.9).

We next consider the problem of when $A^T : F^\beta \to E^\beta$ will imply that $A : E \to F$. For this we always assume that dual spaces carry the strong topology and we write $E^{\beta\beta} = (E^\beta)^\beta$ and $A^{TT} = (A^T)^T$. Then $E^{\beta\beta}$ is a space of X'' valued sequences and A^{TT} consists of linear operators $A''_{ij} : X'' \to Y''$ whose restriction to X is just A_{ij}. In order that the transpose matrix A^{TT} map $E^{\beta\beta}$ into $s(Y'')$, we assume

(**) $(Y'', \sigma(Y'', Y'))$ is sequentially complete.

From Theorem 14.1 and Corollary 14.2, we have

Corollary 14.3. *Assume that $\sigma(F^{\beta\beta}, F^\beta)$ is sequentially complete. If $A^T : F^\beta \to E^\beta$, then $A^{TT} : E^{\beta\beta} \to F^{\beta\beta}$ and $z \cdot A^T y = A^{TT} z \cdot y$ for all $z \in E^{\beta\beta}, y \in F^\beta$. Moreover, A^T is $\sigma(F^\beta, F^{\beta\beta}) - \sigma(E^\beta, E^{\beta\beta})$ continuous and A^{TT} is $\sigma(E^{\beta\beta}, E^\beta) - \sigma(F^{\beta\beta}, F^\beta)$ continuous.*

Proof: Note that the assumption that $\sigma(F^{\beta\beta}, F^\beta)$ is sequentially complete implies condition (**) so the transpose matrix A^{TT} is defined on $E^{\beta\beta}$ and the result follows from Theorem 14.1 and Corollary 14.2.

To consider the problem of when $A^T : F^\beta \to E^\beta$ will imply that $A : E \to F$ we use the second transpose A^{TT}. For this we establish a lemma.

Lemma 14.4. *Let Y be semi-reflexive. If $\sigma(F, F^\beta)$ is sequentially complete, then $F = F^{\beta\beta}$.*

Proof: Let $z \in F^{\beta\beta}$. Since Y is semi-reflexive and $F \supset c_{00}(Y)$, $z^n = (z_1, ... z_n, 0, 0, ...) \in F$, and since $z \in F^{\beta\beta}$, $\{z^n\}$ is $\sigma(F, F^\beta)$ Cauchy. Since $\sigma(F, F^\beta)$ is sequentially complete, $\{z^n\}$ converges to an element of F which must be z.

Note that semi-reflexivity of Y is a necessary condition for the identity $F = F^{\beta\beta}$ to hold. From Corollary 14.3 and Lemma 14.4, we obtain

Corollary 14.5. *Assume that $\sigma(F, F^\beta)$ is sequentially complete and Y is semi-reflexive. If $A^T : F^\beta \to E^\beta$, then $A^{\tilde{T}T} : E^{\beta\beta} \to F$ and $z \cdot A^T y = A^{TT} z \cdot y$ for $z \in E^{\beta\beta}, y \in F^\beta$ so A^{TT} is $\sigma(E^{\beta\beta}, E^\beta) - \sigma(F, F^\beta)$ continuous and A^T is $\sigma(F^\beta, F) - \sigma(E^\beta, E)$ continuous. In particular, $A : E \to F$ and $y \cdot Ax = A^T y \cdot x$ holds for $x \in E, y \in F^\beta$ and A is $\sigma(E, E^\beta) - \sigma(F, F^\beta)$*

continuous. Similar continuity statements hold for any Hellinger-Toeplitz topology w.

Corollary 14.6. *Assume that Y is semi-reflexive and that $\sigma(F, F^\beta)$ and $\sigma(E^\beta, E)$ are sequentially complete. The following are equivalent:*

(i) $A : E \to F$
(ii) $A^T : F^\beta \to E^\beta$
(iii) $A^{TT} : E^{\beta\beta} \to F$.

We now give applications of the automatic continuity results to concrete sequences spaces. Recall that $l^1(X)$ is the space of all absolutely convergent X valued series and if X is quasi-barrelled, the β-dual of $l^1(X)$ is $l^\infty(X'_b)$, the space of all strongly bounded sequences with values in X' (Appendix C.25).

Theorem 14.7. *Let X be barrelled and Y be quasi-barrelled. If $A : l^1(X) \to l^1(Y)$, then $A^T : l^\infty(Y'_b) \to l^\infty(X'_b)$ and A is $\sigma(l^1(X), l^\infty(X'_b)) - \sigma(l^1(Y), l^\infty(Y'_b))$ continuous and A^T is $\sigma(l^\infty(Y'_b), l^1(Y)) - \sigma(l^\infty(X'_b), l^1(X))$ continuous. Similar continuity statements hold for any Hellinger-Toeplitz topology w.*

Proof: Since X is barrelled, condition (*) is satisfied so the transpose map A^T is defined on $l^\infty(Y'_b)$. The space $l^1(X)$ is monotone and $(X', \sigma(X', X))$ is sequentially complete so Corollary 11.18 implies that $(l^\infty(X'_b), \sigma(l^\infty(X'_b), l^1(X)))$ is sequentially complete. Theorem 14.1 and Corollary 14.2 give the result.

We next consider matrices acting between l^∞ spaces. The β-dual of $l^\infty(X)$ is $l^1(X'_b)$ (Appendix C.23).

Theorem 14.8. *Let X be barrelled. If $A : l^\infty(X) \to l^\infty(Y)$, then $A^T : l^1(Y'_b) \to l^1(X'_b)$ and A is $\sigma(l^\infty(X), l^1(X'_b)) - \sigma(l^\infty(Y), l^1(Y'_b))$ continuous and A^T is $\sigma(l^1(Y'_b), l^\infty(Y)) - \sigma(l^1(X'_b), l^\infty(X))$ continuous. Similar continuity statements hold for any Hellinger-Toeplitz topology w.*

Proof: The space $l^\infty(X)$ is monotone and $(X', \sigma(X', X))$ is sequentially complete so Corollary 11.18 implies that $(l^1(X'_b), \sigma(l^1(X'_b), l^\infty(X)))$ is sequentially complete. Theorem 14.1 and Corollary 14.2 give the result.

We give an additional automatic continuity result. The space E has the property I if the injections $x \to e^j \otimes x$ from X into E are continuous for every j.

Theorem 14.9. *Assume that every $T \in E^{\beta Y}$ induces a continuous linear operator belonging to $L(E, Y)$, (E, F) has the Banach-Steinhaus property and F is an AK-space with property I. If $A : E \to F$, then A is continuous.*

Proof: Let R^i be the ith row of A so that $R^i \in L(E, Y)$. By hypothesis $x \to R^j \cdot x \to (R^j \cdot x)e^j$ is continuous from E into F so the operator $A^n : E \to F$ defined by $A^n x = \sum_{j=1}^{n}(R^j \cdot x)e^j$ is continuous. By the AK assumption, $A^n \to A$ pointwise so A is continuous by the Banach-Steinhaus assumption.

Sufficient conditions for each $T \in E^{\beta Y}$ to induce a continuous linear operator are given in Theorem 11.2.

The analogue of Theorem 10.10 also holds with essentially the same proof.

Theorem 14.10. *Let (E, τ) be a K-space with 0-GHP and assume that condition (*) is satisfied. If $A : E \to F$, then A is $\tau - \sigma(F, F^\beta)$ continuous.*

Appendix A
Topological Vector Spaces

In this appendix we will record some of the results pertaining to topological vector spaces (TVS) which will be used throughout the text. For convenience we assume that all vector spaces are real.

A topological vector space (TVS) is a vector space X supplied with a topology τ such that the operations of addition and scalar multiplication are continuous with respect to τ. A subset U of a TVS X is symmetric (balanced) if $x \in U$ implies $-x \in U$ ($x \in U$ implies $tx \in U$ for $|t| \leq 1$). Any TVS has a neighborhood base at 0 which consists of symmetric (balanced, closed) sets. See [Sch], [Sw2] or [Wi] for discussions of TVS.

One other result pertaining to TVS will be used. A quasi-norm on a vector space X is a map $|\cdot| : X \to [0, \infty)$ satisfying $|0| = 0$, $|x + y| \leq |x| + |y|$ and $|x| = |-x|$ for $x, y \in X$, and if $t_k \to t$ in $\mathbb{R}$ and $x_k, x \in X$ with $|x_k - x| \to 0$, then $|t_k x_k - tx| \to 0$. If the quasi-norm satisfies $|x| = 0$ iff $x = 0$, then the quasi-norm is said to be total.

If $|\cdot|$ is a quasi-norm on X, then $d(x, y) = |x - y|$ defines a semi-metric on X which is a metric iff $|\cdot|$ is total. The semi-metric d is translation invariant in the sense that $d(x + z, y + z) = d(x, y)$ for $x, y, z \in X$. The space X is a TVS under the semi-metric d. A useful fact which we will use is that the topology of any TVS is generated by a family of quasi-norms ([BM]). That is, if τ is the vector topology of X, then there exists a family of quasi-norms $\{|\cdot|_a : a \in A\}$ which generate τ in the sense that a net $\{x^\delta\}$ in X converges to 0 with respect to τ iff $|x^\delta - x|_a \to 0$ for every $a \in A$.

A TVS X is locally convex ($LCTVS$) if X has a neighborhood base at 0 consisting of convex sets. Any LCTVS also has a base at 0 consisting of closed, absolutely convex sets. The topology τ of any LCTVS is generated by a family of semi-norms $\{p_a : a \in A\}$ as above. See [Sch], [Sw2] or [Wi] for the basic properties of LCTVS.

We now give a description of polar topologies which will play an important role when we discuss Orlicz-Pettis Theorems. A pair of vector spaces X, X' are said to be in duality if there is a bilinear map $\langle \cdot, \cdot \rangle : X' \times X \to \mathbb{R}$ such that

(i) $\{\langle \cdot, x \rangle : x \in X, x \neq 0\}$ separates the points of X and

(ii) $\{\langle x', \cdot \rangle : x' \in X', x' \neq 0\}$ separates the points of X'.

If X, X' are in duality, the weak topology of X $(X'), \sigma(X, X')$ $(\sigma(X'X))$, is the locally convex vector topology generated by the semi-norms $p(x) = |< x', x >|$, $x' \in X'$ $(p(x') = |< x', x >|), x \in X)$. A subset $A \subset X$ is $\sigma(X, X')$ bounded iff $\sup\{|< x', x >| : x \in A\} < \infty$ for every $x' \in X'$.

Let $\mathcal{A}$ be a family of $\sigma(X'X)$ bounded subsets of X'. For $A \in \mathcal{A}$, set

$$p_A(x) = \sup\{|\langle x, x' \rangle| : x' \in A\}.$$

The semi-norms $\{p_A : A \in \mathcal{A}\}$ generate a locally convex topology $\tau_{\mathcal{A}}$ on X called the polar topology of uniform convergence on $\mathcal{A}$ (for the reason the topology is called a polar topology, see [Sw2] 17). Thus, a net $\{x^\delta\}$ converges to 0 in $\tau_{\mathcal{A}}$ iff $\langle x', x^\delta \rangle \to 0$ uniformly for $x' \in A$ for every $A \in \mathcal{A}$.

We will use the following polar topologies in the text.

(1) The weak topology $\sigma(X, X')$ is generated by the family $\mathcal{A}$ of all finite subsets of X'.

(2) The strong topology of X, denoted by $\beta(X, X')$, is generated by the family of all $\sigma(X', X)$ bounded subsets of X'.

(3) The Mackey topology, denoted by $\tau(X, X')$, is generated by the family of all absolutely convex, $\sigma(X', X)$ compact subsets of X'.

(4) The polar topology generated by the family of all $\sigma(X', X)$ compact subsets of X' is denoted by $\lambda(X, X')$.

(5) A subset $A \subset X'$ is said to be conditionally $\sigma(X', X)$ sequentially compact if every sequence $\{x_j'\} \subset A$ has a subsequence $\{x_{n_j}'\}$ which is $\sigma(X', X)$ Cauchy, i.e., $\lim \langle x_{n_j}', x \rangle$ exists for every $x \in X$. The polar topology generated by the family of conditionally $\sigma(X', X)$ sequentially compact sets is denoted by $\gamma(X, X')$.

The topology $\lambda(X, X')$ was introduced by G. Bennett and Kalton ([BK]) and is obviously stronger than the Mackey topology $\tau(X, X')$; it can be strictly stronger ([K1] 21.4).

Let $w(X, X')$ be a polar topology defined for all dual pairs X, X'. We have the following useful notion introduced by Wilansky ([Wi]).

Definition A.1. The topology $w(\cdot, \cdot)$ is a Hellinger-Toeplitz topology if whenever

$$T : (X, \sigma(X, X')) \to (Y, \sigma(Y, Y'))$$

is linear and continuous, then

$$T : (X, w(X, X')) \to (Y, w(Y, Y'))$$

is continuous.

Wilansky has given a very useful criterion for Hellinger-Toeplitz topologies ([Wi] 11.2.2).

If $T : (X, \sigma(X, X')) \to (Y, \sigma(Y, Y'))$ is linear and continuous, then the adjoint (transpose) operator of T is the linear operator $T' : Y' \to X'$ defined by $\langle T'y', x \rangle = \langle y', Tx \rangle$ for $x \in X, y' \in Y'$. The adjoint T' is $\sigma(Y', Y) - \sigma(X', X)$ continuous.

Let $\mathcal{A}(X', X)$ be a family of $\sigma(X', X)$ bounded subsets which is defined for all dual pairs X, X'. Let $w(X, X')$ be the polar topology generated by the elements of $\mathcal{A}(X', X)$. We have

Theorem A.2. *The topology* $w(X, X')$ *is a Hellinger-Toeplitz topology if whenever* $T : (X, \sigma(X, X')) \to (Y, \sigma(Y, Y'))$ *is linear and continuous, then*

$$T'\mathcal{A}(Y', Y) \subset \mathcal{A}(X', X).$$

Proof: Let $\{x^\delta\}$ be a net in X which converges to 0 in $w(X, X')$. Let $A \in \mathcal{A}(Y', Y)$. Then $\{T'y' : y' \in A\} \in \mathcal{A}(X', X)$ so $\langle y', Tx^\delta \rangle = \langle T'y', x^\delta \rangle \to 0$ uniformly for $y' \in A$. That is, $Tx^\delta \to 0$ in $w(Y, Y')$.

Theorem A.2 clearly implies that the polar topologies given in (1) - (5) are all Hellinger-Toeplitz topologies.

We next consider another notion due to Wilansky which is useful in treating Orlicz-Pettis results.

Definition A.3. Let X be a vector space and σ and τ two vector topologies on X. We say that τ is linked to σ if τ has a neighborhood base at 0 consisting of σ closed sets. [The terminology is that of Wilansky ([Wi] 6.1.9).]

For example, the polar topologies $\beta(X, X'), \tau(X, X'), \gamma(X, X')$ and $\lambda(X, X')$ are linked to the weak topology $\sigma(X, X')$.

Lemma A.4. *Let X be a vector space and σ and τ two vector topologies on X such that τ is linked to σ.*

(i) If $\{x_j\} \subset X$ is τ Cauchy and if $\sigma\text{-lim}\, x_j = x$, then τ -$\lim x_j = x$.
(ii) If (X, σ) is sequentially complete and $\sigma \subset \tau$, then (X, τ) is sequentially complete.

Proof: (i): Let U be a τ neighborhood of 0 which is σ closed. There exists N such that $j, k \geq N$ implies $x_j - x_k \in U$. Since U is τ closed, $x_j - x \in U$ for $j \geq N$.
   (ii) follows from (i).

Remark A.5. It is important that the topologies σ and τ are linked in the Lemma. For example, consider the space c with its weak topology $\sigma(c, l^1)$ and the topology of pointwise convergence p. The series $\sum_j e^j$ is p convergent, the partial sums of the series are $\sigma(c, l^1)$ Cauchy, but the series is not $\sigma(c, l^1)$ convergent [here we are using the pairing between c and l^1 where l^1 is the topological dual of c ([Sw2] 5.12)].

   We now establish a basic lemma.

Lemma A.6. *Let X be a vector space and σ and τ two vector topologies on X such that τ is linked to σ. If every series $\sum_j x_j$ which is σ subseries convergent satisfies $\tau - \lim x_j = 0$, then every series in X which is σ subseries convergent is τ subseries convergent.*

Proof: By the previous lemma it suffices to show that every σ subseries convergent series $\sum_j x_j$ is such that its partial sums $s_n = \sum_{j=1}^n x_j$ form a τ Cauchy sequence. If $\sum_j x_j$ is σ subseries convergent but $\{s_n\}$ is not τ Cauchy, there exists a τ neighborhood of 0, U, and a pairwise disjoint sequence of finite subsets, $\{I_k\}$, of $\mathbb{N}$ such that $\max I_k < \min I_{k+1}$ and $z_k = \sum_{j \in I_k} x_j \notin U$. The series $\sum_k z_k$ is σ subseries convergent, being a subseries of $\sum_j x_j$, so $\tau - \lim z_k = 0$. This contradicts $z_k \notin U$ and establishes the result.

   We next consider topologies for spaces of continuous linear operators. Let X, Y be LCTVS and $L(X, Y)$ the space of all continuous linear operators $T : X \to Y$. Let $\mathcal{A}$ be a family of bounded ($\sigma(X, X')$) subsets of X

and let $\mathcal{Q}$ be the family of all continuous semi-norms on Y. Then $\mathcal{A}$ and $\mathcal{Q}$ generate a locally convex topology on $L(X, Y)$ defined by

$$(*) \quad p_{qA}(T) = \sup\{q(Tx) : x \in A\}, q \in \mathcal{Q}, A \in \mathcal{A}.$$

We denote by $L_{\mathcal{A}}(X, Y)$, $L(X, Y)$ with the locally convex topology generated by the semi-norms in $(*)$. A net $\{T^\delta\}$ in $L(X, Y)$ converges to 0 in $L_{\mathcal{A}}(X, Y)$ iff $T^\delta x \to 0$ uniformly for $x \in A$ for every $A \in \mathcal{A}$ and for this reason, the topology is called the topology of uniform convergence on $\mathcal{A}$ or $\mathcal{A}$-uniform convergence.

We have the following examples which will be considered.

(i) If $\mathcal{A}$ is the family of all finite subsets of X, we denote the topology generated by $\mathcal{A}$ by $L_s(X, Y)$. This is just the topology of pointwise convergence on X and is called the strong operator topology. Thus, a net $\{T^\delta\}$ converges to 0 in $L_s(X, Y)$ iff $T^\delta x \to 0$ in Y for every $x \in X$.

(ii) Let $\mathcal{A}$ again be the family of all finite subsets of X but equip Y with the weak topology $\sigma(Y, Y')$. This topology is called the weak operator topology. Thus, a net $\{T^\delta\}$ in $L(X, Y)$ converges to 0 in the weak operator topology iff for every $x \in X$, $T^\delta x \to 0$ in $\sigma(Y, Y')$ or, equivalently, $\langle y', T^\delta x \rangle \to 0$ for every $x \in X, y' \in Y'$.

(iii) Let $\mathcal{A}$ be the family of all bounded subsets of X. Then the topology generated by $\mathcal{A}$ is denoted by $L_b(X, Y)$. If X and Y are normed spaces, the topology $L_b(X, Y)$ is generated by the operator norm

$$\|T\| = \sup\{\|Tx\| : \|x\| \le 1\}$$

and is called the uniform operator topology.

(iv) Let $\mathcal{A}$ be the family of all precompact subsets of X. Then the topology generated by $\mathcal{A}$ is denoted by $L_{pc}(X, Y)$.

(v) Let $\mathcal{A}$ be the family of all compact subsets of X. Then the topology generated by $\mathcal{A}$ is denoted by $L_c(X, Y)$.

(vi) Let $\mathcal{A}$ be the family of all sequences $\{x_j\} \subset X$ which converge to 0. Then the topology generated by $\mathcal{A}$ is denoted by $L_{\to 0}(X, Y)$.

Appendix B
Scalar Sequence Spaces

In this appendix we will list the sequence spaces and their properties which will be used in the text. If λ is a vector space of (real) sequences containing c_{00}, the space of all sequences which are eventually 0, the β-dual of λ is defined to be

$$\lambda^\beta = \left\{ s = \{s_j\} : \sum_{j=1}^{\infty} s_j t_j = \{s_j\} \cdot \{t_j\} = s \cdot t \text{ converges for every } t = \{t_j\} \in \lambda \right\}.$$

Since $\lambda \supset c_{00}$, the pair λ, λ^β are in duality with respect to the pairing $s \cdot t = \{s_j\} \cdot \{t_j\}$ for $s \in \lambda^\beta, t \in \lambda$.

We now list some of the scalar valued sequence spaces which will be encountered in the text.

- $c_{00} = \{\{t_j\} : t_j = 0 \text{ eventually}\}$
- $c_0 = \{\{t_j\} : \lim t_j = 0\}$
- $c_c = \{\{t_j\} : t_j \text{ is eventually constant}\}$
- $c = \{\{t_j\} : \lim t_j \text{ exists}\}$
- $m_0 = \{\{t_j\} : \text{the range of } \{t_j\} \text{ is finite}\} = span\{\chi_\sigma : \sigma \subset \mathbb{N}\}$
- $l^\infty = \{\{t_j\} : \sup_j\{|t_j|\} = \|\{t_j\}\|_\infty < \infty\}$

All of the sequence spaces above are usually equipped with the sup-norm, $\|\cdot\|_\infty$, defined above.

For $0 < p < 1$,

- $l^p = \{\{t_j\} : \sum_{j=1}^{\infty} |t_j|^p = |\{t_j\}|_p < \infty\}$

The space l^p $(0 < p < 1)$ is usually equipped with the quasi-norm $|\cdot|_p$ which generates the metric $d_p(s,t) = |s-t|_p$ under which it is complete.

For $1 \le p < \infty$,

- $l^p = \{\{t_j\} : (\sum_{j=1}^{\infty} |t_j|^p)^{1/p} = \|\{t_j\}\|_p < \infty\}$

The space l^p is usually equipped with the norm $\|\cdot\|_p$ under which it is a Banach space.

- $bs = \{\{t_j\} : \sup_n \{|\sum_{j=1}^n t_j|\} = \|t\|_{bs} < \infty\}$

The space bs is called the space of bounded series and is usually equipped with the norm $\|\cdot\|_{bs}$ under which it is a Banach space.

- $cs = \{\{t_j\} : \sum_{j=1}^{\infty} t_j \text{ converges}\}$

The space cs is a subspace of bs and is called the space of convergent series; cs is a closed subspace of bs under the norm $\|\cdot\|_{bs}$.

- $bv = \{\{t_j\} : \sum_{j=1}^{\infty} |t_{j+1} - t_j| < \infty\}$

The space bv is called the space of sequences of bounded variation and is a Banach space under the norm $\|\{t_j\}\|_{bv} = \sum_{j=1}^{\infty} |t_{j+1} - t_j| + |t_1|$.

- $bv_0 = bv \cap c_0$

The space bv_0 is a closed subspace of bv.

- $s =$ the space of all real valued sequences.

The space s is a Frechet space under the metric

$$d(s,t) = \sum_{j=1}^{\infty} |s_j - t_j| / 2^j (1 + |s_j - t_j|)$$

of coordinatewise convergence.

We give a list of the β-duals and topological duals of the spaces above. For these, see [HK] and [Bo].

space	β-dual	topological dual
c_{00}	s	l^1
c_0	l^1	l^1
c_c	cs	l^1
c	l^1	l^1
m_0	l^1	ba
l^∞	l^1	ba
$l^p (0 < p < 1)$	l^∞	l^∞
$l^p (1 \le p < \infty)$	$l^q(\frac{1}{p} + \frac{1}{q} = 1)$	$l^q(\frac{1}{p} + \frac{1}{q} = 1)$
bs	bv_0	bv
cs	bv	bv
bv_0	bs	bs
s	c_{00}	c_{00}

We now list some of the properties of sequence spaces which will be encountered in the sequel. Throughout the remainder of this appendix λ will denote a sequence space containing c_{00}. Suppose that λ is equipped with a Hausdorff vector topology.

Definition B.1. The space λ is a K-space if the coordinate functionals $t = \{t_j\} \to t_j$ are continuous from λ into $\mathbb{R}$ for every j. If the K-space λ is a Banach (Frechet) space, λ is called a BK-space (FK-space).

All of the spaces listed above are K-spaces under their natural topologies.

Let e^j be the sequence with a 1 in the jth coordinate and 0 in the other coordinates.

Definition B.2. The K-space λ is an AK-space if the $\{e^j\}$ form a Schauder basis for λ, i.e., if $t = \{t_j\} \in \lambda$, then $t = \lim_n \sum_{j=1}^n t_j e^j$, where the convergence is in λ.

The spaces c_{00}, c_0, l^p $(0 < p < \infty), cs, bv$ and bv_0 are AK-spaces. The spaces m_0 and l^∞ are not AK-spaces.

For each n let $P_n : \lambda \to \lambda$ be the sectional projection (operator) defined by $P_n t = \sum_{j=1}^n t_j e^j = (t_1, ..., t_n, 0, 0, ...)$.

Definition B.3. The K-space λ is an AB-space if $\{P_n t : n \in \mathbb{N}\}$ is bounded for each $t \in \lambda$, i.e., if the $\{P_n\}$ are pointwise bounded on λ.

Definition B.4. The K-space λ has the sections uniformly bounded property (SUB) if $\{P_n t : n \in \mathbb{N}, t \in B\}$ is bounded for every bounded subset B of λ, i.e., if the $\{P_n\}$ are uniformly bounded on bounded subsets of λ.

Definition B.5. The K-space λ has the property SE (sections equicontinuous) if the sectional operators $\{P_n\}$ from λ into λ are equicontinuous.

Obviously, property SUB implies that λ is an AB-space and property SE implies property SUB. If λ is a barrelled AB-space, then λ has property SE and, therefore, SUB ([Sw2], [Wi]). If λ is a metric linear space whose topology is generated by the quasi-norm $|\cdot|$ which satisfies $|P_n t| \leq M |t|$ for some M and all $t \in \lambda$, then λ has property SE; e.g., s, l^∞ and its subspaces, l^p $(0 \leq p < \infty)$, and bs and its subspace cs.

Throughout this text numerous gliding hump properties are employed. We now list these gliding hump properties and give examples of sequence spaces which satisfy the various gliding hump properties. If $\sigma \subset \mathbb{N}$, χ_σ will denote the characteristic function of σ and if $t = \{t_j\}$ is any sequence (scalar or vector), $\chi_\sigma t$ will denote the coordinatewise product of χ_σ and t.

A sequence space λ is *monotone* if $\chi_\sigma t \in \lambda$ for every $\sigma \subset \mathbb{N}$ and $t \in \lambda$. A sequence space λ is *normal (solid)* if $t \in \lambda$ and $|s_j| \leq |t_j|$ implies that $s = \{s_j\} \in \lambda$. Obviously, a normal space is monotone; the space m_0 is monotone but not normal. The spaces c_{00}, c_0, l^p $(0 < p \leq \infty)$ and s are normal whereas c_c, c, bs, cs, bv and bv_0 are not monotone.

An *interval* in $\mathbb{N}$ is a subset of the form

$$[m, n] = \{j \in \mathbb{N} : m \leq j \leq n\},$$

where $m, n \in \mathbb{N}$ with $m \leq n$. A sequence of intervals $\{I_j\}$ is *increasing* if $\max I_j < \min I_{j+1}$ for every j. A sequence of *signs* is a sequence $\{s_j\}$ with $s_j = \pm 1$ for every j.

We begin with 2 gliding hump properties which are algebraic and require no topology on the sequence space λ.

Definition B.6. Let $\Lambda \subset \lambda$. Then Λ has the signed weak gliding hump property (signed-WGHP) if for every $t \in \Lambda$ and every increasing sequence of intervals $\{I_j\}$, there is a subsequence $\{n_j\}$ and a sequence of signs $\{s_j\}$ such that the coordinatewise sum of the series $\sum_{j=1}^\infty s_j \chi_{I_{n_j}} t$ belongs to Λ. If the signs s_j can all be chosen to be equal to 1 for every $t \in \Lambda$, then Λ has the weak gliding hump property (WGHP).

The weak gliding hump property was introduced by Noll ([No]) and the signed weak gliding hump property was introduced by Stuart ([St1],

[St2]). We now give some examples of sequence spaces with WGHP and signed-WGHP.

Example B.7. Any monotone space has WGHP.

We now show that the space cs of convergent series is not monotone but has WGHP.

Example B.8. cs has WGHP but is not monotone. Let $t \in cs$ and $\{I_j\}$ be an increasing sequence of intervals. Since the series $\sum_j t_j$ converges,

$$\chi_{I_j \cap J} \cdot t = \sum_{k \in I_j \cap J} t_k \to 0$$

for any interval J. Pick a subsequence $\{n_j\}$ such that $\left| \chi_{I_{n_j} \cap J} \cdot t \right| < 1/2^j$ for every interval J. Then $\sum_{j=1}^\infty \chi_{I_{n_j}} t \in cs$ since this series satisfies the Cauchy criterion.

If $t = \{(-1)^j/j\}$, then $t \in cs$, but if $\sigma = \{1, 3, 5, ...\}$, then $\chi_\sigma t \notin cs$ so cs is not monotone.

Example B.9. The space c of convergent sequences does not have WGHP. For example, $t = \{1, 1, ...\} \in c$ and if $I_j = \{2j\}$, then $\sum_{j=1}^\infty \chi_{I_{n_j}} t \notin c$ for any subsequence $\{n_j\}$.

We show later in Example B.26 that the space bs of bounded series has signed-WGHP but not WGHP so the inclusion of signs in Definition B.6 is important.

We next establish a general criterion for a space to have WGHP. A TVS X is a $\mathcal{K}$-space if whenever $\{x_j\}$ is a null sequence in X and $\{n_j\}$ is a subsequence, then there is a further subsequence $\{m_j\}$ of $\{n_j\}$ such that the series $\sum_{j=1}^\infty x_{m_j}$ converges in X. For example, any complete metric linear space is a $\mathcal{K}$-space ([Sw1] 3.2.3, there are further examples in this text).

Theorem B.10. *If λ is a $\mathcal{K}$-space and an AK-space, then λ has WGHP.*

Proof: Let $t \in \lambda$ and $\{I_j\}$ be an increasing sequence of intervals. Since $t = \sum_{j=1}^\infty t_j e^j$ converges in λ, $\lim_k \sum_{j \in I_k} t_j e^j = 0$. Since λ is a $\mathcal{K}$-space, there exists $\{n_k\}$ such that $\sum_{k=1}^\infty \sum_{j \in I_{n_k}} t_j e^j$ converges to some $s \in \lambda$. Since λ is a K-space, the series $\sum_{k=1}^\infty \sum_{j \in I_{n_k}} t_j e^j$ also converges to s pointwise.

Theorem B.10 applies to spaces such as c_0, l^p $(0 < p < \infty)$ and cs.

We next consider results related to ideas due to Garling ([Ga]). For this we need the following observation.

Example B.11. Let B_0 be the closed unit ball of bv_0. We show that B_0 has WGHP. Let $t \in B_0$ and $\{I_k\}$ be an increasing sequence of intervals. Then $\sum_{k=1}^{\infty} |t_{k+1} - t_k| \leq 1$ and $\lim t_k = 0$. Choose $\{n_k\}$ such that

$$\sum_{k=1}^{\infty} \sum_{j \in I_{n_k}} |t_{j+1} - t_j| < 1/2$$

and

$$\sum_{k=1}^{\infty} 2 \max\{|t_{\min I_{n_k}}|, |t_{\max I_{n_k}}|\} < 1/2$$

(this is possible since $t \in c_0$, so we can extract a subsequence which belongs to l^1). Then $\chi_{\cup_{k=1}^{\infty} I_{n_k}} t$ has total variation less than or equal to

$$\sum_{k=1}^{\infty} \sum_{j \in I_{n_k}} |t_{j+1} - t_j| + \sum_{k=1}^{\infty} 2 \max\{|t_{\min I_{n_k}}|, |t_{\max I_{n_k}}|\} < 1$$

so $\chi_{\cup_{k=1}^{\infty} I_{n_k}} t \in B_0$.

Definition B.12. (Garling) Let $\mathcal{S}$ be any subset of sequences. The space λ is $\mathcal{S}$ invariant if $\mathcal{S}\lambda = \lambda$.

For example, λ is monotone iff λ is m_0 invariant.

Proposition B.13. *If $\mathcal{S}$ has signed-WGHP (WGHP) and λ is $\mathcal{S}$ invariant, then λ has signed-WGHP (WGHP).*

Proof: For $t \in \lambda$ there exists $u = \{u_k\} \in \mathcal{S}, v \in \lambda$ such that $t = uv$. Let $\{I_k\}$ be an increasing sequence of intervals. By hypothesis there exist a subsequence $\{I_{n_k}\}$ and a sequence of signs $\{s_k\}$ such that $u' = \sum_{k=1}^{\infty} s_k \chi_{I_{n_k}} u \in \mathcal{S}$ (coordinate sum). Then $\{u'_k v_k\} \in \lambda$ and since $\{u'_k v_k\} = \sum_{k=1}^{\infty} s_k \chi_{I_{n_k}} t$, λ has signed-WGHP.

From Example B.11 and Proposition B.14, we have

Corollary B.14. *If λ is B_0 invariant, then λ has WGHP.*

We next consider a notion introduced by Noll ([No]).

Definition B.15. The multiplier space of λ, $M(\lambda)$, is defined to be $\{s : st \in \lambda$ for all $t \in \lambda\}$.

Proposition B.16. *If $M(\lambda)$ has signed-WGHP (WGHP), then λ has signed-WGHP (WGHP).*

Proof: Consider the case of WGHP. The constant sequence e with 1 in each coordinate belongs to $M(\lambda)$. Let $t \in \lambda$ and $\{I_k\}$ be an increasing sequence of intervals. There exist $\{n_k\}$ such that $\sum_{k=1}^{\infty} \chi I_{n_k} e \in M(\lambda)$ so $(\sum_{k=1}^{\infty} \chi I_{n_k} e)t = \sum_{k=1}^{\infty} \chi I_{n_k} t \in \lambda$.

The results above in Propositions B.14 and B.16 are used in Chapter 2 to establish weak sequential completeness for β-duals.

We next consider gliding hump properties which depend on the topology of the sequence space λ. In what follows we assume that λ is a K-space.

Definition B.17. Let $\Lambda \subset \lambda$. Then Λ has the signed strong gliding hump property (signed-SGHP) if for every bounded sequence $\{t^j\} \subset \Lambda$ and every increasing sequence of intervals $\{I_j\}$, there is a subsequence $\{n_j\}$ and a sequence of signs $\{s_j\}$ such that the coordinatewise sum of the series $\sum_{j=1}^{\infty} s_j \chi I_{n_j} t^{n_j} \in \Lambda$. If the signs s_j can be chosen equal to 1 for every $t \in \Lambda$, then Λ is said to have the strong gliding hump property (SGHP).

The strong gliding hump property was introduced by Noll ([No]) and the signed strong gliding hump property was introduced in [Sw4].

Example B.18. The space l^{∞} has SGHP. The spaces l^p $(0 < p < \infty)$ and c_0 do not have SGHP (consider $\{e^j\}$ and $I_j = \{j\}$).

Example B.19. The subset $M_0 = \{\chi_{\sigma} : \sigma \subset \mathbb{N}\} \subset m_0$ has SGHP while the space $m_0 = \mathrm{span} M_0$ does not have SGHP.

We consider smaller subsets of $2^{\mathbb{N}}$ whose characteristic functions have SGHP.

Definition B.20. A family $\mathcal{F}$ of subsets of $\mathbb{N}$ is an FQσ family if $\mathcal{F}$ contains the finite sets and if whenever $\{I_j\}$ is a pairwise disjoint sequence of finite subsets, there is a subsequence $\{I_{n_j}\}$ such that $\cup_{j=1}^{\infty} I_{n_j} \in \mathcal{F}$.

This notion is due to Sember and Samaratanga ([SaSe]). We give an example of an FQσ family which is a proper subset of $2^{\mathbb{N}}$.

Example B.21. Haydon has given an example of an algebra $\mathcal{H}$ of subsets of $\mathbb{N}$ such that $\mathcal{H}$ is an FQσ family but for no infinite $A \subset \mathbb{N}$ do we have $2^A = \{A \cap B : B \in \mathcal{H}\}$ ([Hay]).

Example B.22. If $\mathcal{F}$ is an FQσ family and $\Lambda = \{\chi_\sigma : \sigma \in \mathcal{F}\} \subset m_0$, then Λ has SGHP (if $\{\chi_{\sigma_j}\} \subset \Lambda$ and $\{I_j\}$ is an increasing sequence of intervals, then there exists a subsequence $\{n_j\}$ such that $I = \cup_{j=1}^\infty \sigma_{n_j} \cap I_{n_j} \in \mathcal{F}$ so $\chi_I = \sum_{j=1}^\infty \chi_{I_{n_j}} \chi_{\sigma_{n_j}} \in \Lambda$).

Definition B.23. A family of subsets $\mathcal{F}$ of $\mathbb{N}$ is an IQσ family if $\mathcal{F}$ contains the finite sets and if whenever $\{I_j\}$ is an increasing sequence of intervals, there is a subsequence $\{I_{n_j}\}$ such that $\cup_{j=1}^\infty I_{n_j} \in \mathcal{F}$.

This notion is also due to Sember and Samaratanga ([SaSe]). We give an example of an IQσ family containing $\mathbb{N}$.

Example B.24. Let $\sum_{j=1}^\infty t_j$ be a conditionally convergent scalar series. Put $\mathcal{F} = \{\sigma : \sum_{j \in \sigma} t_j$ converges$\}$. Then $\mathcal{F}$ is an IQσ family containing $\mathbb{N}$. For suppose that $\{I_j\}$ is an increasing sequence of intervals. There exists a subsequence $\{I_{n_j}\}$ such that $\left|\sum_{i \in I_{n_j} \cap J} t_i\right| < 1/2^j$ for every j and every interval J. Then $I = \cup_{j=1}^\infty I_{n_j} \in \mathcal{F}$ since the series $\sum_{j \in I} t_j$ satisfies the Cauchy condition. Note that $\mathbb{N} \in \mathcal{F}$ but $\mathcal{F} \neq 2^\mathbb{N}$ since the series $\sum_j t_j$ is conditionally convergent.

As in Example B.22, we have

Example B.25. Let $\mathcal{F}$ be an IQσ family and $\Lambda = \{\chi_\sigma : \sigma \in \mathcal{F}\} \subset m_0$. Then Λ has SGHP.

We next show that the space bs of bounded series has signed-SGHP but not SGHP.

Example B.26. The space bs has signed-SGHP but not SGHP. Let $\{t^j\} \subset bs$ be bounded and $\{I_j\}$ be an increasing sequence of intervals. Put

$$M = \sup\left\{\left|\sum_{i \in I} t_i^j\right| : j \in \mathbb{N}, I \text{ an interval in } \mathbb{N}\right\} < \infty.$$

Define signs inductively by setting $s_1 = sign\chi_{I_1} \cdot t^1$ and

$$s_{n+1} = -[sign \sum_{k=1}^n s_k\chi_{I_k} \cdot t^k][sign\chi_{I_{n+1}} \cdot t^{n+1}].$$

Put $y = \sum_{k=1}^\infty s_k\chi_{I_k} t^k$. We show $\|y\|_{bs} \leq 2M$. We first show by induction that $\left|\sum_{j=1}^{\max I_n} y_j\right| \leq M$ for every n. For $n = 1$, $\left|\sum_{j=1}^{\max I_1} y_j\right| =$

$\left|\sum_{j\in I_1} s_1 t_j^1\right| \le M$. Suppose the inequality holds for n. Then

$$\left|\sum_{j=1}^{\max I_{n+1}} y_j\right| = \left|\sum_{j=1}^{\max I_n} y_j + \sum_{j\in I_{n+1}} y_j\right| \le M$$

since $\left|\sum_{j\in I_{n+1}} y_j\right| = \left|\sum_{j\in I_{n+1}} s_{n+1} y_j\right| \le M$ and $\left|\sum_{j=1}^{\max I_n} y_j\right| \le M$ and both of these terms have opposite signs. Now for arbitrary n, let $k = k_n$ be the largest integer such that $\max I_k \le n$. Then

$$\left|\sum_{j=1}^{n} y_j\right| = \left|\sum_{j=1}^{\max I_k} y_j + \sum_{j=\min I_{k+1}}^{n} y_j\right| \le \left|\sum_{j=1}^{\max I_k} y_j\right| + \left|\sum_{j=\min I_{k+1}}^{n} s_{k+1} t_j^{k+1}\right| \le 2M$$

so $\|y\|_{bs} \le 2M$ as desired.

Note that bs does not have WGHP (consider $t = \{1, -1, 1, -1, ...\}$ and $I_j = \{2j - 1\}$).

The proof above is essentially that of Stuart who showed that bs has signed-WGHP but not WGHP ([St1], [St2]). Further examples of spaces with SGHP (WGHP) and signed-SGHP (signed-WGHP) are constructed later in this appendix. Note that in the proof of Example B.26 it was not necessary to pass to a subsequence in the definition of signed-SGHP.

Definition B.27. The K-space λ has the zero gliding hump property (0-GHP) if whenever $t^j \to 0$ in λ and $\{I_j\}$ is an increasing sequence of intervals, there is a subsequence $\{n_j\}$ such that the coordinate sum of the series $\sum_{j=1}^{\infty} \chi_{I_{n_j}} t^{n_j}$ belongs to λ.

The 0-GHP was essentially introduced by Lee Peng Yee ([LPY]); see also [LPYS].

We give some examples of spaces with 0-GHP. Recall the section operators $P_n : \lambda \to \lambda$ are defined by $P_n t = \sum_{j=1}^{n} t_j e^j$.

Proposition B.28. *Let λ be a K-space with property SE. Then λ has 0-GHP.*

Proof: Let $t^j \to 0$ in λ and let $\{I_j\}$ be an increasing sequence of intervals. Then $\chi_{I_j} t^j \to 0$ in λ by property SE. Since λ is a K-space, there is a subsequence $\{n_j\}$ such that the subseries $\sum_j \chi_{I_{n_j}} t^{n_j}$ converges to some $t \in \lambda$. Since λ is a K-space, the series $\sum_j \chi_{I_{n_j}} t^{n_j}$ converges coordinatewise to t.

From Proposition B.28, it follows that the spaces l^p $(0 < p \leq \infty)$, s, c, cs, bs and c_0 have 0-GHP. The space c_{00} does not have 0-GHP. Further examples are given later in this appendix and can be found in [Sw1] 12.5.

The gliding hump properties WGHP and 0-GHP are independent; the space c has 0-GHP but not WGHP while the space c_{00} has WGHP but not 0-GHP. We give a simple proposition which relates the two conditions.

Proposition B.29. *If λ is an AK-space with 0-GHP, then λ has WGHP.*

Proof: Let $t \in \lambda$ and $\{I_j\}$ be an increasing sequence of intervals. Since λ has AK, $\chi_{I_j} t \to 0$. By 0-GHP, there is a subsequence $\{n_j\}$ such that $\sum_{j=1}^{\infty} \chi_{I_{n_j}} t \in \lambda$, where the series is coordinatewise convergent since λ is a K-space.

We next define a gliding hump property which is used to establish uniform boundedness principles.

Let μ be sequence space containing c_{00}.

Definition B.30. The K-space λ has the strong μ gliding hump property (strong μ-GHP) if whenever $\{t^j\}$ is a bounded sequence in λ and $\{I_j\}$ is an increasing sequence of intervals, the coordinate sum of the series $\sum_{j=1}^{\infty} u_j \chi_{I_j} t^j$ belongs to λ for every $u = \{u_j\} \in \mu$.

Definition B.31. The K-space λ has the weak μ gliding hump property (weak μ-GHP) if whenever $\{t^j\}$ is a bounded sequence in λ and $\{I_j\}$ is an increasing sequence of intervals, there is a subsequence $\{n_j\}$ such that the coordinate sum of the series $\sum_{j=1}^{\infty} u_j \chi_{I_{n_j}} t^{n_j}$ belongs to λ for every $u = \{u_j\} \in \mu$.

Of course, the difference in the strong μ-GHP and the weak μ-GHP is the necessity to pass to a subsequence in Definition B.31. We refer to the elements $u = \{u_j\} \in \mu$ as multipliers since the coordinates of u multiply the blocks or "humps", $\{\chi_{I_j} t^j\}$, determined by the $\{t^j\}$ and $\{I_j\}$. This is analogous to the situation in the signed-WGHP or signed-SGHP where the humps are multiplied by $\{\pm 1\}$. We give examples of spaces with μ-GHP.

Proposition B.32. *If λ is a locally complete K-space with property SUB, then λ has strong l^1-GHP.*

Proof: Let $\{t^j\} \subset \lambda$ be bounded and $\{I_j\}$ be an increasing sequence of intervals. By SUB, $\{\chi_{I_j} t^j\}$ is bounded in λ so if $u \in l^1$, the series $\sum_{j=1}^{\infty} u_j \chi_{I_j} t^j$ is absolutely convergent in λ and, therefore, converges to an element $t \in \lambda$ by local completeness. Since λ is a K-space, the series $\sum_{j=1}^{\infty} u_j \chi_{I_j} t^j$ is coordinatewise convergent to t.

Example B.33. From Proposition B.32, it follows that the spaces l^p ($0 < p \leq \infty$), s, c, cs, bs and c_0 have strong l^1-GHP.

We also have

Example B.34. The spaces l^∞ and c_0 have strong c_0-GHP; l^p ($0 < p \leq \infty$) has strong l^∞-GHP.

We next give examples of non-complete spaces with weak l^p-GHP. This example requires some properties of integration with respect to finitely additive set functions. We refer the reader to [RR] for a discussion of the integrals.

Example B.35. Let $1 \leq p < \infty$. Let $\mathcal{P}$ be the power set of $\mathbb{N}$ and let $\alpha : \mathcal{P} \to [0, \infty)$ be a finitely additive set function with $\alpha(\{j\}) > 0$ for every j. Let $l^p(\alpha) = L^p(\alpha)$ be the space of all p^{th} power α-integrable functions with the norm $\|f\|_p = (\int_{\mathbb{N}} |f|^p \, d\alpha)^{1/p}$ [see [RR] for details; the assumption that $\alpha(\{j\}) > 0$ makes $l^p(\alpha)$ a K-space]. We show that $l^p(\alpha)$ has weak l^p-GHP. Let $\{f_j\} \subset l^p(\alpha)$ be bounded with $\|f_j\|_p \leq 1$ and $\{I_j\}$ be an increasing sequence of intervals. By Drewnowski's Lemma (Appendix E.2), there is a subsequence $\{n_j\}$ such that α is countably additive on the σ-algebra generated by $\{I_{n_j}\}$. Suppose that $t \in l^p$. Put $f = \sum_{j=1}^{\infty} t_j \chi_{I_{n_j}} f_{n_j}$ [coordinate sum]. We claim that $f \in l^p(\alpha)$ and the series $\sum_{j=1}^{\infty} t_j \chi_{I_{n_j}} f_{n_j}$ converges to f in $l^p(\alpha)$ by using Theorem 4.6.10 of [RR]; this will establish the result. Put $s_n = \sum_{j=1}^{n} t_j \chi_{I_{n_j}} f_{n_j}$ and note that $s_n \to f$ α-hazily [i.e., in α measure] since if $\epsilon > 0$,

$$\alpha(\{j : |s_n(j) - f(j)| \geq \epsilon\}) \leq \alpha(\cup_{j=n+1}^{\infty} I_{n_j}) = \sum_{j=n+1}^{\infty} \alpha(I_{n_j}) \to 0$$

by the countable additivity. Next, $\{s_n\}$ is Cauchy in $l^p(\alpha)$ since

$$\|s_n - s_m\|_p^p = \left\| \sum_{j=m+1}^{n} t_j \chi_{I_{n_j}} f_{n_j} \right\|_p^p \leq \sum_{j=m+1}^{n} |t_j|^p \to 0.$$

It follows that $\{\int |s_n|^p \, d\alpha\}$ is uniformly α-continuous and using [RR] this justifies the claim.

We next consider a gliding hump property called the infinite gliding hump property which has been proven to be useful when considering strong convergence. This property is algebraic and requires no topology on λ.

Definition B.36. The sequence space λ has the infinite gliding hump property (∞-GHP) if whenever $t \in \lambda$ and $\{I_j\}$ is an increasing sequence of intervals, there exist a subsequence $\{n_j\}$ and $a_{n_j} > 0, a_{n_j} \to \infty$ such that every subsequence of $\{n_j\}$ has a further subsequence $\{p_j\}$ such that the coordinate sum of the series $\sum_{j=1}^{\infty} a_{p_j} \chi_{I_{p_j}} t \in \lambda$.

The term "infinite" gliding hump property is used to suggest that the humps $\{\chi_{I_j} t\}$ are multiplied by a sequence of scalars which tend to infinity. We now give examples of sequence spaces with ∞-GHP. For this, we introduce another property for a sequence space.

Definition B.37. The space λ is c_0-factorable (called c_0-invariant by Garling ([Ga])) if $t \in \lambda$ implies that there exist $s \in c_0, u \in \lambda$ with $t = su$ [coordinate product].

Proposition B.38. *If λ is normal and c_0-factorable, then λ has ∞-GHP.*

Proof: Let $t \in \lambda$ with $t = su$, $s \in c_0, u \in \lambda$ and let $\{I_j\}$ be an increasing sequence of intervals. Pick an increasing sequence $\{n_j\}$ such that $\sup\{|s_i| : i \in I_{n_j}\} = b_{n_j} > 0$ (if this choice is not possible there is nothing to do). Note that $b_{n_j} \to 0$ so $a_{n_j} = 1/b_{n_j} \to \infty$. Define $v_j = s_j a_{n_k}$ if $j \in I_{n_k}$ and $v_j = 0$ otherwise; then $v \in l^{\infty}$ so $vu \in \lambda$ since λ is normal. We have $\sum_{j=1}^{\infty}(vu)_j e^j = \sum_{k=1}^{\infty} a_{n_k} \chi_{I_{n_k}} t \in \lambda$. Since the same argument can be applied to any subsequence, λ has ∞-GHP.

We now give some examples of spaces which satisfy the conditions of Proposition B.38.

Example B.39. The space c_0 is normal and c_0-factorable so has ∞-GHP.

Example B.40. Let $0 < p < \infty$. Let $t \in l^p$. Pick an increasing sequence $\{n_j\}$ with $n_0 = 0$ such that $\sum_{i=n_j+1}^{n_{j+1}} |t_j|^p < 1/2^{j(p+1)}$ and set $I_j = [n_j + 1, n_{j+1}]$, $s = \sum_{j=1}^{\infty} 2^{-j} \chi_{I_j}$, $u = \sum_{j=1}^{\infty} 2^j \chi_{I_j} t$. Then $t = su$ so l^p is c_0-factorable and is obviously normal so Proposition B.38 applies.

Example B.41. As in Example B.40, it can be shown that the spaces $d = \{t : \sup\{|t_j|^{1/j} < \infty\}$ and $\delta = \{t : \lim |t_j|^{1/j} = 0\}$ are c_0-factorable and are both clearly normal so Proposition B.38 applies [see [KG] for these spaces].

We give examples of non-normal spaces with ∞-GHP.

Proposition B.42. *If λ is a Banach AK-space, then λ has ∞-GHP.*

Proof: Let $t \in \lambda$ and $\{I_j\}$ be an increasing sequence of intervals. Choose a subsequence $\{n_j\}$ such that $\left\|\sum_{i \in I_{n_j} \cap J} t_j e^j\right\| < 1/j2^j$ for any interval J. Consider $s = \sum_{j=1}^{\infty} j\chi_{I_{n_j}} t$. If J is any interval contained in the interval $[\min I_{n_j}, \infty)$, then $\left\|\sum_{j \in J} s_j e^j\right\| = \left\|\sum_{i=j}^{\infty} \sum_{k \in I_{n_j} \cap J} it_k e^k\right\| \leq \sum_{i=j}^{\infty} 1/2^j = 2^{-j+1}$ so the partial sums of the series generated by s are Cauchy and, therefore, convergent. Hence, $s \in \lambda$ and λ has ∞-GHP since λ is a K-space.

Example B.43. For example, it follows from Proposition B.42 that the non-normal space cs of convergent series has ∞-GHP. Likewise, bv_0 has ∞-GHP.

Example B.44. The spaces l^{∞}, m_0, bs and bv do not have ∞-GHP.

To this point the only example of a space with signed-SGHP (signed-WGHP) is bs. We will now describe a method which can be used to construct more sequence spaces with these and other gliding hump properties ([BSS]).

Let $A = [a_{ij}]$ be an infinite matrix with scalar entries. We use A and λ to generate a further sequence space. The *matrix domain* of A and λ is defined to be

$$\lambda_A = \{t = \{t_j\} : At \in \lambda\}.$$

Thus, A is a linear map from $\lambda_A \to \lambda$. Some of the familiar sequence spaces can be generated by this procedure. In particular, c_A and $c_{0A} = (c_0)_A$ are the spaces of all sequences which are A-summable and A-summable to 0, respectively ([Bo]).

Example B.45. Let $B = [b_{ij}]$ be the matrix with $b_{ij} = 1$ for $j \leq i$ and $b_{ij} = 0$ otherwise. Then $l_B^{\infty} = bv$, the space of sequences with bounded variation, and $c_B = cs$, the space of convergent series.

Example B.46. Let $B_1 = [b_{ij}]$ be the matrix with $b_{ii} = 1, b_{i+1,i} = -1$ and $b_{ij} = 0$ otherwise. Then $l_{B_1}^1 = bv$.

Example B.47. Let $C = [c_{ij}]$ be the Cesaro matrix , $c_{ij} = 1/i$ for $1 \leq j \leq i$ and $c_{ij} = 0$ otherwise. Then l_C^{∞} is the space of sequences $\{t_j\}$ with bounded averages, $\sup_j \left|\sum_{i=1}^{j} t_i/j\right| < \infty$.

Let $\{d_j\}$ be a scalar sequence and let D be the diagonal matrix with the $\{d_j\}$ as the entries down the diagonal.

Theorem B.48. *If λ has WGHP (signed-WGHP), then λ_D has WGHP (signed-WGHP).*

Proof: Let $t \in \lambda_D$ and $\{I_j\}$ be an increasing sequence of intervals. Since $Dt \in \lambda$, there is a subsequence $\{n_j\}$ such that

$$v = \sum_{j=1}^{\infty} \chi_{I_{n_j}} Dt \in \lambda.$$

Then

$$u = \sum_{j=1}^{\infty} \chi_{I_{n_j}} t \in \lambda_D$$

since $Du = v$.

The proof for the signed-WGHP case is similar.

Corollary B.49. *bs_D has signed-WGHP for any diagonal matrix D.*

If each $d_j \neq 0$, the spaces bs and bs_D are algebraically isomorphic, but, in general, the spaces may have very different topological properties depending on the growth of the sequence $\{d_j\}$. For example, if $d_j \to \infty$ and $t \in bs_D$, then $t \in c_0$ so $bs_D \subset c_0$ in this case. On the other hand, if $d_j \to 0$ and $d_j \neq 0$ for all j, then

$$u = (1/d_1, -1/d_2, 1/d_3, ...) \in bs_D$$

and $\lim u_j = \infty$ so $u \notin l^{\infty}$ in this case. Also, if $d_j = 0$ for some j, then the jth coordinate of any element in bs_D can be arbitrarily large so bs_D will be very different from bs. Thus, the spaces bs_D furnish a large class of sequence spaces with signed-WGHP.

We next consider the construction above when the sequence space λ is a K-space. Assume that λ is a K-space whose locally convex topology is generated by a family of semi-norms $\mathcal{P}$. We give the space λ_A the locally convex topology generated by the semi-norms

$$p_A(t) = p(At) \ for \ p \in \mathcal{P} \ and \ p_k(t) = |t_k|, \ k \in \mathbb{N}.$$

Since only triangular matrices A will be considered below, this agrees with the usual topology defined on λ_A ([Wi2] 4.3.12). Note that λ_A is a K-space and $A : \lambda_A \to \lambda$ is a linear, continuous operator.

Theorem B.50. *If λ has SGHP (signed-SGHP), then λ_D has SGHP (signed-SGHP) for any diagonal matrix D.*

Proof: Let $\{t^j\}$ be bounded in λ_D and $\{I_j\}$ be an increasing sequence of intervals. Since $D : \lambda_D \to \lambda$ is continuous, $\{Dt^j\}$ is bounded in λ. There exists a subsequence $\{n_j\}$ such that the coordinate sum of the series $v = \sum_{j=1}^{\infty} \chi_{I_{n_j}} Dt^{n_j} \in \lambda$. Then the coordinate sum of the series $u = \sum_{j=1}^{\infty} \chi_{I_{n_j}} t^{n_j} \in \lambda_D$ since $Du = v$.

The other case is treated similarly.

Corollary B.51. *l_D^{∞} has SGHP and bs_D has signed-SGHP for any diagonal matrix D.*

As noted earlier the spaces l_D^{∞} furnish examples of spaces of a different nature with SGHP. The only spaces other than l^{∞} with SGHP seem to have been constructed by Noll ([No]).

Theorem B.52. *If λ has 0-GHP, then λ_D has 0-GHP.*

Proof: Let $t^j \to 0$ in λ_D and $\{I_j\}$ be an increasing sequence of intervals. Since $D : \lambda_D \to \lambda$ is continuous, $Dt^j \to 0$ in λ. There exists a subsequence $\{n_j\}$ such that the coordinate sum of the series $v = \sum_{j=1}^{\infty} \chi_{I_{n_j}} Dt^{n_j} \in \lambda$. Then the coordinate sum of the series $u = \sum_{j=1}^{\infty} \chi_{I_{n_j}} t^{n_j} \in \lambda_D$ since $Du = v$.

Again Theorem B.52 furnishes a large number of examples of spaces with 0-GHP.

Let μ be a sequence space containing c_{00}. As in Theorems B.48, B.50 and B.52, we have

Theorem B.53. *If λ has strong μ-GHP (weak μ-GHP), then λ_D has strong μ-GHP (weak μ-GHP).*

Further examples of sequence spaces λ_A with various gliding hump properties for matrices which are not diagonal matrices are given in [BSS].

Appendix C
Vector Valued Sequence Spaces

In this appendix we give a list of vector valued sequence spaces, review the gliding hump properties of sequence spaces and give examples of the sequence spaces satisfying various gliding hump properties.

Let X be a TVS. We give a list of X valued sequence spaces and their natural topologies.

- $c_{00}(X)$: all X valued sequences which are eventually 0.
- $c_0(X)$: all X valued sequences which converge to 0.
- $c_c(X)$: all X valued sequences which are eventually constant.
- $c(X)$: all X valued sequences which are convergent.
- $m_0(X)$: all X valued sequences with finite range.
- $l^\infty(X)$: all X valued sequences which are bounded.

Let X be an LCTVS whose topology is generated by the family of semi-norms $\mathcal{X}$. The natural topology of all the spaces above is generated by the semi-norms

$$q_\infty(\{x_j\}) = \sup_j q(x_j), \ q \in \mathcal{X}.$$

There is a significant difference between the scalar and vector case for spaces of bounded sequences. In the scalar case m_0 is dense in l^∞ with respect to $\|\cdot\|_\infty$. However, when X is an infinite dimensional Banach space, $m_0(X)$ is not dense in $l^\infty(X)$. For suppose that X is an infinite dimensional Banach space. By Riesz's Lemma ([Sw2] 7.6), There exists a sequence $\{x_j\}$ such that $\|x_i - x_j\| \geq 1$ for $i \neq j$ with $\|x_j\| = 1$ for all j. Then $x = \{x_j\} \in l^\infty(X)$. However, if $m_0(X)$ is dense in $l^\infty(X)$, then $\{x_j : j \in \mathbb{N}\}$ has a finite ϵ-net for every $\epsilon > 0$ and is, therefore, precompact which is clearly not the case.

Let $0 < p < \infty$. Then

- $l^p(X)$: all X valued sequences such that $\sum_{j=1}^{\infty} q(x_j)^p < \infty, q \in \mathcal{X}$.

If $1 \le p < \infty$, the topology of $l^p(X)$ is generated by the semi-norms

$$q_p(\{x_j\}) = \left(\sum_{j=1}^{\infty} q(x_j)^p \right)^{1/p} , \ q \in \mathcal{X}.$$

If $0 < p < 1$, the topology of $l^p(X)$ is generated by the quasi-norms

$$q_p(\{x_j\}) = \sum_{j=1}^{\infty} q(x_j)^p, \ q \in \mathcal{X}.$$

- $BS(X)$: all X valued sequences $\{x_j\}$ satisfying $\{\{x_j\} : \{\sum_{j=1}^{n} x_j\}_n$ is bounded$\}$.
- $CS(X)$: all X valued sequences $\{x_j\}$ satisfying $\{\{x_j\} : \sum_j x_j$ is Cauchy$\}$.

If X is the scalar field, then $bs = BS(X)$ and $cs = CS(X)$. We define a locally convex topology on $BS(X) \supset CS(X)$ by the semi-norms

$$q'(\{x_j\}) = \sup \left\{ q \left(\sum_{j \in I} x_j \right) : I \text{ a finite interval} \right\} , q \in \mathcal{X}.$$

- $BV(X)$: all X valued sequences satisfying $\sum_{j=1}^{\infty} q(x_{j+1} - x_j) < \infty, q \in \mathcal{X}$.
- $BV_0(X) = BV(X) \cap c_0(X)$.

If X is the scalar field, then $bv = BV(X)$ and $bv_0 = BV_0(X)$. These spaces are topologized by the semi-norms

$$q\hat{\ }(\{x_j\}) = q(x_1) + \sum_{j=1}^{\infty} q(x_{j+1} - x_j), q \in \mathcal{X}.$$

- $s(X)$: all X valued sequences.

We now describe a method of constructing vector valued sequence spaces from scalar valued sequence spaces. Let ν be a scalar valued sequence space which is normal and a K-space under a locally convex topology generated by the semi-norms $\mathcal{N}$. If $t = \{t_j\} \in \nu$, we write $|t| = \{|t_j|\}$; note $|t| \in \nu$ since ν is normal. We will also consider a monotone property for ν. The sequence space ν satisfies condition

(M) if there is a family of semi-norms $\mathcal{N}$ generating the topology of ν which is such that $q(s) \leq q(t)$ whenever $s, t \in \nu$ and $|s| \leq |t|$ and $q \in \mathcal{N}$.

Define

$$\nu\{X\} = \{\{x_j\} : x_j \in X, \{q(x_j)\} \in \nu \text{ for all } q \in \mathcal{X}\}.$$

Since ν is normal, $\nu\{X\}$ is a vector space. We supply $\nu\{X\}$ with the locally convex topology generated by the semi-norms

$$(\#) \quad \pi_{q,p}(x) = q(\{p(x_j)\}), p \in \mathcal{X}, q \in \mathcal{N}.$$

Thus, we have $c_{00}(X) = c_{00}\{X\}, c_0(X) = c_0\{X\}$ and $l^p(X) = l^p\{X\}$ for $0 < p \leq \infty$.

We extend some of the topological properties of scalar sequence spaces to vector valued sequence spaces.

Let E be a vector space of X valued sequences equipped with a locally convex Hausdorff topology.

Definition C.1. The space E is a K-space if the maps $x = \{x_j\} \to x_j$ from E into X are continuous for every $j \in \mathbb{N}$.

Since ν is a K-space, then $\nu\{X\}$ is a K-space.

If $z \in X$, then $e^j \otimes z$ will denote the sequence with z in the jth coordinate and 0 in the other coordinates. For every $n \in \mathbb{N}$, the sectional operator (projection) is defined to be the map $P_n : E \to E$ defined by

$$P_n(x) = \sum_{j=1}^{n} e^j \otimes x_j = (x_1, ..., x_n, 0, ...).$$

Definition C.2. The K-space E is an AK-space if for every $x \in E$, we have $x = \sum_{j=1}^{\infty} e^j \otimes x_j$, with convergence in E.

Definition C.3. The K-space E is an AB-space if $\{P_n x\}$ is bounded for every $x \in E$.

Definition C.4. The K-space E has the sections uniformly bounded property (SUB) if $\{P_n x : n \in \mathbb{N}, x \in B\}$ is bounded for every bounded set $B \subset E$.

Definition C.5. The K-space E has the property SE (sections equicontinuous) if the sectional operators $\{P_n\}$ are equicontinuous.

The spaces $\nu\{X\}$ supply a large number of spaces with the properties defined above.

Proposition C.6. *Consider the following properties for* $\nu\{X\}$:

(i) *If* ν *is an AK-space, then* $\nu\{X\}$ *is an AK-space.*
(ii) *If* ν *is an AB-space, then* $\nu\{X\}$ *is an AB-space.*
(iii) *If* ν *has property SUB, then* $\nu\{X\}$ *has property SUB.*
(iv) *If* ν *has property SE, then* $\nu\{X\}$ *has property SE.*
(v) *If* X *is sequentially complete,* ν *is sequentially complete, has property* (M) *and the topology of* ν *is linked to the topology of coordinatewise convergence, then* $\nu\{X\}$ *is sequentially complete.*

Proof: We prove (i); the other proofs of (ii)-(iv) are similar. Let $x \in E$ and let $\pi_{q,p}$ be a basic semi-norm for $\nu\{X\}$. Then

$$\pi_{q,p}\left(\sum_{j=n}^{\infty} e^j \otimes x_j\right) = q\left(\sum_{j=n}^{\infty} p(x_j)e^j\right) \to 0$$

since $\{p(x_j)\} \in \nu$, so the result follows.

(v): Suppose $\{x^k\}$ is Cauchy in $\nu\{X\}$. Since X is sequentially complete, $\lim_k x_j^k = x_j$ exists for every j. Set $x = \{x_j\}$. Let $p \in \mathcal{X}$ and $q \in \mathcal{N}$. By (M),

$$q(\{p(x_j^k)\}_j - \{p(x_j^l)\}_j) \leq q(\{p(x_j^k - x_j^l)\}_j)$$

so $\{p(x_j^k)\}_j$ is a Cauchy sequence in ν. Since ν is sequentially complete, there exists $u \in \nu$ such that $\{p(x_j^k)\} \to u$ as $k \to \infty$. Also, $\lim_k p(x_j^k) = p(x_j)$ for each j so $u = \{p(x_j)\} \in \nu$ and, therefore, $x = \{x_j\} \in \nu\{X\}$. Since $\{x^k\}$ is Cauchy in ν and $x^k \to x \in \nu\{X\}$ in the topology of coordinatewise convergence, $x^k \to x$ in ν since the topology of ν is linked to the topology of coordinatewise convergence (Appendix A.4).

From (v) above, we have

Corollary C.7. *If* X *is sequentially complete, then* $l^p(X)$ *and* $c_0(X)$ *are sequentially complete for* $1 \leq p \leq \infty$.

We now give statements of various gliding hump properties for vector valued sequence spaces. These statements are straightforward generalizations of the corresponding properties for scalar sequence spaces (see

Appendix B). If $\sigma \subset \mathbb{N}$ and $x = \{x_j\}$ is an X valued sequence, the coordinatewise product of χ_σ and x is denoted by $\chi_\sigma x$.

Definition C.8. The space E has the signed weak gliding hump property (signed-WGHP) if for every $x \in E$ and every increasing sequence of intervals $\{I_j\}$, there exist a subsequence $\{n_j\}$ and a sequence of signs $\{s_j\}$ such that the coordinatewise sum of the series $\sum_{j=1}^{\infty} s_j \chi_{I_{n_j}} x \in E$. If the signs above can be chosen to be equal to 1 for every $x \in E$, then E has the weak gliding hump property (WGHP).

The space E is *monotone* if $\chi_\sigma x \in E$ for every $\sigma \subset \mathbb{N}$ and $x \in E$. Any monotone space has WGHP. The spaces $\nu\{X\}$ are all monotone since ν is assumed to be normal so $\nu\{X\}$ always has WGHP. The proof in Example 8 of Appendix B shows that the non-monotone space $CS(X)$ has WGHP (use the semi-norms $|\cdot|$ which generates the topology of X in place of absolute value).

Definition C.9. The K-space E has the signed strong gliding hump property (signed-SGHP) if for every bounded sequence $\{x^j\}$ from E and every increasing sequence of intervals $\{I_j\}$, there exist a subsequence $\{n_j\}$ and a sequence of signs $\{s_j\}$ such that the coordinatewise sum of the series $\sum_{j=1}^{\infty} s_j \chi_{I_{n_j}} x^{n_j} \in E$. If the signs above can be chosen equal to 1 for every $x \in E$, then E has the strong gliding hump property (SGHP).

The space $l^{\infty}(X)$ has SGHP.

Proposition C.10. *If ν has SGHP and if X is normed, then $\nu\{X\}$ has SGHP.*

Proof: Let $\{x^j\}$ be bounded in $\nu\{X\}$ and $\{I_j\}$ be an increasing sequence of intervals. Then $\{\{\left\|x_i^j\right\|\}_i : j \in \mathbb{N}\}$ is bounded in ν so there exists a subsequence $\{n_j\}$ such that the coordinate sum $\sum_{j=1}^{\infty} \chi_{I_{n_j}} \{\left\|x_i^{n_j}\right\|\}_i \in \nu$ so $\sum_{j=1}^{\infty} \chi_{I_{n_j}} x^{n_j} \in \nu\{X\}$.

Definition C.11. The K-space E has the zero gliding hump property (0-GHP) if for every null sequence $\{x^j\}$ in E and every increasing sequence of intervals $\{I_j\}$, there is a subsequence $\{n_j\}$ such that the coordinatewise sum of the series $\sum_{j=1}^{\infty} \chi_{I_{n_j}} x^{n_j} \in E$.

As in Proposition C.10, we have

Proposition C.12. *If ν has 0-GHP and if X is normed, then $\nu\{X\}$ has 0-GHP.*

Let μ be a scalar sequence space containing c_{00}.

Definition C.13. The K-space E has the strong μ gliding hump property (strong μ-GHP) if whenever $\{x^j\}$ is a bounded sequence in E and $\{I_j\}$ is an increasing sequence of intervals, then for every $u = \{u_j\} \in \mu$ the coordinatewise sum of the series $\sum_{j=1}^\infty u_j \chi_{I_j} x^j \in E$.

Definition C.14. The K-space E has the weak μ gliding hump property (weak μ-GHP) if whenever $\{x^j\}$ is a bounded sequence in E and $\{I_j\}$ is an increasing sequence of intervals, then there is a subsequence $\{n_j\}$ such that for every $u = \{u_j\} \in \mu$ the coordinatewise sum of the series $\sum_{j=1}^\infty u_j \chi_{I_{n_j}} x^{n_j} \in E$.

Proposition C.15. *If ν has strong μ-GHP, then $\nu\{X\}$ has strong μ-GHP.*

Proof: Let $\{x^j\}$ be bounded in $\nu\{X\}$ and let $\{I_j\}$ be an increasing sequence of intervals. Let $u = \{u_j\} \in \mu$ and set $x = \sum_{j=1}^\infty u_j \chi_{I_j} x^j$ (coordinate sum). Let $p \in \mathcal{X}$ and note that $p(x(\cdot)) = \sum_{j=1}^\infty |u_j| \chi_{I_j} p(x^j(\cdot))$, where $x(\cdot)$ is the function $j \to x_j$. Now $\{\{p(x_i^j)\}_i : j \in \mathbb{N}\}$ is bounded in ν by definition. By strong μ-GHP, $\{p(x_j)\} \in \nu$ so $x \in \nu\{X\}$.

As in Propositions C.10 and C.12, we have

Proposition C.16. *If ν has weak μ-GHP and X is normed, then $\nu\{X\}$ has weak μ-GHP.*

We give an example of a non-monotone space with strong l^1-GHP.

Example C.17. $CS(X)$ has strong l^1-GHP. Suppose that $\{x^j\}$ is bounded in $CS(X)$ and $\{I_j\}$ is an increasing sequence of intervals. Let $u = \{u_j\} \in l^1$. Put $x = \sum_{j=1}^\infty u_j \chi_{I_j} x^j$. Let $\epsilon > 0$ and $p \in \mathcal{X}$ and set $M = \sup\{p(\sum_{i \in I} x_i^j) : I$ a finite interval, $j \in \mathbb{N}\}$. Pick N such that $\sum_{j=N}^\infty |u_j| < \epsilon$. Suppose I is a finite interval such that $\min I > N$. Then $p(\sum_{j \in I} x_j) \le \sum_{j=N}^\infty |u_j| M < \epsilon M$ so $x \in CS(X)$.

As in the example above, $BS(X)$ also has strong l^1-GHP.

We next extend the ∞-GHP to vector valued sequence spaces (Appendix B.36).

Definition C.18. The space E has the infinite gliding hump property (∞-GHP) if whenever $x \in E$ and $\{I_j\}$ is an increasing sequence of intervals, there exist a subsequence $\{n_j\}$ and $a_{n_j} > 0, a_{n_j} \to \infty$ such that every subsequence of $\{n_j\}$ has a further subsequence $\{p_j\}$ such that the coordinatewise sum of the series $\sum_{j=1}^{\infty} a_{p_j} \chi_{I_{p_j}} \in E$.

As in Propositions C.10 and C.12, we have

Proposition C.19. *If X is normed and ν has ∞-GHP, then $\nu\{X\}$ has ∞-GHP.*

In particular, if X is normed, then the spaces $c_0(X)$ and $l^p(X)$ $(0 < p < \infty)$ have ∞-GHP (Examples B.39 and B.40 of Appendix B). As in Proposition B.42 of Appendix B, we also have

Proposition C.20. *If E is a Banach AK-space, then E has ∞-GHP.*

Let Y be a TVS. The β-dual of E with respect to Y is defined to be

$$E^{\beta Y} = \left\{ \{T_j\} : T_j \in L(X, Y), \sum_{j=1}^{\infty} T_j x_j \text{ converges for every } \{x_j\} \in E \right\}.$$

Thus, $E^{\beta Y}$ consists of all operator valued series in $L(X, Y)$ which are E multiplier convergent. If Y is the scalar field, we write $E^{\beta \mathbb{R}} = E^{\beta}$; in this case, E, E^{β} form a dual pair under the pairing

$$x' \cdot x = \sum_{j=1}^{\infty} \langle x'_j, x_j \rangle, x = \{x_j\} \in E, x' = \{x'_j\} \in E^{\beta}.$$

We give several examples of β-duals.

Example C.21. Let X be a normed space and assume that the dual space X' is equipped with the dual norm. Then $\nu\{X\}^{\beta} = \nu^{\beta}\{X'\}$. First, let $\{x'_j\} \in \nu^{\beta}\{X'\}$ and suppose $\{x_j\} \in \nu\{X\}$. Then $\{\|x_j\|\} \in \nu$ and $\{\|x'_j\|\} \in \nu^{\beta}$. Hence, $\sum_{j=1}^{\infty} |\langle x'_j, x_j \rangle| \leq \sum_{j=1}^{\infty} \|x'_j\| \|x_j\| < \infty$ so $\{x'_j\} \in \nu\{X\}^{\beta}$ and $\nu^{\beta}\{X'\} \subset \nu\{X\}^{\beta}$. Next, let $\{x'_j\} \in \nu\{X\}^{\beta}$. If $\{x'_j\} \notin \nu^{\beta}\{X'\}$, then $\{\|x'_j\|\} \notin \nu^{\beta}$. Since ν is normal, there exists $t \in \nu$ such that $\sum_{j=1}^{\infty} |t_j| \|x'_j\| = \infty$. For each j pick $x_j \in X$ such that $\|x_j\| = 1$ and $\|x'_j\| < |\langle x'_j, x_j \rangle| + 1/2^j$. Then $\{t_j x_j\} \in \nu\{X\}$ and $\sum_{j=1}^{\infty} |t_j| |\langle x'_j, x_j \rangle| = \infty$. Thus, $\{x'_j\} \notin \nu\{X\}^{\beta}$ so $\nu\{X\}^{\beta} \subset \nu^{\beta}\{X'\}$ and equality follows.

Let X'_b be the dual space of X equipped with the strong topology $\beta(X', X)$.

Proposition C.22. $BS(X)^{\beta} = BV_0(X'_b)$.

Proof: Let $y \in BS(X)^\beta$. To show that $y_j \to 0$ strongly, it suffices to show that $\langle y_j, x_j \rangle \to 0$ for every bounded sequence $\{x_j\} \subset X$. If $x_0 = 0$, then $\{x_j - x_{j-1}\} \in BS(X)$ so $\sum_{j=1}^\infty \langle y_j, x_j - x_{j-1} \rangle$ converges, and we have that $\lim_j \langle y_j, x_j - x_{j-1} \rangle = 0$ for every bounded sequence $\{x_j\}$. This implies that $\lim \langle y_j, x_j \rangle = 0$ for every bounded sequence $\{x_j\}$ [define a bounded sequence $\{z_j\}$ by $0, x_1, 0, x_3, 0, ...$; then the sequence $\{\langle y_j, z_{j+1} - z_j \rangle\}$ contains the sequence $\{\langle y_{2j+1}, x_{2j+1} \rangle\}$ as a subsequence so $\lim \langle y_{2j+1}, x_{2j+1} \rangle = 0$ and similarly, $\lim \langle y_{2j}, x_{2j} \rangle = 0$ so $\lim \langle y_j, x_j \rangle = 0$]. Thus, $y \in c_0(X)$.

Let $\{x_j\}$ be bounded in X and set $z_j = x_{j+1} - x_j$. Then $\{z_j\} \in BS(X)$ and $\sum_{j=1}^\infty \langle y_j, z_j \rangle$ converges. Now

$$(*) \quad \sum_{j=1}^n \langle y_j, z_j \rangle = \sum_{j=1}^n \langle y_j, x_{j+1} - x_j \rangle = \sum_{j=1}^n \langle y_j - y_{j+1}, x_j \rangle - \langle y_n, x_n \rangle.$$

By what was established above, $\langle y_n, x_n \rangle \to 0$ so $\sum_{j=1}^\infty \langle y_j - y_{j+1}, x_j \rangle$ converges absolutely for every bounded sequence $\{x_j\}$ by $(*)$. This implies that $\sum_{j=1}^\infty (y_j - y_{j+1})$ is absolutely convergent in X_b' [if $B \subset X$ is bounded, for every j pick $x_j \in B$ such that

$$p_B(y_j - y_{j+1}) = \sup\{|\langle y_j - y_{j+1}, x \rangle| : x \in B\} < |\langle y_j - y_{j+1}, x_j \rangle| + 1/2^j$$

so $\sum_{j=1}^\infty p_B(y_j - y_{j+1}) < \infty$]. Thus, $y \in BV_0(X_b')$.

Next, let $y \in BV_0(X_b')$ and $x \in BS(X)$. Then $\{s_n = \sum_{j=1}^n x_j\}$ is bounded so $\sum_{j=1}^\infty \langle y_j - y_{j+1}, s_j \rangle$ converges absolutely. Now

$$(**) \quad \sum_{j=1}^n \langle y_j, x_j \rangle = \sum_{j=1}^{n-1} \langle y_j - y_{j+1}, s_j \rangle + \langle y_n, s_n \rangle.$$

Since $y_n \to 0$ strongly, $\langle y_n, s_n \rangle \to 0$ so $(**)$ implies that $\sum_{j=1}^\infty \langle y_j, x_j \rangle$ converges. That is, $y \in BS(X)^\beta$ so $BV_0(X_b') \subset BS(X)^\beta$.

Proposition C.23. $l^\infty(X)^\beta = l^1(X_b')$.

Proof: Let $y \in l^\infty(X)^\beta$. Suppose that $\{x_j\} \subset X$ is bounded and $t \in l^\infty$. Then $\{t_j x_j\} \in l^\infty(X)$ so $\sum_{j=1}^\infty |\langle y_j, t_j x_j \rangle| = \sum_{j=1}^\infty |t_j| |\langle y_j, x_j \rangle| < \infty$. Thus, $\{\langle y_j, x_j \rangle\} \in l^1$ or $\sum_{j=1}^\infty |\langle y_j, x_j \rangle| < \infty$. Hence, $y \in l^1(X_b')$ [see the argument in Proposition C.22] and $l^\infty(X)^\beta \subset l^1(X_b')$.

Let $y \in l^1(X_b')$. Suppose that $\{x_j\} \in l^\infty(X)$ and set $B = \{x_j : j \in \mathbb{N}\}$, $p_B(x') = \sup\{|\langle x', x \rangle| : x \in B\}$. Then $\sum_{j=1}^\infty |\langle y_j, x_j \rangle| \leq \sum_{j=1}^\infty p_B(y_j) < \infty$ so $y \in l^\infty(X)^\beta$ and $l^1(X_b') \subset l^\infty(X)^\beta$.

Similarly, we have

Proposition C.24. $c_0(X)^\beta = l^1(X_b')$.

Let $l_e^\infty(X')$ be the sequences in $l^\infty(X_b')$ which are equicontinuous (such sequences are always strongly bounded ([Sw2] 19.5)).

Proposition C.25. *We have the following relationships:*

(i) $l_e^\infty(X') \subset l^1(X)^\beta$.
(ii) $l^1(X)^\beta \subset l^\infty(X_b')$.
(iii) *If X is quasi-barrelled, then $l^1(X)^\beta = l^\infty(X_b')$.*

Proof: (i): Let $y = \{y_j\} \in l_e^\infty(X')$. Then there exists a continuous semi-norm p on X such that $|\langle y_j, x \rangle| \leq p(x)$ for all $x \in X, j \in \mathbb{N}$. Let $\{x_j\} \in l^1(X)$. Since $\sum_{j=1}^\infty |\langle y_j, x_j \rangle| \leq \sum_{j=1}^\infty p(x_j) < \infty$, $y \in l^1(X)^\beta$ and (i) holds.

(ii): Let $y \in l^1(X)^\beta$ and let $\{x_j\}$ be bounded. If $t \in l^1$, then $\{t_j x_j\} \in l^1(X)$ so $\sum_{j=1}^\infty |\langle y_j, t_j x_j \rangle| = \sum_{j=1}^\infty |t_j| \, |\langle y_j, x_j \rangle| < \infty$. Hence, $\{\langle y_j, x_j \rangle\} \in l^\infty$ which implies that $\{y_j\}$ is strongly bounded [if $B \subset X$ is bounded, then for every j there exists $x_j \in B$ such that $p_B(y_j) = \sup\{|\langle y_j, x \rangle| : x \in B\} \leq |\langle y_j, x_j \rangle| + 1$ so $\{p_B(y_j)\}$ is bounded]. Therefore, (ii) holds.

(iii) follows from the definition of quasi-barrelled and (i) and (ii) ([Sw2] 19.13, [Wi] 10.1.11).

Finally, we compute one example of an operator valued β-dual. Below when X and Y are normed spaces, the dual spaces are always assumed to be equipped with the dual norm.

Proposition C.26. *Let X and Y be normed spaces with Y complete. Let $1 < p < \infty$ and $\frac{1}{p} + \frac{1}{q} = 1$. Then $\{T_j\} \in l^p(X)^\beta$ iff $\{\{T_j' y'\} : y' \in Y', \|y'\| \leq 1\}$ is bounded in $l^q(X')$.*

Proof: Suppose $\{\{T_j' y'\} : y' \in Y', \|y'\| \leq 1\}$ is bounded in $l^q(X')$ and M is a bound for the norms of the elements in this set. Then

$$\left\| \sum_{j=m}^n T_j x_j \right\| = \sup\left\{ \left| \sum_{j=m}^n \langle y', T_j x_j \rangle \right| : \|y'\| \leq 1 \right\} \leq \sup_{\|y'\| \leq 1} \sum_{j=m}^n \|T_j' y'\| \, \|x_j\|$$

$$\leq \left(\sum_{j=m}^n \|x_j\|^p \right)^{1/p} \sup_{\|y'\| \leq 1} \left\{ \left(\sum_{j=m}^n \|T_j' y'\|^q \right)^{1/q} \right. \leq M \left(\sum_{j=m}^n \|x_j\|^p \right)^{1/p}$$

which implies that $\sum_j T_j x_j$ is Cauchy in Y and, therefore, converges.

Suppose $\{T_j\} \in l^p(X)^\beta$. First, we claim that $\{T_j' y'\} \in l^q(X')$ for every $y' \in Y'$. Suppose that this condition fails to hold. Then $\sum_{j=1}^\infty \|T_j' y'\|^q = \infty$ for some $y' \in Y'$. Then there exists $t \in l^p$ such that $\sum_{j=1}^\infty \|T_j' y'\| \, |t_j| = \infty$.

For every j there exists $x_j \in X, \|x_j\| \leq 1$ with $\langle y', T_j x_j \rangle \geq \|T'_j y'\| / 2$. Then $\sum_{j=1}^{\infty} |t_j| |\langle y', T_j x_j \rangle| = \infty$. Since $\{t_j x_j\} \in l^p(X)$, $\{T_j\} \notin l^p(X)^\beta$.

Now $l^q(X')$ is complete (Corollary C.7) and the linear map $y' \rightarrow \{T'_j y'\}$ from Y' into $l^q(X')$ has a closed graph. Hence, from the Closed Graph Theorem the map is continuous and the result follows.

Other examples of β-duals for vector valued sequence spaces can be found in [FP2], [Fo], [GKR] and [Ros].

Appendix D
The Antosik-Mikusinski Matrix Theorems

In this appendix we will present two versions of the Antosik-Mikusinski Matrix Theorems. These matrix theorems have proven to be very useful in treating applications in functional analysis and measure theory where gliding hump techniques are employed (see [Sw1] for more versions of the matrix theorem and applications). These theorems are used at various points in the text in gliding hump proofs.

Let X be a (Hausdorff) TVS. We begin with a simple lemma.

Lemma D.1. Let $x_{ij} \in X$ for $i, j \in \mathbb{N}$. If $\lim_i x_{ij} = 0$ for every j and $\lim_j x_{ij} = 0$ for every i and if $\{U_k\}$ is a sequence of neighborhoods of 0 in X, then there exists an increasing sequence $\{p_i\}$ such that $x_{p_i p_j} \in U_j$ for $j > i$.

Proof: Set $p_1 = 1$. There exists $p_2 > p_1$ such that $x_{ip_1} \in U_2$, $x_{p_1 j} \in U_2$ for $i, j \geq p_2$. Then there exists $p_3 > p_2$ such that $x_{ip_1}, x_{ip_2}, x_{p_1 j}, x_{p_2 j} \in U$ for $i, j \geq p_3$. Now just continue the construction.

We now establish our version of the Antosik-Mikusinski Matrix Theorem.

Theorem D.2. *(Antosik-Mikusinski)* Let $x_{ij} \in X$ for $i, j \in \mathbb{N}$. Suppose

(I) $\lim_i x_{ij} = x_j$ exists for each j and

(II) for each increasing sequence of positive integers $\{m_j\}$ there is a subsequence $\{n_j\}$ of $\{m_j\}$ such that $\{\sum_{j=1}^{\infty} x_{in_j}\}_i$ is Cauchy.

Then $\lim_i x_{ij} = x_j$ uniformly for $j \in \mathbb{N}$. In particular,
$$\lim_i \lim_j x_{ij} = \lim_j \lim_i x_{ij} = 0 \text{ and } \lim_i x_{ii} = 0.$$

Proof: If the conclusion fails, there is a closed, symmetric neighborhood U_0 of 0 and increasing sequences of positive integers $\{m_k\}$ and $\{n_k\}$ such that $x_{m_k n_k} - x_{n_k} \notin U_0$ for all k. Pick a closed, symmetric neighborhood U_1 of 0 such that $U_1 + U_1 \subseteq U_0$ and set $i_1 = m_1, j_1 = n_1$. Since $x_{i_1 j_1} - x_{j_1} = (x_{i_1 j_1} - x_{ij_1}) + (x_{ij_1} - x_{j_1})$, there exists i_0 such that $x_{i_1 j_1} - x_{ij_1} \notin U_1$ for $i \geq i_0$. Choose k_0 such that

$$m_{k_0} > \max\{i_1, i_0\}, n_{k_0} > j_1 \text{ and set } i_2 = m_{k_0}, \; j_2 = n_{k_0}.$$

Then $x_{i_1 j_1} - x_{i_2 j_1} \notin U_1$ and $x_{i_2 j_2} - x_{j_2} \notin U_0$. Proceeding in this manner produces increasing sequences $\{i_k\}, \{j_k\}$ such that $x_{i_k j_k} - x_{j_k} \notin U_0$ and $x_{i_k j_k} - x_{i_{k+1} j_k} \notin U_1$. For convenience, set $z_{k,l} = x_{i_k j_l} - x_{i_{k+1} j_l}$ so $z_{k,k} \notin U_1$.

Choose a sequence of closed, symmetric neighborhoods of 0, $\{U_n\}$, such that $U_n + U_n \subseteq U_{n-1}$ for $n \geq 1$. Note that

$$U_3 + U_4 + \cdots + U_m = \sum_{j=3}^{m} U_j \subseteq U_2$$

for each $m \geq 3$. By (I) and (II), $\lim_k z_{kl} = 0$ for each l and $\lim_l z_{kl} = 0$ for each k so by Lemma D.1 there is an increasing sequence of positive integers $\{p_k\}$ such that $z_{p_k p_l}, z_{p_l p_k} \in U_{k+2}$ for $k > l$. By (II), $\{p_k\}$ has a subsequence $\{q_k\}$ such that $\{\sum_{k=1}^{\infty} x_{i q_k}\}_{i=1}^{\infty}$ is Cauchy so

$$\lim_k \sum_{l=1}^{\infty} z_{q_k q_l} = 0.$$

Thus, there exists k_0 such that $\sum_{l=1}^{\infty} z_{q_{k_0} q_l} \in U_2$. Then for $m > k_0$,

$$\sum_{l=1, l\neq k_0}^{m} z_{q_{k_0} q_l} = \sum_{l=1}^{k_0-1} z_{q_{k_0} q_l} + \sum_{l=k_0+1}^{m} z_{q_{k_0} q_l} \in \sum_{l=1}^{k_0-1} U_{k_0+2}$$

$$+ \sum_{l=k_0+1}^{m} U_{l+2} \subseteq \sum_{l=3}^{m+2} U_l \subseteq U_2$$

so $z_{k_0} = \sum_{l=1, l\neq k_0}^{\infty} z_{q_{k_0} q_l} \in U_2$. Thus,

$$z_{q_{k_0} q_{k_0}} = \sum_{l=1}^{\infty} z_{q_{k_0} q_l} - z_{k_0} \in U_2 + U_2 \subseteq U_1$$

This is a contradiction and establishes the result.

A matrix $[x_{ij}]$ satisfying conditions (I) and (II) of Theorem D.2 is called a $\mathcal{K}$-*matrix*. [The appellation "$\mathcal{K}$" here refers to the Katowice branch of the Mathematical Institute of the Polish Academy of Science where the matrix theorems and applications were developed by Antosik, Mikusinski and other members of the institute.]

At other points in the text we will also require another version of the matrix theorem which was developed by Stuart ([St1], [St2]) to treat weak sequential completeness of β-duals.

Theorem D.3. *(Stuart) Let $x_{ij} \in X$ for $i, j \in \mathbb{N}$. Suppose*

(I) $\lim_i x_{ij} = x_j$ exists for all j and

(II) for each increasing sequence of positive integers $\{m_j\}$ there is a subsequence $\{n_j\}$ and a choice of signs $s_j \in \{-1, 1\}$ such that $\{\sum_{j=1}^{\infty} s_j x_{in_j}\}_{i=1}^{\infty}$ is Cauchy.

Then $\lim_i x_{ij} = x_j$ uniformly for $j \in \mathbb{N}$. In particular,

$$\lim_i \lim_j x_{ij} = \lim_j \lim_i x_{ij} = 0 \quad and \quad \lim_i x_{ii} = 0.$$

Proof: If the conclusion fails, there is a closed, symmetric neighborhood of $0, U_0$, and increasing sequences of positive integers $\{m_k\}$ and $\{n_k\}$ such that $x_{m_k n_k} - x_{n_k} \notin U_0$ for all k. Pick a closed, symmetric neighborhood of $0, U_1$, such that $U_1 + U_1 \subset U_0$ and set $i_1 = m_1, j_1 = n_1$. Since

$$x_{i_1 j_1} - x_{j_1} = (x_{i_1 j_1} - x_{ij_1}) + (x_{ij_1} - x_{j_1}),$$

there exists i_0 such that $x_{i_1 j_1} - x_{ij_1} \notin U_1$ for $i \geq i_0$. Choose k_0 such that $m_{k_0} > \max\{i_1, i_0\}, n_{k_0} > j_1$ and set $i_2 = m_{k_0}, j_2 = n_{k_0}$. Then $x_{i_1 j_1} - x_{i_2 j_1} \notin U_1$ and $x_{i_2 j_2} - x_{j_2} \notin U_0$. Proceeding in this manner produces increasing sequences $\{i_k\}$ and $\{j_k\}$ such that $x_{i_k j_k} - x_{j_k} \notin U_0$ and $x_{i_k j_k} - x_{i_{k+1} j_k} \notin U_1$. For convenience, set $z_{kl} = x_{i_k j_l} - x_{i_{k+1} j_l}$ so $z_{kk} \notin U_1$.

Choose a sequence of closed, symmetric neighborhoods of 0, $\{U_n\}$, such that $U_n + U_n \subset U_{n-1}$ for $n \geq 1$. Note that

$$U_3 + U_4 + \dots + U_m = \sum_{j=3}^{m} U_j \subset U_2 \text{ for each } m \geq 3.$$

By (I), $\lim_k z_{kl} = 0$ for each l and by (II), $\lim_l z_{kl} = 0$ for each k so by Lemma D.1 there is an increasing sequence of positive integers $\{p_k\}$ such that $z_{p_k p_l}, z_{p_l p_k} \in U_{k+2}$ for $k > l$. By (II) there is a subsequence $\{q_k\}$ of $\{p_k\}$ and a choice of signs s_k such that

$$\{\sum_{k=1}^{\infty} s_k x_{iq_k}\}_{i=1}^{\infty}$$

is Cauchy so

$$\lim_k \sum_{l=1}^{\infty} s_l z_{q_k q_l} = 0.$$

Thus, there exists k_0 such that

$$\sum_{l=1}^{\infty} s_l z_{q_{k_0} q_l} \in U_2.$$

Then for $m > k_0$,

$$\sum_{l=1, l \neq k_0}^{m} s_l z_{q_{k_0} q_l} = \sum_{l=1}^{k_0 - 1} s_l z_{q_{k_0} q_l} + \sum_{l=k_0+1}^{m} s_l z_{q_{k_0} q_l} \in \sum_{l=1}^{k_0-1} U_{k_0+2} + \sum_{l=k_0+1}^{m} U_l \subset U_2$$

so

$$z_{k_0} = \sum_{l=1, l \neq k_0}^{\infty} s_l z_{q_{k_0} q_l} \in U_2.$$

Thus,

$$s_{k_0} z_{q_{k_0} q_{k_0}} = \sum_{l=1}^{\infty} s_l z_{q_{k_0} q_l} - z_{k_0} \in U_2 + U_2 \subset U_1$$

since U_1 is symmetric

$$z_{q_{k_0} q_{k_0}} \in U_1$$

as well. This is a contradiction.

A matrix which satisfies conditions (I) and (II) of Theorem D.3 will be called a *signed $\mathcal{K}$-matrix* and Theorem D.3 will be referred to as the signed version of the Antosik-Mikusinski Matrix Theorem.

We give an example of a matrix which is a signed $\mathcal{K}$-matrix but is not a $\mathcal{K}$-matrix.

Example D.4. Let $t = \{t_j\} \in bs$, the space of bounded series, and let X be bs equipped with the topology of coordinatewise convergence, $\sigma(bs, c_{00})$ [Appendix A]. Define a matrix $M = [m_{ij}]$ with entries from X by $m_{ij} = e^j$. Then no row of M has a subseries which converges in X so M is not a $\mathcal{K}$-matrix. However, given any subsequence $\{n_j\}$ the series $\sum_{j=1}^{\infty} (-1)^j e^{n_j}$ converges in X so M is a signed $\mathcal{K}$-matrix.

Other refinements and comments on the matrix theorems can be found in [Sw1] 2.2. The text [Sw1] contains numerous applications of the matrix theorems to topics in topological vector spaces, measure and integration theory and sequence spaces.

Appendix E
Drewnowski's Lemma

In this appendix we establish a remarkable result of Drewnowski which asserts that a strongly bounded, finitely additive set function defined on a σ-algebra is in some sense not "too far" from being countably additive ([Dr]). This result is very useful in treating finitely additive set functions.

Let Σ be a σ-algebra of subsets of a set S, X be a TVS whose topology is generated by the quasi-norm $|\cdot|$ and let $\mu : \Sigma \to X$ be finitely additive and strongly bounded. (Recall μ is strongly bounded if $\mu(E_j) \to 0$ whenever $\{E_j\}$ is a pairwise disjoint sequence from Σ.) For $E \in \Sigma$, set

$$\mu'(E) = \sup\{|\mu(A)| : A \subset E, A \in \Sigma\};$$

μ' is called the *submeasure majorant* of μ and μ' is also strongly bounded in the sense that $\mu'(E_j) \to 0$ whenever $\{E_j\}$ is a pairwise disjoint sequence from Σ.

Lemma E.1. *(Drewnowski) If* $\mu : \Sigma \to X$ *is finitely additive and strongly bounded and* $\{E_j\}$ *is a pairwise disjoint sequence from* Σ, *then* $\{E_j\}$ *has a subsequence* $\{E_{n_j}\}$ *such that* μ *is countably additive on the* σ-*algebra generated by* $\{E_{n_j}\}$.

Proof: Partition $\mathbb{N}$ into a pairwise disjoint sequence of infinite sets $\{K_j^1\}_{j=1}^\infty$. By the strong additivity of μ', $\mu'(\cup_{j \in K_i^1} E_j) \to 0$ as $i \to \infty$ so there exists i such that $\mu'(\cup_{j \in K_i^1} E_j) < 1/2$. Set $N_1 = K_i^1$ and $n_1 = \inf N_1$. Now partition $N_1 \setminus \{n_1\}$ into a pairwise disjoint sequence of infinite subsets $\{K_j^2\}_{j=1}^\infty$. As above there exists i such that $\mu'(\cup_{j \in K_i^2} E_j) < 1/2^2$. Let $N_2 = K_i^2$ and $n_2 = \inf N_2$. Note $N_2 \subset N_1$ and $n_2 > n_1$. Continuing this construction produces a subsequence $n_j \uparrow \infty$ and a sequence of infinite subsets of $\mathbb{N}$, $\{N_j\}$, such that $N_{j+1} \subset N_j$ and $\mu'(\cup_{i \in N_j} E_i) < 1/2^j$. If Σ_0 is the σ-algebra generated by $\{E_{n_j}\}$, then μ is countably additive on Σ_0.

We also have a version of Drewnowski's Lemma for a sequence of finitely additive set functions.

Corollary E.2. *Let $\mu_i : \Sigma \to X$ be finitely additive and strongly bounded for each $i \in \mathbb{N}$. If $\{E_j\}$ is a pairwise disjoint sequence from Σ, then there is a subsequence $\{E_{n_j}\}$ such that each μ_i is countably additive on the σ-algebra generated by $\{E_{n_j}\}$.*

Proof: Define a quasi-norm $|\cdot|'$ on $X^{\mathbb{N}}$ by

$$|x|' = |\{x_i\}|' = \sum_{i=1}^{\infty} |x_i| \,/((1 + |x_i|)2^i).$$

Define $\mu : \Sigma \to X^{\mathbb{N}}$ by $\mu(E) = \{\mu_i(E)\}$. Then μ is finitely additive and strongly bounded with respect to $|\cdot|'$ so by Lemma 1 there is a subsequence $\{E_{n_j}\}$ such that μ is countably additive on the σ-algebra Σ_0 generated by $\{E_{n_j}\}$. Thus, each μ_i is countably additive on Σ_0.

References

[AP1] A. Aizpura and J. Perez-Fernandez, Spaces of s-bounded multiplier convergent series, Acta Math. Hunjar., 87(2000), 135 - 146.

[AP2] A. Aizpura and J. Perez-Fernandez, Sequence space associated to a series in Banach space, Indian J. Pure Appl. Math., 33(2002), 1317-1329.

[A] P. Antosik, On interchange of limits, Generalized Functions, Convergence Strucctures and Their Applications, Plenum Press, N.Y., 1988, p. 367 - 374.

[Ap] T. Apostol, Mathematical Analysis, Addison-Wesley, Reading, 1975.

[Ba] S. Banach, Theorie des Operations Lineaires, Warsaw, 1932.

[Bar] R. Bartle, A general bilinear vector integral, Studia Math., 15(1956), 337-352.

[Bs] B. Basit, On a Theorem of Gelfand and a new proof of the Orlicz-Pettis Theorem, Rend. Inst. Matem. Univ. di Trieste, 18(1986), 159 -162.

[Bt] J. Batt, Applications of the Orlicz-Pettis Theorem to operator-valued measures and compact and weakly compact linear transformations on the space of continuous functions, Revue Roum. Math. Pures Appl., 14(1969), 907 - 945.

[Be] G. Bennett, Some inclusion theorems for sequence spaces, Pacific J. Math. 46(1973), 17 - 30.

[BK] G. Bennett and N. Kalton, FK spaces containing c_0, Duke Math. J., 39(1972), 561 - 582.

[BP] C. Bessaga and A. Pelczynski, On Bases and Unconditional Convergence of series in Banach Space, Studia Math., 17(1958), 151-164.

[Bo] J. Boos, Classical and Modern Methods in Summability, Oxford University Press, Oxford, 2000.

[BSS] J. Boos, C. Stuart and C. Swartz, Gliding Hump Properties and

Matrix Domains, Analysis Math., 30(2004), 243 - 257.

[BW]   Qingying Bu and Cong Xin Wu, Unconditionally Convergent Series of Operators on Banach Spaces, J. Math. Anal. Appl., 207(1997), 291 - 299.

[BM]   J. Burzyk and P. Mikusinski, On Normability of Semigroups, Bull. Polon. Acad. Sci., 28(1980), 33-35.

[Day] M. Day, Normed Linear Spaces, Springer-Verlag, Berlin, 1962.

[DeS] J. DePree and C. Swartz, Introduction to Real Analysis, Wiley, N.Y., 1987.

[Die] P. Dierolf, Theorems of Orlicz-Pettis type for locally convex spaces, Man. Math., 20(1977), 73 - 94.

[DF]   J. Diestel and F. Faires, On Vector Measures, Trans. Amer. Math. Soc., 198(1974), 253 - 271.

[DU]   J. Diestel and J. Uhl, Vector Measures, Amer. Math. Soc. Surveys #15, Providence, 1977.

[Din] N. Dinculeanu, Weak Compactness and Uniform Convergence of Operators in Space of Bochner Integrable Functions, J. Math. Anal. Appl., 1090(1985), 372 - 387.

[Dr]   L. Drewnowski, Equivalence of Brooks-Jewett, Vitali-Hahn-Saks and Nikodym Theorems, Bull. Acad. Polon. Sci., 20(1972), 725-731.

[DS]   N. Dunford and J. Schwartz, Linear Operators I, Interscience, N.Y., 1958.

[FL]   W. Filter and I. Labudu, Essays on the Orlicz - Pettis Theorem I, Real. Anal. Exch., 16(1990/91), 393 - 403.

[FP]   M. Florencio and P. Paul, A Note on λ- Multiplier Convergent Series, Casopis Pro. Post. Mat. 113(1988), 421 - 428.

[FP2] M. Florencio and P. Paul, Barrelledness conditions on certain vector valued sequence spaces, Arch. Math., 48(1987), 153 - 164.

[Fo]   J. Fourie, Barrelledness conditions on generalized sequence spaces, South African J. Sci., 84(1988), 346 - 348.

[Ga]   D.J.H. Garling, The β − and γ− duality of sequence spaces, Proc. Camb. Phil. Soc., 63(1967), 963 - 981.

[Gi]   D.P. Giesy, A Finite-valued Finitely Additive Unbounded Measure, Amer. Math. Monthly, 77(1970), 508-510.

[GDS] H.G. Garnir, M. DeWilde and J. Schmets, Analyse Fontionnelle I, Birkhauser, Basel, 1968.

[GR]   W. Graves, Proceedings of the Conference on Integration, Topology, and Geometry in Linear Spaces, Amer. Math. Soc., Providence, 1980.

[GKR] M. Gupta, P.K. Kamthan and K.L.N. Rao, Duality in Certain

Generalized Kothe Sequence Spaces, Bull. Inst. Math. Acad. Sinica, 5(1977), 285 - 298.

[Ha] H. Hahn, Uber Folgen Linearen Operationen, Monatsch. fur Math. und Phys. 32(1922), 1 - 88.

[Hay] R. Haydon, A non-reflexive Grothendieck space that does not contain l^∞, Isreal J. Math, 40(1981), 65-73.

[HT] E. Hellinger and O. Toeplitz, Grundlagen fur eine Theorie den unendlichen Matrizen, Math. Ann., 69(1910), 289 - 330.

[Ho] J. Howard, The Comparison of an Unconditionally Converging Operator, Studia Math., 33(1969), 295 - 298.

[Ka] N. J. Kalton, Spaces of Compact Operators, Math. Ann., 208(1974), 267 - 278.

[Ka2] N. Kalton, Subseries Convergence in Topological Groups and Vector Spaces, Isreal J. Math., 10(1971), 402 - 412.

[Ka3] N.J. Kalton, The Orlicz - Pettis Theorem, Contemporary Math., Amer. Math. Soc., Providence, 1980.

[KG] P.K. Kamthan and M. Gupta, Sequence Spaces and Series, Marcel Dekker, N.Y., 1981.

[K1] G. Köthe, Topological Vector Spaces I, Springer - Verlag, Berlin, 1969.

[K2] G. Köthe, Topological Vector Spaces II, Springer-Verlag, Berlin, 1979.

[LW] E. Lacey and R.J. Whitley, Conditions under which all the Bounded Linear Maps are Compact, Math. Ann., 158 (1965), 1-5.

[LPY] Lee Peng Yee, Sequence Spaces and the Gliding Hump Property, Southeast Asia Bull. Math., Special Issue (1993), 65 - 72.

[LPYS] Lee Peng Yee and C. Swartz, Continuity of Superposition Operators on Sequence Spaces, New Zealand J. Math., 24(1995), 41-52.

[LB] R. Li and Q. Bu, Locally Convex Spaces Containing no Copy of c_0, J. Math. Anal. Appl., 172(1993), 205-211.

[LS] R. Li and C. Swartz, A Nonlinear Schur Theorem, Acta Sci. Math., 58(1993), 497 - 508.

[LT] J. Lindenstrauss and L. Tzafriri, Classical Banach Spaces I, Springer-Verlag, Berlin, 1977.

[MA] S.D. Madrigal and J.M.B. Arrese, Local Completeness and Series, Simon Stevin, 65(1991), 331 - 335.

[Mc] C. W. McArthur, On a theorem of Orlicz and Pettis, Pacific J. Math. 22(1967), 297 - 303.

[MR] C. McArthur and J. Rutherford, Some Applications of an Inequality in Locally Convex Spaces, Trans. Amer. Math. Soc., 137(1969), 115 -123.

[No] D. Noll, Sequential Completeness and Spaces with the Gliding Humps Property, Manuscripta Math., 66(1990), 237 - 252.

[Or] W. Orlicz, Beiträge zur Theorie der Orthogonalent Wichlungen II, Studia Math., 1(1929), 241 - 255.

[PBA] J. Perez - Fernandez, F. Benitez - Trujillo and A. Aizpuru, Characterizations of completeness of normed spaces through weakly unconditionally Cauchy series, Czech. Math. J., 50(2000), 889-896.

[Pe] B. J. Pettis, On Integration in Vector Spaces, Trans. Amer. Math. Soc., 44(1938), 277 - 304.

[Pl] A. Pelczynski, On Strictly singular and Strictly cosingular Operators, Bull. Acad. Polon. Sci., 13 (1965), 31 - 36.

[RR] K.P.S. Rao and M. Rao, Theory of Charges, Academic Press, N. Y., 1983.

[Ro1] A. Robertson, Unconditional Convergence and the Vitali-Hahn-Saks Theorem, Bull. Soc. Math, France, Supp;. Mem. 31-32 (1972), 335 - 341.

[Ro2] A. Robertson, On Unconditional Convergence in Topological Vector Spaces, Proc. Royal Soc. Edinburgh, 68(1969), 145-157.

[Rol] S. Rolewicz, Metric Linear Spaces, Polish Sci. Publ., Warsaw, 1972.

[Ros] R. Rosier, Dual Spaces of Certain Vector Sequence Spaces, Pacific J. Math., 46(1973), 487 - 501.

[RS] W. Ruckle and S. Saxon, Generalized Sectional Convergence, J. Math. Anal. Appl., 193(1995), 680 - 705.

[SaSe] R. Samaratunga and J. Sember, Summability and Substructures of 2^N, Southeast Asia Math. Bull., 66(1990), 237 - 252.

[Sm] W. Schachermeyer, On some classical measure-theoretic theorems for non-sigma-complete Boolean algebras, Dissert. Math., Warsaw, 1982.

[Sch] H. H. Schaefer, Topological Vector Spaces, MacMillan, N.Y., 1966.

[Sr] J. Schur, Über lineare Tranformation in der Theorie die unendlichen Reihen, J. Reine Angew. Math., 151(1920), 79 - 111.

[Sti] W. J. Stiles, On Subseries Convergence in F-spaces, Israel J. Math., 8(1970), 53 - 56.

[St1] C. Stuart, Weak Sequential Completeness in Sequence Spaces, Ph.D. Dissertation, New Mexico State University, 1993.

[St2] C. Stuart, Weak Sequential Completeness of β-duals, Rocky Mt. Math. J., 26(1996), 1559 -1568.

[St3]   C. Stuart, Interchanging the limit in a double series, Southeast Asia Bull. Math., 18(1994), 81 - 84.

[SS]   C. Stuart and C. Swartz, A Projection Property and Weak Sequential Completeness of α-duals, Collect. Math.1, 43(1992), 177 - 185.

[Sw1]   C. Swartz, Infinite Matrices and the Gliding Hump, World Sci. Publ., Singapore, 1996.

[Sw2]   C. Swartz, An Introduction to Functional Analysis, Marcel Dekker, N.Y., 1992.

[Sw3]   C. Swartz, Measure Integration and Function Spaces, World Sci., Pub., Singapore, 1994.

[Sw4]   C. Swartz, Orlicz-Pettis Theorems for Multiplier Convergent Operator Valued Series, Proy. J. Math., 23(2004), 61-72.

[Sw5]   C. Swartz, Subseries Convergence in Spaces with Schander Basis, Proc. Amer. Math. Soc.,129(1995), 455-457.

[Th]   G.E.F. Thomas, L'integration par rapport a une mesure de Radon vectorielle, Ann. Inst. Fourier, 20(1970), 55 - 191.

[Thr]   B.L. Thorp, Sequential Evaluation Convergence, J. London. Math. Soc., 44(1969), 201-209.

[Tw]   I. Tweddle, Unconditional Convergence and Vector-valued Measures, J. London Math. Soc., 2(1970), 603 - 610.

[Wi]   A. Wilansky, Modern Methods in Topological Vector Spaces, McGraw - Hill, N.Y., 1978.

[Wi2]   A. Wilansky, Summability through Functional Analysis, North Holland, Amsterdam, 1984.

[Wu]   Wu, Junde, The compact sets in the infinite matrix topological algebras, Acta Math. Sinica, to appear.

[WL]   Wu Junde and Lu Shijie, A Summation Theorem and its Applications, J. Math. Anal. Appl., 257(2001), 29 - 38.

[Y]   K. Yosida, Functional Analysis, Springer - Verlag, N.Y., 1966.

Index